TEMPERATE CROP SCIENCE AND BREEDING

Ecological and Genetic Studies

TEMPERATE CROP SCIENCE AND BREEDING

Ecological and Genetic Studies

Edited by
Sarra A. Bekuzarova, DSc
Nina A. Bome, DSc
Anatoly I. Opalko, PhD
Larissa I. Weisfeld, PhD

Reviewer and Advisory Board Members:
Gennady E. Zaikov, DSc
A. K. Haghi, PhD

APPLE ACADEMIC PRESS

Apple Academic Press Inc. | Apple Academic Press Inc.
3333 Mistwell Crescent | 9 Spinnaker Way
Oakville, ON L6L 0A2 | Waretown, NJ 08758
Canada | USA

©2016 by Apple Academic Press, Inc.

First issued in paperback 2021

Exclusive worldwide distribution by CRC Press, a member of Taylor & Francis Group
No claim to original U.S. Government works

ISBN 13: 978-1-77463-570-4 (pbk)
ISBN 13: 978-1-77188-225-5 (hbk)

Library and Archives Canada Cataloguing in Publication

Temperate crop science and breeding : ecological and genetic studies/edited by Sarra A. Bekuzarova, DSc, Nina A. Bome, DSc, Anatoly I. Opalko, PhD, Larissa I. Weisfeld, PhD; reviewer and advisory board members: Gennady E. Zaikov, DSc A. K. Haghi, PhD.

Includes bibliographical references and index.
Issued in print and electronic formats.
ISBN 978-1-77188-225-5 (hardcover).--ISBN 978-1-77188-229-3 (pdf)
1. Agricultural productivity. 2. Agricultural ecology. 3. Plant breeding. I. Bekuzarova, Sarra A., author, editor II. Bome, Nina A., author, editor III. Opalko, Anatoly I., author, editor IV. Weisfeld, Larissa I., editor

S494.5.P75T44 2016   338.1'6   C2016-901109-7   C2016-901110-0

Library of Congress Cataloging-in-Publication Data

Names: Bekuzarova, Sarra A., editor. | Bome, Nina A., editor. | Opalko, Anatoly I., editor. | Weisfeld, Larissa I., editor.
Title: Temperate crop science and breeding : ecological and genetic studies / editors: Sarra A. Bekuzarova, Nina A. Bome, Anatoly I. Opalko, Larissa I. Weisfeld ; reviewer and advisory board members: Gennady E. Zaikov, A.K. Haghi.
Description: 1st ed. | Waretown, NJ : Apple Academic Press, [2016] | Includes bibliographical references and index.
Identifiers: LCCN 2016006421 (print) | LCCN 2016007007 (ebook) | ISBN 9781771882255 (hardcover : alk. paper) | ISBN 9781771882293 ()
Subjects: LCSH: Agricultural productivity. | Agricultural ecology. | Plant breeding.
Classification: LCC S494.5.P75 T39 2016 (print) | LCC S494.5.P75 (ebook) | DDC 338.1/6--dc23
LC record available at http://lccn.loc.gov/2016006421

Apple Academic Press also publishes its books in a variety of electronic formats. Some content that appears in print may not be available in electronic format. For information about Apple Academic Press products, visit our website at **www.appleacademicpress.com** and the CRC Press website at **www.crcpress.com**

CONTENTS

     Differentiated Fertilizer Application 475

     Rafail A. Afanas'ev

27. Gas Discharge Visualization of Selection Samples
     of *Trifoliumpratense L.* 491

     Victoria A. Belyayeva

28. Application *Galega orientalis* Lam. for Solving Problems
     of Reduction the Cost of Forage 505

     Igor Y. Kuznetsov

29. The Effect of Aromatic Plants on the Incidence and the Development
     of Malignant Tumors... 527

     Valery N. Erokhin, Tamara A. Misharina, Elena B. Burlakova,
     and Anna V. Krementsova

     Glossary *535*

     Index.. *557*

LIST OF CONTRIBUTORS

R. A. Afanas'ev
DSc in Agriculture, Chief of Lab., Professor, Pryanishnikov All-Russian Scientific Research Institute of Agrochemistry, Pryanishnikov St., 31a, Moscow, 127550, Russia, Mobile: 89191040585, 84999764757; E-mail: Rafail-afanasev@mail.ru

C. S. Andronovich
Post-graduated Student, Department of Breeding and Genetics, Belorussian State Agricultural Academy, d. 5, Michurin St., Gorki, Mogilev region, 213407, Republic Belarus, Tel.: +375445998381; E-mail: andronovich.88@mail.ru

M. N. Avramenko
PhD in Agriculture, Senior Lecturer, Department of Breeding and Genetics, Belorussian State Agricultural Academy, d. 5, Michurin St., Gorki, Mogilev region, 213407, Republic Belarus, Tel.: +375293378064; E-mail: avramenko_77@mail.ru

G. A. Batalova
DSc in Agriculture, Professor, Head of Department of Oats Breeding, N.V. Rudnitsky Zonal North-East Agricultural Research Institute, 166a, Lenin St., Kirov, 610007 Russia; Professor of cathedra of Ecology and Zoology; Vyatka State Agricultural Academy, d. 133 October □ʊ., Kirov, 610017, Russia, Tel.: +79128231553; E-mail: g.batalova@mail.ru

S. A. Bekusarova
DSc in Agriculture, Honored the Inventor of Russian Federation, Professor, Gorsky State Agrarian University, d. 37, Kirov St., Vladikavkaz, Republic of North Ossetia Alania 362040, Russia, +7(8672)362040; North-Caucasian Research Institute of Mountain and Foothill Agriculture, Republic of North Ossetia Alania, Suburban region, Mikhailovskoye vil., d. 1, Williams St., 363110, Russia; E-mail: bekos37@mail.ru

V. A. Belyayeva
PhD in Agriculture, Senior Scientist, Institute for Biomedical Research of Vladikavkaz Scientific Centre of Russian Academy of Sciences and RNO-Alania Government, d. 40, Pushkinskaya St., Vladikavkaz, Republic North Ossetia—Alania, 362019, Russia, Tel.: +79064944493; E-mail: pursh@inbox.ru

L. A. Bobyleva
PhD in Biology, Researcher, Institute of Biomedical Research of Vladikavkaz Scientific Center of the Russian Academy of Science, Vladikavkaz, RNO-Alania,; Associate Professor, North Ossetian State University, d. 44–46, Vatutina St., Vladikavkaz, RNO_Alania, 362025, Russia, Tel.: +7(8672)531304; E-mail: medgenetika435@yandex.ru

A. Ja. Bome
PhD in Agriculture, Senior Research Associate at the Tyumen Basing Point of the N.I. Vavilov All-Russia Research Institute of Plant Growing, d. 42–44, Bol'shaya Morskaya St., St. Petersburg, 190000, Russia, Tel.: +7(812)3142234; E-mail: office@vir.nv.ru

N. A. Bome
DSc in Agriculture, Professor, Head of the Department of Botany, Biotechnology and Landscape Architecture, Institute of Biology of the Tyumen State University, d. 10, Semakova St., Tyumen, 625003, Russia, Tel.: +7(3452)464061, +7(912)9236177; E-mail: bomena@mail.ru

O. A. Boyko
PhD in Agriculture, National University of Life and Environmental Sciences of Ukraine, Head of the Vegetable Physiology, Biochemistry and Bioenergetics ☐ηαιρ, d. 15 Heroyiv Oborony St., Kyiv, 03041 Ukraine, Tel.: +380974752714; E-mail: boets2008@ukr.net

M. O. Bublyk
DSc in Agriculture, Professor, Executive Director, Institute of Horticulture of NAAS of Ukraine, d. 23 Sadova St., Kyiv, 03027, Ukraine, Tel.: +380445266548; E-mail: mbublyk@mail.ru

E. B. Burlakova
DSc in Biology, Professor, Head of Department of the kinetics of chemical and biological processes, N.M. Emanuel Institute of Biochemical Physics, Russian Academy of Sciences, d. 4, Kosygin St., Moscow, 119334, Russia, Tel.: 84959397179; E-mail: chembio@sky.chph.ras.ru

V. I. Bushuyeva
DSc in Agriculture, Professor, Associate Professor of the Department of Breeding and Genetics, Belorussian State Agricultural Academy, d. 5, Michurin St., Gorki, 213407, Republic Belarus, Tel.: +3750223379674, +375296910383; E-mail: vibush@mail.ru

L. V. Chopikashvili
DSc in Biology, Professor, Institute of Biomedical Research of Vladikavkaz Scientific Center of the Russian Academy of Science, Head of Medical Genetics Department; d. 47, Pushkinskaya St., Vladikavkaz, RNO-Alania, 362025, Russia; North Ossetian State University, Head of the Department of Zoology, d. 44–46, Vatutin St., Vladikavkaz, RNO-Alania, 362025, Russia, Tel.: +7(8672)531304, e-mail: medgenetika435@yandex.ru

G. A. Chorna
Researcher, Institute of Horticulture of NAAS of Ukraine, d. 23, Sadova St., Kyiv, 03027, Ukraine, Tel.: +380445266542; E-mail: chg3@i.ua

M. T. Dzodzikova
DSc in Biology, Senior Researcher, Academician of the International Academy of Ecology and Life Safety, d. 1, Chabahan Basiev St., Alagir, Republic of North Ossetia-Alania, 363000, Russia, Tel.: +7(8672)550697, +7(918)8224269; E-mail: Dzodzikova_m@mail.ru

V. N. Erokhin
PhD in Chemical, Senior Scientist, N.M. Emanuel Institute of Biochemical Physics, Russian Academy of Sciences, d. 4, Kosygin St., Moscow, 119334, Russia, Tel.: +7(495)9397178; E-mail: valery@sky.chph.ras.ru

L. A. Fryziuk
Researcher, Institute of Horticulture of NAAS of Ukraine, d. 23, Sadova Str., Kyiv, 03027, Ukraine, Tel.: +380445266542; E-mail: lufri@ukr.net

I. S. Kosenko
DSc in Biology, Full Professor, Director, National Dendrological Park "Sofiyivka" of NAS of Ukraine, d. 12-a, Kyivska St., Uman, Cherkassy region, 20300, Ukraine, +380975081246; E-mail: sofievka@ck.ukrtel.net

O. N. Kovaleva
PhD in Agriculture, Senior Researcher of Genetic Resources of Oats, Rye, Barley, N.I. Vavilov All-Russia Research Institute of Plant Growing, d. 42–44, Bol'shaya Morskaya St., St. Petersburg, 190000, Russia, Tel.: +7(812)3142234; E-mail: o.kovaleva@vir.nw.ru

G. P. Khubaeva
PhD in Technical Sciences, North-Caucasian Mining and Metallurgical Institute, d.4, Nikolaev St., Vladikavkaz, RNO-Alania, 362020, Russia; E-mail: lady.almana@mail.ru

A. V. Krementsova
PhD in Chemical, Senior Scientist, N.M. Emanuel Institute of Biochemical Physics, Russian Academy of Sciences, d. 4, Kosygin St., Moscow, 119334, Russia, Tel.: +7(495)9397178; E-mail: valery@sky. chph.ras.ru

A. A. Kuzemko
DSc in Biology, Chief Researcher, National Dendrological Park "Sofiyivka" of National Academy of Sciences of Ukraine, St. Kyivska, d. 12a, Uman, 20300, Ukraine, Tel.: +380979193987; E-mail: anya_meadow@mail.ru

I. Yu. Kuznetsov
PhD in Agriculture, Associate Professor of the Chair of Plant Growing, Forages Production and Horticulture, Bashkir State Agrarian University, d. 34, 50 years of October St., Ufa, Bashkortostan Republic, 450001, Russia, Tel.: +7(9050)039426; E-mail: kuznecov_igor74@mail.ru

L. M. Levchuk
Researcher, Institute of Horticulture of NAAS of Ukraine, d. 23, Sadova St., Kyiv, 03027, Ukraine Tel.: +380445266542; E-mail: l.levchuk@ukr.net

E. M. Lisitsyn
DSc in Biology, Assistant Professor, Head of Department of Plant Edaphic Resistance, N.V. Rudnitsky Zonal North-East Agricultural Research Institute, d. 166a, Lenin St., Kirov, 610007 Russia; Professor of Cathedra of Ecology and Zoology, Vyatka State Agricultural Academy, d. 133 October A℧., Kirov, 610017, Russia, +7(912)3649822; E-mail: edaphic@mail.ru

G. E. Merzlaya
DSc in Agriculture, Professor, Head of Laboratory, D.N. Pryanishnikov All-Russia Research and Development Institute of Agrochemistry, d. 31A, Pryanishnikov St., Moscow, 127550, Russia, Tel.: +7(962)3694197; E-mail: lab.organic@mail.ru

T. A. Misharina
DSc in Chemistry, Head of Department, N.M. Emanuel Institute of Biochemical Physics, Russian Academy of Sciences, d. 4, Kosygin St., Moscow, 119334, Russia, Tel.: +7(495)9397343; E-mail: tmish@rambler.ru

O. I. Mykychuk
Researcher, Prydnistrovska Research Station of Horticulture of Institute of Horticulture of the National Academy of Agrarian Sciences of Ukraine, d. 1, Yablunivska St., Godyliv, Storozhynets district, Chernivtsi region, 59052, Ukraine, Tel.: +380372243707; E-mail: to_olgamu@mail.ru

R. A. Nazyrov
Student, Institute of Biology of the Tyumen State University, Tyumen State University, Department of Botany, Biotechnology and Landscape Architecture, d. 3, Pirogova St., Tyumen, 625045, Russia; E-mail: rvl5577@mail.ru

A. I. Opalko
PhD in Agriculture, Full Professor, Head of the Physiology, Genetics, Plant Breeding and Biotechnology Division in National Dendrological Park "Sofiyivka" of NAS of Ukraine, d. 12-□, Kyivska St., Uman, Cherkassy region, 20300, Ukraine; and Professor of the Genetics, Plant Breeding and Biotechnology Chair in Uman National University of Horticulture, d. 1, Institutska St., Uman, Cherkassy region, 20305, Ukraine, Tel.: +380506116881; E-mail: opalko_a@ukr.net

O. A. Opalko
PhD in Agriculture, Associate Professor, Senior Researcher of the Physiology, Genetics, Plant Breeding and Biotechnology Division in National Dendrological Park "Sofiyivka" of NAS of Ukraine, d. 12-a, Kyivska St., Uman, Cherkassy region, 20300, Ukraine, Tel.: +380664569116; E-mail: opalko_o@ ukr.net

E. G. Pukhaeva

Junior Researcher, Institute of Biomedical Research of Vladikavkaz Scientific Center of the Russian Academy of Science, d. 40, Pushkinskaya St., RNO-Alania, Vladikavkaz, 362025, Russia, Tel.: +7(8672) 53–13–04, E-mail: medgenetika435@yandex.ru

E. I. Ripberger

PhD, Department of Botany, Biotechnology and Landscape Architecture, Tyumen State University, d. 1, Semakova St., Tyumen, 625003, Russia, Tel.: +7(345)2464061, +7(345)2468169; E-mail: lena-umka@yandex.ru

F. K. Rurua

Junior Researcher, Institute of Biomedical Research of Vladikavkaz Scientific Center of the Russian Academy of Science, d. 40, Pushkinskaya St., RNO-Alania, Vladikavkaz, 362025, Russia, +7(8672)531304, E-mail: medgenetika435@yandex.ru

I. T. Samova

Specialist rank 1 of Forestry Committee, d. 25, Iristonskaya St., Vladikavkaz, RNO-Alania. 362020, Russia, Tel.: +7(928)4940085; E-mail: pochta@leskom-15.ru

Yu. A. Semenishchenkov

PhD in Biology, Associate professor, Department of Botany, I.G. Petrovsky Bryansk State University, d.14, Bezhitskaya St., Bryansk, 241036, Russia, Tel.: +7(483)2666834, +7(905)1000390, E-mail: yuricek@yandex.ru

T. K. Sheshegova

DSc in Agriculture, Assistant Professor, Head of Laboratory of Plant Protection, N.V. Rudnitsky Zonal North-East Agricultural Research Institute, d. 166a, Lenin St., Kirov, 610007, Russia, +7(912)7376344; E-mail: niish-sv@mail.ru

L. N. Shikhova

DSc in Agriculture, Associated professor, Head of Cathedra of Ecology and Zoology, Vyatka State Agricultural Academy, d. 133 October, Kirov, 610017, Russia, +7(912)7213758; E-mail: shikhola-l@mail.ru

O. N. Shupletsova

PhD in Biology, Assistant professor, Senior Researcher, Laboratory of Plant Genetics, N.V. Rudnitsky Zonal North-East Agricultural Research Institute, d. 166a, Lenin St., Kirov, 610007, Russia, Tel.: +7(962)8934754; E-mail: olga.shuplecova@mail.ru

I. N. Shchennikova

PhD in Agriculture, Associated Professor, Head of Department of Barley Breeding, N.V. Rudnitsky Zonal North-East Agricultural Research Institute, d. 166a, Lenin St., Kirov, 610007, Russia, +7(912)7376344; E-mail: i.shchennikova@mail.ru

S. V. Skupnevskiy

PhD in Biology, Junior Researcher, Institute of Biomedical Research of Vladikavkaz Scientific Center of the Russian Academy of Science, d. 40, Pushkinskaya St., RNO-Alania, Vladikavkaz, 362025, Russia; Associate Professor, North Ossetian State University, d. 44–46, Vatutina St., Vladikavkaz, RNO-Alania, 362025, Russia, Tel.: +7(8672)531304; E-mail: dreammas@yandex.ru

M. O. Smirnov

PhD in Biology, Senior Researcher, D.N. Pryanishnikov All-Russian Scientific Research Institute of Agrochemistry, d. 31A, Pryanishnikov St., Moscow, 127550, Russia, Tel.: +7(905)7966323; E-mail: User53530@yandex.ru

K. E. Sokaev
DSc in Agriculture, Station of Agrochemical Service Station "North-Ossetia," d. 36, Sadonskia St., Vladikavkaz, RNO-Alania, 362013, Russia, Tel.: +7(8672)761279; E-mail: agrohim-15@mail.ru

L. B. Sokolova
DSc in Biology, Professor, Gorsky State Agrarian University, d. 37, Kirov St., Vladikavkaz, RNO-Alania, 362040, Russia, Tel.: +7672563422, +7(8672)530142; E-mail: agrofak1918@ yandex.ru

G. A. Tarasenko
MPhil in Biology, Junior Researcher, Department of Reproductive Biology of Plants, National Dendrological Park "Sofiyivka" of NAS of Ukraine, d. 12-a, Kyivska St., Uman, Cherkassy region, 20300, Ukraine, Tel.: +380954226463; E-mail: vernyuk_galina@mail15.com

N. V. Tetyannikov
Graduate student, Institute of Biology of the Tyumen State University, Tyumen State University, Department of Botany, Biotechnology and Landscape Architecture, d. 3, Pirogova St, Tyumen, 625045, Russia; E-mail: kolyannn@yandex.ru

T. I. Tsidaeva
DSc in Medicine, Deputy Minister, Ministry of Health RNO-Alania, d. 9a, Borodinskaya St., Vladikavkaz, 362025, Russia; Professor, Head of the Department of Obstetrics and Gynecology, Northern Ossetian State Academy of Medicine, d. 40, Pushkinskaya St., Vladikavkaz, RNO-Alania, 362025, Russia, Tel.: +7(8672)403894; E-mail: minzdrav@osetia.ru

L. I. Weisfeld
Senior Research, N.M. Emanuel Institute of Biochemical Physics of Russian Academy of Sciences, d. 4, Kosygin St., Moscow, 119334, Russia, Tel.: +7(916)2278685; E-mail: liv11@yandex.ru

I. L. Zamorska
PhD in Agriculture, Associate Professor, Department of Technology Storage and Processing of Fruits and Vegetables, Uman National University of Horticulture, 535 flat, 2 International St., Uman, Cherkassy region, 20305, Ukraine, Tel.: +380661479983; E-mail: zil1976@mail.ru

LIST OF ABBREVIATIONS

ANOVA	analysis of variances
ApMV	apple mosaic virus
BCP	bioclimatic potential
BEA	beef-extract agar
BiOECOFUNGE-1	antipathogenic biopreparations based on mushrooms components and carries of plants
BSAA	Belorussian state agricultural academy
BVG_i	breeding value of a genotype
CA	congenital anomalies
CC	critical concentration of every element
Cd	cadmium
cGy	cantiGrey
ChA	chromosomal aberrations
cM	cantimol
cm	centimeter
CMD	congenital malformations of development
CO	carbon monoxide
Co	cobalt
CO_2	carbon dioxide
Cs	cesium
Cu	copper
CV	coefficient of variation
cwt	centner, hundredweight
DAS-eA	double antibody sandwich-enzyme-linked immunosorbent assay
DIECA	sodium dietildytiocarbomat
DSS	dry soluble substances
DUS	distinguishability uniformity stability
e.g.	exempli gratia (lat.)
EIV	ellenberg indicator values
ELISA	enzyme-linked immunosorbent assay

EQ.	equalent
Etc.	et cetĕra (lat.)
eV	electronvolt
FAO	Food and Agriculture Organization of the United Nations
FC	form coefficient
FDR	field drought resistance
GAA	general adaptive ability of a genotype, characterizes average value of a trait in various environmental conditions
GDV	gas discharge visualization
GENAN	genetic analysis (computer software)
GPS	Global Positioning System
HC	hydrothermal coefficient
Hd	soil moisture
$HgCl_2$	mercury (II) chloride
HTC	hydrothermal coefficient
i.e.	Id est (lat.)
K	potassium
LAR	leaf area ratio, the ratio of total leaf area to stem or twig mass
LSD_{05}	least statistical distinction = least significant difference at $p<0.05$
LWR	leaf weight ratio, ratio of leaf mass to total plant mass
MAC	maximum allowable concentrations
Mo	molybdenum
MPa	megapaskal
MPC	maximum permissible concentration
MS	Murashige–Skoog nutritive medium
N	nitrogen
NAAS	National Academy of Agrarian Sciences
NaOCl	sodium hypochlorite
NAS	National Academy of Science
$NaS_2CN(C_2H_5)_2$	sodium diethyldithiocarbamate
$NaSO_3$	sodium sulfur

NDP	National Dendrological Park
NFE	nitrogen-free extract
NPK	fertilizer: nitrogen, phosphorus, and potassium
NPP	net photosynthesis productivity
NSC	National Science Centre
OAR	obstetric anamnestic record
P	phosphorus
PABA	para-aminobenzoic acid
PAGe	polyacrylamide gel
Pb	lead
PEG	polyethylene glycol (used as osmotic)
pH	hydrogen ion concentration
PNRSV	prunus necrotic ringspot virus
QTL	quantitative trait locus
RA	regenerant obtained of aluminum-acid media
RAC	roughly allowable concentration of every element
Rc	soil reaction
RFLP	restriction fragment length polymorphism
RNA-asa	ribonucleasa
RNO – Alania	Republic of North Ossetia – Alania
ROS	Reactive Oxygen Species
RRG	Relative Root Growth
RTI	root tolerance index, is counted as average root length in test treatment divided by root length in control treatment
SA	spontaneous abortion
SAA	specific adaptive ability of a genotype, characterizes a deviation from *GAA* in the exact environment
SB	stillborn
SLA	specific leaf area, one-sided area of a fresh leaf, divided by its oven-dry mass two-way
SNP	single nucleotide polymorphisms
SOD	superoxide dismutase
Sr	strontium

SSD	smallest significant difference
two-way ANOVA	ANalysis Of VAriance between groups with two factors
UPOV	International Union for the Protection of New Varieties of Plants = distinctness, homogeneity, stability = distinctness, uniformity, stability
VIR	N.I. Vavilov Research Institute of Plant Industry
WPI	index of water pollution
WSFO	concentration of water-soluble fractions of oil
X-ray exposure	X-ray irradiation, X-irradiating
Zn	zinc

LIST OF SYMBOLS

$\sqrt{H_1/D}$	average degree of dominance by all heterozygous loci
°X	degree Celsius
2n	diploid chromosomal set
$AgNO_3$	silver nitrate
Al	aluminum
Al^{+++}	ion of aluminum
$Al_a Al_a$	dominant alleles of aluminum resistance in oats
$al_a al_a$	recessive alleles of aluminum resistance in oats
b_i	linear regression coefficient
Bq/kg: kBq/m²	coefficient of transition of radionuclide from soil to plants
c/ha	center/ha, hundredweight
$C_2H_5(OH)$	ethyl alcohol
$C_6H_8O_6$	ascorbic acid
$C_9H_9HgNaO_2S$	(Thiomersal) mercury((o-carboxyphenyl)thio) ethyl sodium salt
Ca^{++}	ion of calcium
F	component of variability reflecting a direction of dominance on the average on a number
F_1, F_2, F_3, F_4	generations of organisms from first to fourth and so on
g	gram
G_0	phase 0 of the mitotic cycle
G_1	presyntetic phase of the mitotic cycle
G_2	postsyntetic phase of the mitotic cycle
H^+	hydrogen ion
H_1, H_2	components of variability caused by dominante effects

ha	hectare – is area unit that equal to 10,000 square meters
kBq/m^2	kilo Becquerel per square meter
KCl	potassium chloride
K_d	phytocoenosis destruction index
kg	kilogram
Kg/ha	kilogram/hectare
KH_2PO_4	potassium dihydrogen phosphate
km	kilometer
l	liter
mg	milligram
Mg/kg	milligram/kilogram
Mg^{++}	ion magnesium
mkg	microgram
mkl	microliter
ml	milliliter
Mcg	microgram
ml_1–ml_0	difference between average mean for parents and average
mln	million
Mm	micromoles
mm	millimeter
mm per year	millimeters per year – atmospheric fallouts amount
M	gram-molecule
$M_1 - M_7$	generations of mutants from first to seventh
n	haploid chromosomal set, chromosome complement
NH_4^+-N	ammoniacal form of nitrogen fertilizer
NH_4NO_3	ammonium nitrate
NH_4NO_3	mixed form of nitrogen fertilizer
NO_3^-N	nitric form of nitrogen fertilizer
$P_{30}K_{30}$	fertilizer: phosphors, potassium
psc.	pieces
r	coefficient of pair correlation
R	Roentgen

R_4 and R_5	fourth and fifth generation of regenerants
S	phase synthesis of DNA of the mitotic cycle
S_{gi}	relative stability of genotype
sm	centimeter
Sx	mean arithmetic error
Sx, % –	the relative mean arithmetic error
t	ton – is mass unit, that equal to 1000 pounds in the units
t/ha	ton/hectare
t_{05}	criterion of Student
th. t	thousand tones
Tr	nutrient content in the soil
ts	technical system
urc	units of regeneration coefficient
V	coefficient of variation
Xav	the average value of the characteristic
Xmax	maximum value of the characteristic
Xmin	minimum value of the characteristic
y-axis	axis of ordinates
Y_{dr}	yield in drought conditions
Y_{fav}	yield in favorable conditions of growth
$Y_{min}, Y_{max}, Y_{aver}$	minimum, maximum and average productivity of a variety

ABOUT THE EDITORS

Sarra A. Bekuzarova, DSc in agriculture, is head of the Laboratory at Plant Breeding of Fodder Crops at the North Caucasus of Institute of Mountain and Foothill Agriculture of the Republic of North Ossetia-Alania. She is also a professor at Gorsky State University of Agriculture, Vladikavkaz, Republic of North Ossetia-Alania, Russia, as well as a professor at L. N. Kosta Khetagurov North-Ossetia State University, Vladikavkaz, Republic of North Ossetia-Alania, Russia.

She is also a prolific author, researcher, and lecturer, and has received the Medal of Popova. She is a corresponding member of the Russian Academy of Natural Sciences as well as a member of the International Academy of Authors of Scientific Discoveries and Inventions, the International Academy of Sciences and Ecology, All-Russian Academy of Non-traditional and Rare Plants, and the International Academy of Agrarian Education, among others. She is a member of the editorial boards of several scientific journals and co-edited the books *Ecological Consequences of Increasing Crop Productivity: Plant Breeding and Biotic Diversity*, and also *Biological Systems, Biodiversity, and Stability of Plant Communities.*

Nina Anatolievna Bome, DSc in agriculture, is professor and head of the Department of Botany, Biotechnology and Landscape Architecture at the Institute of Biology at the Tyumen State University, Tyumen, Russia. She is the author of monographs, articles, schoolbooks, and patents, and she is a lecturer. She is the director and founder of the Scientific School for Young Specialists. She is the author of about 300 publications. She participates in long-term Russian and international programs.

Her main field of interest concerns basic problems of adaptive potential of cultivated crops, mutagenesis, possibility of conservation, enhancing biodiversity of plants, methods of evaluation of plants' resistance to the phytopatogens and other unfavorable environmental factors, and genetic resources of cultivated plants in the extreme conditions of the Western Siberia. She is a co-editor of the books *Ecological Consequences of Increasing Crop Productivity: Plant Breeding* and *also Biotic Diversity* and *Biological Systems, Biodiversity, and Stability of Plant Communities.*

Anatoly Iv. Opalko, PhD, is a professor and head of the Physiology, Genetics, Plant Breeding and Biotechnology Division at the National Dendrological Park "Sofiyivka" of the National Academy of Sciences of Ukraine, Uman, Cherkassy region, Ukraine, and a professor and Genetics, Plant Breeding and Biotechnology Chair in Uman National University of Horticulture, Uman, Ukraine.

He is also the head of the Cherkassy Regional Branch of the Vavilov Society of Geneticists and Breeders of Ukraine. He is also a prolific author, researcher, and lecturer. He has received several awards for his work, including the badge of honor for "Excellence in Agricultural Education" and the badge of honor of the National Academy of Sciences of Ukraine for professional achievement. He is member of many professional organizations and on the editorial boards of the Ukrainian biological and agricultural science journals. In 2014, he was a co-editor of the books *Ecological Consequences of Increasing Crop Productivity: Plant Breeding and Biotic Diversity* and also *Biological Systems, Biodiversity, and Stability of Plant Communities.*

Larissa I. Weisfeld, PhD, is a senior researcher at the N.M. Emanuel Institute of Biochemical Physics, Russian Academy of Sciences in Moscow, Russia, and a member of the N. I. Vavilov Society of Geneticists and Breeders. She is the author of about 300 publications in scientific journals, patents, and conference proceedings, as well as the co-author of a work on three new cultivars of winter wheat.

Her research interests concern the basic problems of chemical mutagenesis, cytogenetic, and the other ecological problems. She has worked as a scientific editor at the publishing house Nauka (Moscow) and of the journals *Genetics* and *Ontogenesis*. In 2014, she was a co-editor of the books *Ecological Consequences of Increasing Crop Productivity: Plant Breeding and Biotic Diversity* and also *Biological Systems, Biodiversity, and Stability of Plant Communities.*

PREFACE

Survival of the individuals of *Homo sapiens* L., as well as of the species as a whole, since prehistoric times was conditioned during millenniums by the success in hunting on wild animals and in gathering of edible parts of wild plants. Only a little more than 10 thousands year ago, when humans in different parts of the world started cultivation of plants and domestication of animals, our ancestors considerably decreased their dependence on fortuitousness due to consequent *agricultural revolution spread*, which allowed producing more food with smaller physical costs. Namely primitive *breeding*, as *selection* of the best specimens from extremely *heterogeneous* populations of wild animals and wild plants, unconsciously applied by ancient humans, provided possibilities of the *animal husbandry* and *field-husbandry* development in the remote past and, when taking more advanced modern forms, provided food supply of continuously increasing human population of the planet.

However, although a field, fully sowed with agricultural crops, gives the possibility to feed more people then forest, where edible plants occur separately, and a herd of *cattle* can provide more mouth to feed than can bag permanently nomadic hunter, further spread of Agricultural revolution brought up to the mankind new challenges unknown scent agriculture.

The possibility of population food supply under considerably smaller amount of farm workers provided labor resources to the mankind for the development of industry and a set of branch not related with the production of material goods. At the same time, enormous plants and factories, megalopolises, as well as giant orchards, enormous cattle-breeding farms and monocultural fields of several thousands hectares, the continuous cultivation of corn, sunflower, rape or other highly remunerative culture became source of permanent pollution of human habitat. Because of human economic activity the environment is changing more and more under permanently increasing anthropogenic impact, which reaches threatening scale in the conditions of the new globalization wave in 21st century, which swept over much of the developing world and many countries, which kept until recent time traditional agriculture. The planetary ecosystem, which was formed and evolved very slowly during centuries, is now exposed to the destruction in the past unknown.

The pages of this book are devoted to the analysis of processes affecting atmospheric, water, soil, mineral and other natural resources, their effect on human gene pool, as well as to the search of agricultural methods in stress conditions of pollution under rational use of recent achievements in plant breeding.

—*Anatoly I. Opalko*

INTRODUCTION

This new collection covers a wide variety of research on the ecological aspects of crops growing under stress conditions due to atmospheric changes and pollution and the impact on both plant and human health. The book provides research that will help to find ways to overcome adverse abiotic environmental factors and unfavorable anthropogenic pressures on crop plants, which also eventually impact human health.

This book is divided into six parts, united by common ideas: to finding ways to overcoming as of adverse abiotic environmental factors as and unfavorable anthropogenic pressures on crop plants and eventually on human health.

Science as the special kind of human activity has its own features that attract the intellectual elite of society. Science is characterized by continuity in reception, processing and generalization of knowledge. Therefore, scientific schools are formed; pupils continue work of the teachers. So, the chapters given by the scientist are already having popularity, which theoretically generalize the previous experience, and original experimental works of their colleges and pupils—post-graduate students, students, etc. These scientists work in the different institutes, different cities and countries, but the desire unites them to receive new knowledge and to share them with all interested people unites them.

Science develops both at regional and national levels. At the same time, regional scientific discoveries are part of a global science. As well as Vysotsky's songs attracted Finnish, Swedish, French performers and listeners, so scientific achievements can be understood and applied in different countries.

The authors invested maximum efforts in order to make the collection presented the appreciable contribution to development of a biological science.

Geneticists and breeders are creating new cultivars and hybrids of crops; thus, greatly expanding the range of source material.

Readers are invited to the results of studies from leading experts in the fields of biology, genetics, breeding of crops, taking into account of climatic and environmental changes.

The main agricultural crops like cereals, fodder crops, and horticultural plants are studied in various ecological and climatic conditions.

The book presents the works of leading scientists from different regions of Russia, Ukraine, and Belarussia that were carried out in contrasting environmental conditions. These works focuses on the impact of human activities on the environment, health, and status of the gene pool of the population in modern conditions.

Plant communities, interaction plant–soil–plant, ways of using plants as anticancer drugs and other important problems of nature management are examined.

Part I is titled "Plant Breeding Under Adverse Conditions of Acid Soils" and consists of seven chapters. The research of the ecological aspects of crops growing on acid soils of European-North Russia are presented in part I. In studies of the North-East Agricultural Research Institute and Vyatka State Agricultural Academy, Kirov, Russia, great attention is paid to generalization of studies conducted in various soil-climatic conditions of the country. The authors' opinions on different discussion problems are shown. These problems include seasonal and profile dynamics of elements of soil acidity; role of genetic and agronomical approaches in improving of plant productivity; using of methods of classic breeding; and biotechnology in creation of stress-tolerant cereal cultivars. Comprehensive analysis of the genetics and breeding of cereal crops was obtained with participation of specialists in the field of soil science (L. N. Shikhova, DSc), phytopathology (T. K. Sheshegova, DSc), plant physiology (E. M. Lisitsyn, DSc), tissue culture (O. N. Shupletsova, PhD), and plant breeding (G. A. Batalova DSc; I. N. Shchennikova, PhD). In the presented articles, the authors' opinions on different discussion problems are shown. These problems include seasonal and profile dynamics of elements of soil acidity; role of genetic and agronomical approaches in improving of plant productivity; and using of methods of classic breeding and biotechnology in creation of stress-tolerant cereal cultivars.

Part II is titled "Horticultural Crop Science" and consists of four chapters. Global trends on the market of horticultural production are characterized by stable growth of unsatisfied demand, which formed as a result of increasing consumption of fruits and berries, first of all in the states of European Union, Northern America, Japan, and other developed countries of the northern hemisphere. Now considerable changes in the nutrition structure in favor of fruits and berries is taking place in these countries. Apples, plums, apricots, raisins, nuts and other fruits and berries are more and more introduced not only as fruit addition to traditional vegetable salads, but are also added to soups, meat and fish dishes and are used in the baking not only of pastry and buns, but also of rye bread. Consequently, the states of European Union, Northern America, and Japan, in spite of high volumes of domestic production, belong to the biggest importers of fruits and berries in the world.

The analysis of fruits and berries production supply of the population of Ukraine shows its considerable deficit. Under science-based annual consumption norm, which amounts in the climatic conditions of the temperate climate zone 82 kg fruits and berries per capita of Ukraine population, their average production is 29–42 kg only. In contrast, in Spain this indicator exceeds 400 kg, in Italy and Moldova it approaches to 300 kg, and in Greece it exceeds 300 kg fruits and berries per capita. Even higher deficit is in berries, which production per capita in Ukraine is 3.4 kg under physiological norm of consumption for this region about 10 kg. At that a high inter-regional diversity of consumption level of horticultural production is observed in the country.

Annually Ukraine imports 15–20 kg of fruits and berries per capita, which provides together with domestic production about 70% of mean demand. Of that, under two million tons of total yield of horticultural products in Ukraine the portion of berries does not exceed 1.5–2.0%, whereas in the neighboring Poland, where total yield of fruits and berries reaches almost three and half million tons, namely berries, amount to 15–20%. The soil and climatic conditions of Ukraine are much more favorable for horticulture than in Poland or other neighboring countries. Consequently, it is necessary to perform in Ukraine suitable organizing and technological arrangements in order to overcome the deficit of horticultural production. It is, in particular, a question of deepening of zonal specialization and increasing of state protectionism or both, domestic production and scientific, first of all selection and genetic, studies aimed to the increase of the anthropoadaptive potential of the whole horticulture and in particular the anthropoadaptability of new cultivars.

Consequently, key elements of the strategy and particular details of horticulture improvement are developed by researchers of universities and academic research institutes concerning biological peculiarities of new pear cultivars (*Pyrus communis* L.) in the Ukrainian Pridnestrovya; the amino acid composition of strawberries fruits (*Fragaria ananassa* Duch.); the viral diseases of the representatives of the genus *Corylus* in the ecological conditions of the National Dendrological Park "Sofiyivka" of NAS of Ukraine and the production biotechnology of improved planting material, as well as the phylogenetic connections between representatives of the genus *Amelanchier* Medik., grown in Ukraine as an initial material for the horticultural plant breeding.

Part III is titled "Ecological Peculiarities of the Foothills of the Northern Caucasus: Cytogenetic Anomalies of the Local Human Population" and consists of four chapters. It contains the original works of the specialists of the Republic of North Ossetia–Alania. The Republic, located in the northern part of the Main Caucasian Range, is one of the most densely populated regions of the Russian Federation. The climate here is continental due to the weak influence of the seas. The vegetation period lasts from May to October, which promotes agriculture. However, zoning mountain and foothill areas creates difficulties for agriculture and the cultivation of forage grasses. Confined space contributes to the accumulation in the soil, air, and in plants of heavy metals of hazardous industries. Analysis of genetic changes in humans confirms this.

There is a need for conducting environmentally sound agriculture. Grasses and leguminous plants play a significant role in solving this problem. They have a significant impact on the preservation and restoration of soil fertility and are the most efficient source of cheap highly nourishing fodders for livestock.

In the review paper "Introduction of Clover Species (*Trifolium* L.) in the North Caucasus," it was studied that wild species of clover in contrasting environmental conditions on different heights of mountains (600, 800, 1300, 1600, and 2000 m above sea level).

The important role for environment is played by the water supply in the region. These data in detail are presented in the chapter "Sources of

Fresh and Mineral Water in North Ossetia—Alania." The current climatic and industrial conditions have increased the level of environmental pollution with heavy metals. The following chapter, "Detoxification of Soils Contaminated with Heavy Metals," describes biological methods of evaluation of heavy metal in the soil and air and proposes a method of detoxification of the soil using the clays and zeolites. In "Genetic Health of the Human Population as a Reflection of the Environment: Cytogenetic Analysis," it was studied the cytogenetic and demographic aspect of monitoring the population living in conditions of high anthropogenic pressure. Environmental pollution by heavy metals affects the genetic health of the population. A study of the correlation of mutational load in the population with birthrate dynamics in 1996–2000 and 2008–2012 have shown that the higher the frequency of spontaneous abortions and preterm births, the less likely the birth of children with congenital anomalies and vice versa.

Part IV is titled "Phenogenetic Studies of Cultivated Plants and Biological Properties of the Seeds" and consists of four chapters. This part is dedicated to one fundamental task of farming—preserving and expanding biodiversity of cultivated plants in difficult soil and climatic conditions of Western Siberia. In the Department of Botany, Biotechnology and Landscape Architecture of Tyumen State University jointly with Tyumen Strong point of N.I. Vavilov All-Russian Research Institute of Plant Industry, forming and storaging the valuable collections of cultivated plants are carried out. According to the results of the study in the Tyumen region, world collection of barley from VIR possesses valuable characters for breeding ("Ecological and Biological Study of Collection of Genus *Hordeum* L."). A new breeding material of spring wheat, possessing the wide adaptive capacities, also was created. Field germination and viability of plants during the growing season served as an indicator of ecological plasticity of hybrid of spring wheat. In this case, locally adapted cultivars and the best examples of the world collection were involved for hybridization. In the chapter, "Reaction of Collection Samples of Barley (*Hordeum* L.) and Oats (*Avena* L.) on Chloride Salinization," the authors have shown the resistance to the chloride stress on criteria of the ability of seeds to germinate and the variability of parameters of plantlets in the lab. The selection of salt-tolerant specimens of the oats and the barley are considered as valuable source material for breeding and genetic programs. The chapter "Resistance to the Impact of Biotic and Abiotic Factors of the Environment" studies the field seed germination and biological resistance of the parent cultivar plants and hybrids F_1–F_4 of the soft spring wheat in sharply changing climatic conditions and cultivar of soil types of the Western Siberian Lowland. The hybrid forms were studied within four years (2010–2013). From them were singled out samples having a wider adaptive capability according to the indexes of the field seed germination and biological resistance of the plants. The chapter "Comparative Trials of Variety Samples of Eastern Galega (*Galega orientalis* Lam.)" presents the characteristics of samples of *Galega orientalis* on morphological and economically valuable traits in the competitive test. It was found that the studied samples of *Galega orientalis* differed significantly from each other in color of flowers, leaves, seeds and other morphological characteristics. According to the results of a comprehensive evaluation of the economically useful traits, cultivar samples SEG-7, SEG-10, and SEG-12 were characterized by higher rates.

Part V is titled "Anthropogenic Pressure on Environmental and Plant Diversity" and consists of four chapters. The chapter "Plant Response to Oil Contamination in Simulated Condition" deals with response of perennial gramineous (awnless brome, red fescue) and leguminous (red clover) grasses to influence hydrocarbons at different stages of ontogenesis in laboratory and field conditions. The topic of oil pollution is addressed in the article. The observations show that treatment of seeds by hydrocarbons of oil soil pollution can result in both growth inhibition and in the stimulation of growth depending on the reactant concentration and from plant species. The chapter "Influence of Anthropogenic Pressure on Environmental Characteristics of Meadow Habitats in the Forest and Forest-Steppe Zones" deals with the use of modern methods of ecological and geobotanical studies. It was revealed that human pressure changes the basic environmental characteristics of mesic grasslands in a Forest and Forest-Steppe zones of Ukraine. The general trends of changes are decreasing of soil moisture, increasing of soil reaction, and rise of nutrient content in the soil. These patterns should be considered in organization of environmental management and monitoring of natural grasslands, for the development of optimal regimes of grazing, mowing, and recreation.

The chapter "Botanico-Geographical Zoning of the Upper Dnieper Basin on the Base of the J. Braun-Blanquet Vegetation Classification Approach" is an important theoretical direction of modern botany and, in particular, geobotany. Approaches to the allocation of the main botanico-geographical units are constantly being improved. In European countries, the widespread J. Braun-Blanquet approach for vegetation classification today is increasingly used also in zoning. In this regard the presented work on the using of syntaxonomy for the aims of zoning of the Russian part Upper Dnieper basin is very actual.

Part VI is titled "Methods of Evaluation of the Quantitative and Qualitative Characters of Selection Samples" and consists of six chapters. The Pryanishnikov All-Russian Scientific Research Institute of Agrochemistry is the scientific-methodical center of the geographical network of experiments with Russian fertilizers. The collection includes the most significant, at the level of discoveries, scientific works on the study of investigation and regulation of the substance circulation in ecosystems and agrosystems. These scientific works were done in recent years by well-known Russian agrochemists. Studying precision agriculture allowed the identification of previously unknown statistical and agrochemical regularities in the variation of within-field soil fertility ("Accounting Within-Field Variability of Soil Fertility to Optimize Differentiated Fertilizer Application"). They can serve as a theoretical basis for the development of efficient technologies of differentiated fertilizer application, taking into account the heterogeneity of the soil cover. In general, studies display this work on the level of scientific discovery. In another chapter, "Transformation of Mobile Phosphorus in the Soils of Agroecosystems During Prolonged Trials," some new regularities of transformation of phosphorus in soils under long-term interaction between fertilizer and soil are shown. These regularities permit the prediction of the content of mobile phosphorus in the soil for the long term and determine the need for agricultural crops in phosphate fertilizers. The chapter "Sustainability of Agrocenoses in the Use of Fertilizers on the Basis of Sewage Sludge" discusses the theoretical and practical aspects of rational use in biological systems domestic wastes of communal services. It is shown that municipal waste is on one hand a source of environmental pollution and on the other raw materials for the production of valuable fertilizer funds. Application of processed waste as organic mineral fertilizer assists in the closing of a significant portion of the small biological cycle, and helps to protect the environment from contamination of biological wastes.

Particular interest from a bioecological point of view attracts the work (in Part V) "Dynamics of the Floristic Diversity of Meadows as a Stability factor of Herbaceous Ecosystems," revealing in a historical perspective the nature of the interaction in the soil-plant-animal system. The role of vegetation in the formation of soil cover was noticed by Leonardo da Vinci; however, the linkages between all elements of this ecosystem have not previously been considered. This article shows that the usual, at first sight, processes of changing the floristic composition of herbaceous ecocenoses were caused by the historical interaction in the system of plant–animal, and ultimately aimed at conservation of soil as the basis for the existence of plants and animals on land part of our planet. For the first time the author revealed the role of weeds, for example, uneatable plants, as planetary protection function of soils from pasture soil erosion. He also explains that the emergence of weeds on arable land is a protective reaction of nature against the violation of the integrity of grassy ecosystems. In another chapter, "Gas Discharge Visualization of Selection Samples of *Trifolium pratense* L.," a new physical method of gas discharge visualization was developed. The plants, leaf blades of which have a high intensity of luminescence, differ with largest percentage of sugars. The GDV-bioelectrography allows in short term to produce a selection of samples of red clover by sugar content, as well as to assess the impact of X-rays on the vitality of clover plants derived from irradiated seeds. The chapter "Application *Galega orientalis* Lam. for Solving Problems of Reduction the Cost of Forage" has proposed a method of cultivation of valuable fodder crop *Galega otientalis* Lam. to improve the gustatory quality of green mass and improve the quality of harvested forage. *Galega otientalis* Lam. should be cultivated in a mixture with components of cereals and legumes. And it should take into account the timetable for cleaning cover crop, and should take into account the need for adding to the soil of certain mineral fertilizers. In "The Effect on the Incidence and the Development of Malignant Tumors" chapter, the effect of different doses of savory essential oil on the development of spontaneous leukemia was studied on mice. The drug efficiency was determined from the survival curves, animal life spans, and the incidence of leukemia. The savory essential oil in low doses added with drinking water (150 ng/mL) or with feed (2.5 µg/g) increased the average lifetime of mice by 20–35%. The low doses of essential oil from this aromatic plant seems promising as a prophylactic agents.

The articles in this volume are from the following scientific institutions:

- Bashkir State Agrarian University, Ufa, Republic of Bashkortostan, Russia;
- Belorussian State Agricultural Academy, Gorki, Republic of Belarus; Bryansk State University, Bryansk, Russia;
- Gorsky State Agrarian University, Vladikavkaz, Republic of North Ossetia (RNO–Alania), Russia;
- Institute of Biomedical Research of Vladikavkaz Scientific Center of the Russian Academy of Sciences and the Government of the Republic of North Ossetia–Alania, Vladikavkaz, RNO–Alania, Russia;
- Institute of biomedical research of Vladikavkaz Scientific Center of the Russian Academy of Science, RNO–Alania, Russia;
- North Ossetian State Nature Reserve, Chabahan, RNO–Alania, Russia; North-Caucasian Mining and Metallurgical Institute, Vladikavkaz, RNO–Alania, Russia;
- Station of Agrochemical Service "Northy Ossetia," Vladikavkaz, RNO–Alania, Russia;
- National Dendrological Park "Sofiyivka" of NAS of Ukraine; Uman National University of Horticulture, Uman, Cherkasy region, Ukraine;
- National University of Life and Environmental Sciences of Ukraine;
- Institute of Horticulture of the National Academy of Agrarian Sciences of Ukraine, Kyiv, Ukraine, Storozhynets district, Chernivtsi region, Ukraine;
- Emanuel Institute of Biochemical Physics, Russian Academy of Sciences, Moscow, Russia;

- North East Agricultural Research Institute, Kirov, Russia;
- Vyatka State Agricultural Academy, Kirov, Russia;
- Pryanishnikov All-Russian Scientific Research Institute of Agrochemistry, Moscow, Russia;
- Tyumen State University, Tyumen, Russia;
- N.I. Vavilov Research Institute of Plant Industry, St. Petersburg, Russia.

    —*Anatoly I. Opalko, Larissa I. Weisfeld, and Gennady E. Zaikov*

PART I

PLANT BREEDING UNDER ADVERSE CONDITIONS OF ACID SOILS

CHAPTER 1

BREEDING OF GRAIN CROPS IN EXTREME CLIMATIC CONDITIONS

GALINA A. BATALOVA, IRINA N. SHCHENNIKOVA, and
EUGENE M. LISITSYN

CONTENTS

ABSTRACT

The zone of activity of the North-East breeding center are characterized by a complex of the adverse ecological factors caused by low natural fertility of widespread podzolic soils, a variation of temperatures and non-uniformity of distribution of precipitations. On the other hand successes of breeding of varieties of intensive type considerably lowered their resistance to stressful ecological factors. Efficiency of breeding is provided, along with studying of questions of genetics of quantitative and qualitative traits, with use of selective and stressful backgrounds; with a network of ecological test for territories of Volga-Vyatka region. It allows receiving varieties with high plasticity and stability of the genotype, providing formation of yield stable on years and territories under conditions of stressful agriculture. The combination of limiting (provocative) and favorable backgrounds, laboratory and greenhouse experiments has allowed to obtain varieties of cereals characterized by tolerance and/or resistance to soil acidity and to drought: oats Faust, Dens, Krechet, Gunter, Eclips, Sapsan, and Avatars; barley Dina, Ecolog, Lel, Novichok, Pamiaty Rodinoy, Rodnik Prikamia, and Tandem. These varieties are created with use of methods as traditional breeding (hybridization, selection) and bio-technologies. The special attention in biotechnological programs is given to a combination of high potential efficiency of varieties and ability to resist to action of abiotic and biotic stressors. Methods of cellular selection are developed for reception of an initial material of barley and oats resistant against a drought and toxicity of aluminum on acid soils.

1.1   INTRODUCTION

Instability of manufacture of an oats and barley in Volga-Vyatka economic region of the Russian Federation is related with extremeness of natural-environmental conditions of agriculture on a considerable part of territory, unstable distribution of precipitations and heat on years and territory.

Successes of breeding of varieties of intensive type in last 30 years, unfortunately, have considerably lowered their resistance to stressful ecological factors that is expressed in instability of grain productivity. The spectrum of early ripening varieties and varieties of a fodder direction cultivated for green forage was simultaneously reduced in various soil-climatic territories of Russia. It specifies in necessity of expansion of researches on breeding of cereal of various groups of ripeness and directions of use taking into account quality of grain and of dry matter. Along with it, the scientific basis of a technological complex of manufacture of stable on years high crop of biologically high-grade production, ways of increase of sowing and yield properties of seeds and qualitative seeds of covered and naked varieties of oats is of great importance.

1.2 MAIN PROBLEMS OF AGRICULTURE AT NORTHEAST OF EUROPEAN PART OF RUSSIA

The zone of activity of the North-East breeding center includes the Kirov and Nizhniy Novgorod regions, the Perm Kray and republics of Mordovia, Mary El, Chuvash, Udmurt which territories are characterized by a complex of the adverse ecological factors caused by low natural fertility of widespread podzolic soils, a variation of temperatures and non-uniformity of distribution of precipitations, both in growth season, and on region territory. The average long-term temperature of air in a zone of activity of the breeding center varies from 0.6°C in the north of the Kirov region to 4.1°C in the south of Republic of Mordovia.

1.3 TYPE OF SOILS OF VOLGA-VYATKA ECONOMIC REGION

Four soil sub bands: The podzolic, sod-podzolic, gray forest-steppe and chernozem soils pass through all extensive territory of Volga-Vyatka economic region. The river Volga divides economic region into Left-bank and Right-bank parts. Sod-podzolic soils on sandy, sandy-loam, and loamy parent material of soil which are characterized by fragile structure, swell, have acid reaction, and humus content no more than 2–3% (Table 1.1) are extended basically in forest-covered and more damp Left bank. The Right bank differs with the best soils and is strongly plowed up. Here are extended sod-podzolic, gray forest and chernozem soils which are formed on coating loess-like loams and clays, contain about 3 to 8% of humus, and capacity of humus horizon reaches 20–25 sm [1].

TABLE 1.1   Structure of a Soil Cover of Territory of Activity of the North-East Breeding Center (% of Total Area of Arable Land)

Republic, Kray, region	Soil types		
	Podzolic and sod-podzolic	Grey forest	Podzolized and leached chernozems
Republic of Mordovia	8.2	42.2	43.5
Republic of Mari El	79.6	6.3	—
Chuvash Republic	25.0	55.0	18.0
Udmurt Republic	82.0	16.7	—
Permsky Kray	75.0	2.7	0.34
Kirov region	80.0	9.0	—
Nizhny Novgorod region	50.0	32.0	18.0

Sod-podzolic soils differ with superficial arable layer frequently less than 16–17 sm; at gray forest and light gray soils capacity of humus horizon and an arable layer often coincides and fluctuates within 18–24 sm. At dark gray soils capacity of humus horizon reaches 35–45 sm; at podzolized and leached chernozems – 60–75 sm.

In easy and sandy soils the content of humus fluctuates from 0.5 to 1.0%, and nitrogen – from 0.06 to 0.08%. In sod-podzolic soils of loamy mechanical structure the content of humus makes 1.5–2.0%, and the total nitrogen 0.09–0.12%. In gray forest-steppe soils the content of humus fluctuates within 1.2–3.0%, nitrogen – from 0.08 to 0.16%, and in dark gray soil – accordingly within 3.0–8.0% and 0.16–0.44%. The content of humus in podzolized and leached chernozems reaches 5–10% in an arable layer, and nitrogen – 0.20–0.50% [2].

1.4   AGRO-CLIMATIC AND GEOGRAPHICAL FEATURES OF THE REGION

The climate of the Kirov region where the North-East Agricultural Research Institute (North-east ARI) settles down is characterized by the continentality accruing in east and southeast directions, and sharpness of seasonal transitions, with long, multi-snow, cold winter [3]. Cyclones and anticyclones bring into area the Arctic air from the north, moderate sea and continental air – from the west and the east. Along with others climate-forming factors (solar radiation, character of a spreading surface) it causes moderate-continental climate with long, both multisnow and cold winter and with moderately warm summer [4].

The area is located in the North-East of the European part of the Russian Federation between 56° and 61° of northern width, in Predural, and occupies 120.8 thousand km^2 from which 2.6 million hectares makes the arable land [5]. The area is extended from the north on the south on 570 km, from the west on the east – 440 km. The big extent of area from the north on the south causes distinctions as in solar energy inflow and a temperature mode. The annual radiating balance fluctuates from 22 kcal/sm^2 in the north to 25 kcal/sm^2 in the south, mid-annual temperature – from 0.6°C to 2.76°C accordingly. The quantity of an atmospheric precipitation decreases from the Northwest for the South-East [6]. The area is in a zone of sufficient humidifying, however loss of precipitations on months and their distribution on territory is unequal, and 75–80% drop out during the warm period of year. The amount of precipitation decreases in a direction from the north to the south. The sum of precipitations per year makes 400–500 mm in the extreme south and 550–625 mm in the Northwest and the north; during growth season, accordingly, 250–300 and 320–400 mm. The high danger to agriculture of the region is represented by the droughty periods in two decades and more, marked on the average one time in five years. On the average for the warm period it is observed 20–35 droughty days, in separate years 30–60 days happen without a rain [7].

The growth season makes 155–170 days (from April, 20–29th till September, 26th – October, 8th) that is 40–60 days exceeded the period necessary for cultivation of spring grain crops. The sum of daily average temperatures of air between transition dates through 10°C in spring and in fall makes 1550–2175°C.

Territories of Kirov region are divided into three agro-climatic latitudinal zones considerably differing under natural and climatic factors: northern, central and southern.

Northern agro-climatic zone is the coldest and damp. The sum of active temperatures makes 1700°C. The mid-annual temperature of air makes 1.5–2°C. Number of frost-free days is 192–203 [5]. The period of active growth of agricultural crops consists 105–115 days. The zone territory almost completely is in a strip of superfluous humidifying.

The central agro-climatic zone is characterized by moderately warm and damp climate. The greatest quantity of precipitations is in the central part of a zone. The sum of active temperatures changes from 1700 to 1900°C. Duration of active growth of plants – 116–120 days. The central zone can be divided into the western and east areas characterized by various moisture supplies. In the western area the amount of precipitation in period of active growth of plants is sufficient, but changes on years. East area is characterized by non-uniform drop of precipitations during growth season. Dry winds and droughts are observed.

The southern agro-climatic zone of the Kirov region is on the first place on security with heat and on the last – on security with moisture; the sum of daily average temperatures above 10°C equals 1900°C. The period of active growth of agricultural crops makes 126–135 days [8].

Soil cover of the Kirov region is motley; poor podzolic and sod-podzolic soils (83% of all areas) of various mechanical structures prevail. They differ by raised acidity, the low content of humus and low capacity of compost horizon meet in all three subbands [9]. The average content of humus – 2.17%, and soils with the content of humus less than 2.1% occupy 954.1 thousand hectare (44.8%). Security of soils of area with micronutrients is low. In many areas there is a negative balance of humus, there is an irreversible process of soils de-humification. Granulometric soil structure is basically heavy, very dense, badly air – and water-permeable. In the majority they require radical improvement: liming and regular entering of organic and mineral fertilizers [10].

Real podzolic soils occupy northern areas covering middle taiga subband and northern part of a South taiga subband. Sod-podzolic soils dominate in the central part and in the south of a zone. In a southern zone of area there are more fertile light gray forest soils (9%), besides in small amounts (1–4%) in area there are sod-podzolic gley and gleyic, sod-gley, sod-carbonate, and gray forest soils [11].

The analysis of quality of farmlands shows that the steady tendency to active degradation of the soil cover caused by absence of effective measures on preservation and reproduction of soils fertility is observed everywhere in territory of the Kirov region. By results of last cycle of agrochemical inspection acid soils occupy 72.7% (1548.1 thousand hectares) of the arable land areas; 530.4 thousand hectares or 24.9% of arable land have low content of mobile phosphorus and 511.5 thousand hectares or 24% of arable land have low content of exchange potassium that limits their efficiency. The cited data characterizes fertility of soils of the region as low, corresponding to natural fertility of sod-podzolic soils [11].

The presented data testifies to different level of natural fertility of various soils of Volga-Vyatka economic region that predetermines (along with weather conditions) a corresponding set of cultures and structure of areas under crops, specificity of cultivation of agricultural plants and ways of increase of soil fertility. As a whole a climate and soils of Volga-Vyatka region correspond to agro-biological requirements of cereals to growth conditions and are favorable enough for cultivation of oats and barley for seeds, the food and fodder purposes.

Efficiency of breeding is provided, along with studying of questions of genetics of quantitative and qualitative traits, with use of selective and stressful backgrounds; with a network of ecological test for territories of Volga-Vyatka region. It allows to receive varieties with high plasticity and stability of the genotype, providing formation of yield stable on years and territories under conditions of stressful agriculture: the growth period is short and insufficiently provided with the sum of effective temperatures; low fertility and high acidity of soils of podzolic type; drought display during the various periods of plant growth; return of colds and frosts during growth season (before middle of June and after second half of August); non-uniform distribution of precipitations.

1.5 POSSIBILITIES AND SUCCESSES OF CREATION OF ADAPTIVE VARIETIES OF AGRICULTURAL CROPS IN NORTH-EAST BREEDING CENTER

1.5.1 SOME HISTORICAL FACTS

In the end of twentieth to beginning of twenty-first century oats and barley breeding in the North-East breeding center has received a new orientation. Along with productivity, resistance against pests and precocity works are spent on creation of adaptive varieties resistant and/or tolerant to edaphic stresses providing reception of economically defensible yield of qualitative production. The great attention is given to working out of theoretical bases of breeding; studying of features of biology of flowering and photosynthetic activity of oats and barley in favorable and stressful conditions; working off of methods of hybridization, principles of selection of initial material for crossing. For creation of an initial material, along with hybridization and selection, the biotechnology method is applied – reception of regenerants on rigid aluminum selective environments at selection on acid-resistance and osmotic – on drought resistance.

Working out of laboratory express methods of screening of varieties on acid- and drought resistances, differentiations of an initial and selection material, to allocation of forms contrast on stability and an estimation of efficiency of laboratory methods in field conditions is carried out. The estimation method on the root index is used in selection allowing not only to differentiate correctly selection lines, but also to conduct selection of a perspective material with obtaining of seed progeny for further breeding work on stress resistance. Target selection of oats grain for processing, including naked forms develops; breeding of fodder oats and varieties of a universal direction of use is renewed.

Level of productivity of agricultural crops is genetically determined trait; however potential possibility of a variety to give a real crop depends from soil-environmental conditions of plants growth and level of resistance of a variety to stressful ecological factors of environment. Edaphic stress caused by ionic toxicity of aluminum and manganese, related with low pH, for example, soil acidity is count as the most important economic and ecological stresses. The share of such soils all over the world makes about 40% [12]. In structure of acid soils of the Kirov region very strongly acid (pH less than 4.0), strong acid (pH 4.1–4.5) and average acid soils (pH 4.6–5.0) consist 1012.8 thousand hectares [13].

1.5.2 SOIL ACIDITY

The major factor defining toxicity of acid sod-podzolic soils of the European part of Russia is high level of the content of mobile (exchangeable) ions of trivalent aluminum. Toxicity of Al^{3+} is the leading factor reducing efficiency of plants on 67% of all acid soils [14]. Aluminum interferes with active absorption of phosphorus, competes to calcium, and inhibits division and elongation of cells of absorbing organs. The size of root system thus decreases its ability to absorb water and nutrients decreases.

1.5.3 PHOTOSYNTHESIS

The lack of nutrients directly and indirectly influences photosynthesis. In this relation studying of influence of acidity of sod-podzolic soils on the content of chlorophyll a and b, carotenoids in leaves of the upper layer and final productivity of plants [15, 16] is of interest. Edaphic stress of acid soils made essential impact on change of the area of flag and second leaves, the total area of leaves. The flag leaf has been most subject to negative influence of soil acidity. Depression of the area of flag leaf under conditions of stress in comparison with neutral soil background has made in our studies 38.5–72.9%, of second leaf 19.4–63.9%, and the total area of leaves 18.8–63.5%. Depression of chlorophyll a content in the conditions of stress has made for flag leaf 17.8%, for second leaf – 41.7%; of chlorophyll b 36.7% and 61.3%, of carotenoids 7.7% and 33.9%; for the sum of a chlorophyll $a + b$ 22.1% and 45.9% accordingly. Content of carotenoids and ratio of chlorophyll/carotenoids in flag leaf has rendered the greatest influence on productivity of oats plants in the conditions of stress ($r = 0.77$ and $r = 0.80$ accordingly); chlorophyll a and chlorophyll b contents in flag ($r = 0.69$ and $r = 0.78$) and second ($r = 0.53$ and $r = 0.78$) leaves. Similar influence of pigment content in flag and second leaves is noted on number of grains per panicle.

1.5.4 ALUMINUM RESISTANCE

Despite the fact that selection on resistance to biotic and abiotic factors, as a rule, leads to decrease in potential productivity in non-stressful environmental conditions, creation of varieties with a combination of the given traits is obviously possible. The researches spent on the large set of oats of world collection of All Russia Institute of Plant Industry (VIR, St. Petersburg, Russia) have shown that level of aluminum resistance does not depend on a place of origin of a variety, and obtaining of resistant forms is possible among samples of any origin [17]. Breeding on drought resistance becomes complicated with absence of the sources combining high productivity and stress-resistance [17]. The analysis of a genetic variability on studied trait and search of new genetic resources for carrying over of desirable trait is represented actual.

1.5.5 RESULTS BREEDING OF OATS

The combination of limiting (provocative) and favorable backgrounds, laboratory and greenhouse experiments has allowed to obtain in 1992 the selection lines of oats characterized by tolerance and/or resistance to soil acidity and to drought. High-yielding (8.5–9.0 t/ha) varieties Faust (tolerant to soil acidity and to drought) and Dens (early-maturing and drought-resistant) have been transferred to State Test in 1999 and in 2002 are included in the State Register. After that the plastic variety Krechet combining productivity up to 9.1 t/ha with resistance to toxicity of acid soils caused by aluminum ions has been created and admitted to use in agricultural industry since 2005. Varieties Gunter and Eclips are characterized with high grain (up to 11.2 t/ha) and productivity fodder are suitable for cultivation on grain-hay technologies and are included in the State Register in 2007 and 2012, accordingly. In 2009, variety Butsefal of a universal direction of use is transferred to the State Test: it is high-yielding on grain (up to 8.7 t/ha) and on dry matter (up to 10.6 t/ha); adaptive, capable to form high stable grain yields at various ecological-geographical points; with high nature of the grain; poorly defeat with loose smut; resistant against damage by the Swedish fly. In 2012 works are finished on breeding of adaptive to biotic and abiotic ecological factors covered varieties of oats Sapsan and Avatars of a universal direction of use (on grain and green mass).

1.5.6 INHERITANCE OF VALUABLE TRAITS

Despite a variety of methods of breeding, along with selection and hybridization we use mutagenesis, post-genome technologies, genetic transformation; a basis of success is knowledge and understanding of inheritance and preservation in progeny of the traits defining productivity of a variety.

Insufficient knowledge of genetics of quantitative and qualitative traits in selection leads to necessity of a considerable quantity of crossings. At crossing of the individuals differing on quantitative traits, dominance of trait of one of parents is not always observed, and in second generation of hybrids there is not accurate segregation on a small number of phenotypically different classes. It complicates carrying out of screening after crossing as genetic variability intertwines with the ecological in progeny. The success of selection depends both on effect of action of genes, and on character and degree of inheritance of trait of interest [18].

1.5.7 TRANSGRESSIONS

The transgression phenomenon is considered important in the selection when at crossing of the organisms different from each other on quantitative expression of a certain trait in hybrid progenies there are stable forms with much stronger or weaker expression of a corresponding trait than it was in initial parental forms. It occurs when one or both parental forms do not possess extreme degree of expression of a trait, which the given genetic system can give.

In connection with the progress in breeding of grain crops the set of highly productive forms, which can be used as components of crossing for reception of transgressions on the basis of high grain efficiency or on separate components of efficiency, has considerably extended. It also complicates somewhat the problem of a choice of the best component. For the most economic and exact selection of components of crossing application of quantitative methods of an estimation of traits of an initial material is necessary [19].

1.5.8 THE USE OF BIOTECHNOLOGY IN BREEDING

For increase of efficiency of selection of cereal crops introduction of the new biotechnological methods allowing to design the genotypes on the basis of cellular engineering is necessary. One of such methods in creation of new forms of plants is reception of somaclonal variants in callus culture. Somaclonal changes arise in process of cultivation of isolated cells and tissues as a result of mutations, a various expression of genes, and somatic crossing-over [20, 21].

The variability arising in vitro is not always adaptive: separate traits can change as towards increase, and fall of values in comparison with an initial variety [22]. Somaclonal variability gets adaptive advantages, if selection is possible to carrying out in in vitro system. For selection of resistant forms at cellular level selective media are used which simulating natural stressful conditions provide an expression of a trait of resistance and give the chance to select the necessary variants.

1.5.9   SUCCESSES IN BREEDING OAT AND BARLEY

Complex researches on studying of an initial material on breeding valuable traits, obtaining of sources and donors, an estimation of combinational and adaptable possibilities of a material involved in selection provide high efficiency of researches in the field of barley and oats breeding in the North-East Agricultural research Institute (Kirov, Russia). During activity of the breeding center it is created more than 100 varieties of different agricultural crops which most part was successfully cultivated, in due time, on fields of Russia and the countries of the former Soviet Union. Now 188 varieties of spring barley are included in the State Register of protected selection achievements of the Russian Federations admitted in manufacture and 109 varieties of oats; some of them are bred in North-East Agricultural research Institute: 8 barley varieties (Dzhin, Dina, Ecolog, Lel, Novichok, Pamiaty Rodinoy, Rodnik Prikamia, and Tandem) and 10 oats varieties (covered oats Argamak, Dens, Eclips, Fakir, Faust, Gunter, Krechet, and Teremok, naked oats – Persheron and Vyatsky).

These varieties are created with use of methods as traditional breeding (hybridization, selection) and biotechnologies. The special attention in biotechnological programs is given to a combination of high potential efficiency of varieties and ability to resist to action of abiotic and biotic stressors. Methods of cellular selection are developed for reception of an initial material of barley and oats resistant against a drought and toxicity of aluminum on acid soils. Methods are improved of creation of rigid selective nutrient mediums with pH 3.8 and concentration of ions of aluminum up to 40 mg/L. The structure of organogenic media is modified that has allowed to achieve reception in high quantity of plant-regenerants of barley and oats resistant against aluminum.

The first aluminum-tolerant barley variety Novichok is created with use of selective medium and earlier received variety – dihaploid Duet. On acid soils with pH 4.0 and the content of aluminum up to 7.9 mg/100 g of soils variety Novichok exceeds the standard on 10.9% at productivity of 5.6 t/ha. The variety has field resistance to black and covered smut, is characterized with high tillering capacity. Dihaploid variety Tandem is received with use of wild bulbous barley as haplo-producer. Varieties Ecolog, Pamiaty Rodinoy, and Rodnik Prikamia are created which have high productivity with a complex of economic valuable traits.

KEYWORDS

- additive effects
- alleles
- correlation
- dominance
- hybrids
- oats

REFERENCES

1. Program of breeding activity of breeding center of North-East Agricultural research Institute till 1990, Kirov: North-East Agricultural Institute, 1976, 117 p. [in Russian].
2. System of agricultural industry in Volga-Vyatka zone. Crop farming and plant industry. Kirov: Vyatskoe Book Publishers, 1976, 352 p. [in Russian].
3. Molodkin, V. N. Fertility of arable soils of Kirov region as at 01.01.2007. Basic direction of improvement of crop farming system of Kirov region. Kirov: North-East Agricultural Institute, 2007, 91–94. [in Russian].
4. Frenkel, M. O. Climate. Nature, economy, and ecology of Kirov region. Kirov: Vyatka State Humanitarian University, 1996, 115–135. [in Russian].
5. Tyulin, V. V., Roslyakov, N. P. Soil resources and rational use of soils in Kirov region. Intensification of agrarian industry in Kirov region. Perm: Perm Book House, 1980, 3–10. [in Russian].

6. Ecological safety of a region (Kirov region at the turn of the century). Kirov: Vyatka, 2001, 416 p. [in Russian].

7. Tyulin, V. V., Kopysov, I.Ya. Soil estimation and effective use in North-East of Non-Chernozem Zone. Kirov: Vyatka State Agricultural Academy, 1994, 161 p. [in Russian].

8. Agro-climatic characteristic of Kirov region. Kirov: Zonal hydro-meteoobservatory, 1970, 36 p. [in Russian].

9. Shikhova, L. N., Egoshina, T. L. Heavy metals in soils and plants in Northeast of European part of Russia. Kirov: North-East Agricultural Institute, 2004, 262 p. [in Russian].

10. Lisitsyn, E. M., Batalova, G. A., Shchennikova, I. N. Dynamics of sowing areas and productivity of barley and oats in different regions of European Russia having acid sod-podzolic soils. Creation of varieties of oats and barley for acid soils. Theory and practice. Palmarium Academic Publishing, Saarbrucken, Germany, 2012, 11–28. [in Russian].

11. On state of environment of Kirov region in 2007 (Regional report). Ed. Perestoronin, V. P. Kirov: Triada plus Ltd., 2008, 51–52. [in Russian].

12. Delhaize, E., Ryan, P. R., Hebb, D. M., Yamamoto, Y., Sasaki, T., Matsumoto, H. Engineering high-level aluminum tolerance in barley with the *ALMT1* gene. Proc. Natl. Acad. Sci. USA. 2004, 15249–15254.

13. On state of environment of Kirov region in 2011: Regional report. [Ed. A. V. Albegova]. Kirov: Staraya Vyatka Publisher Ltd. (Old Vyatka – in Rus.), 2011, 185 p. [in Russian].

14. Eswaran, H., Reich, P., Beinroth, F. Global distribution of soils with acidity. Brazilian Soil Science Society. 1997, 159–164.

15. Batalova, G. A. Oats, technology of cultivation and breeding. Kirov: North-East Agricultural Institute, 2000, 206 p. [in Russian].

16. Gubanova, A. S., Batalova, G. A. Photosynthetic activity of oats in condition of edaphic stress of sod-podzolic acid soils. Business. Science. Ecology of native land: problems and ways of its decision. Kirov: Vesi Publ., 2013, 102–104. [in Russian].

17. Batalova, G. A., Lisitsyn, E. M., Rusakova, I. I. Biology and genetics of oats. Kirov: North-East Agricultural Institute, 2008, 456 p. [in Russian].

18. Boroevitch, S. Principles and methods of plant breeding. Moscow: Kolos, 1984, 344 p. [in Russian].

19. Fedin, M. A., Silis, D.Ya., Smiryaev, A. V. Statistical methods of genetic analysis. Moscow: Kolos, 1980, 207 p. [in Russian].

20. Larkin, P. J., Scowcroft, W. R. Somaclonal variation – a noval source of variability from cell culture for plant improvement. Theor. Appl. Genet. 1981, Vol. 60. 197–214.

21. Karp, A. Somaclonal variation as a tool for crop improvement. Euphytica. 1995, Vol. 85. 295–302.

22. Shayakhmetov, I. F. Somatic embryogenesis and breeding of cereal crops. Ufa: Bashkir State University press, 1999, 166 p. [in Russian].

CHAPTER 2

GENETICS OF QUANTITATIVE TRAITS OF PRODUCTIVITY AND QUALITIES OF GRAIN OF OAT *AVENA SATIVA* L.

GALINA A. BATALOVA and EUGENE M. LISITSYN

CONTENTS

ABSTRACT

For hybrids from crossing of oats covered varieties Argamak, E-1643, Ulov (Russia), Freja, Petra (Sweden) with naked oat variety Torch (Canada) it is established that additive effects prevailed in F_1 in the genetic control of traits "plant height," "number of spikelets per panicle," "number of grains per panicle," "grain mass per panicle" and "1000 grains mass"; effects of overdominance prevailed in F_2, except of a trait "plant height". Additive effects prevailed in the control of all traits in F_3 that testifies to possibility of effective screening of lines with breeding valuable properties of genotypes. Covered hybrids Freja x Ulov, E-1643 x Ulov, E-1643 x Argamak, Ulov x Argamak, and naked hybrids Freja x Torch, Torch x Freja, Torch x Argamak had the best combination of some breeding valuable traits. The greatest number of transgressive forms has been segregated in combinations with participation of varieties E-1643, Ulov, Petra, and Freja. Significantly high mass of grain per panicle (productivity) in comparison with parental naked variety Torch (1.26–0.91 g) was observed in the first – the third generations of naked hybrids Freja x Torch (2.64–1.56 g), Torch x Freja (1.65–1.44), and Torch x Argamak (2.20–1.35).

2.1   INTRODUCTION

Considerable part of breeding valuable traits of plants is quantitative. Their hereditary distinctions are caused by interaction of several pairs the polymeric genes, thus each gene makes essential impact on development of the given trait or quantitative traits are more strongly dependent on external factors, than qualitative traits. Selection on quantitative traits in early generations of hybrid populations is complicated because genetic variability in progeny is combined with ecological; in F_2, for example, there is no accurate segregation on small number of phenotypic different classes. The success of selection depends both from effect of action of genes and from character and degree of inheritance of those trait on which selection was led [1].

There are some approaches to the description of inheritance of quantitative traits. The *phenomenological* approach unites set of methods and techniques of the description of inheritance, which do not contain enough accurate data on the mechanism of inheritance of a trait. Methods of mathematical statistics allow to estimate possibilities of selection and to predict its results not less correctly than it can be made by means of indexes of heritability and genetic correlations [2].

Principles of Mendel's genetics are use widely enough at the description of inheritance of quantitative traits [3, 4]. Thus the clear boundary between quantitative and qualitative traits does not exist they differ among themselves both on character of a variation, and on phonotypic variability within a specie [5–7]. Simple quantitative signs concern: "duration of the period germination – leaf-tube formation," "plant height," "length of an ear," "number of spikelets per ear," "1000 grains mass," etc., variability at which does not exceed 10% [8, 9]. The *genetic analysis* assumes reception of such knowledge of trait genetics, as number of genes and their alleles, localization of these genes in linkage groups, survival rate of genotypes, a way of phenotypic realizations of a genotype, interaction of genes, and environmental modification of a trait.

Genetic-and-biometric approach is widespread enough at studying of inheritance of quantitative traits too; it is related with the analysis of components of expansion of phenotypic and genotypic variance of a trait and their using in selection. Indexes of heritability [10] characterize a share of genetically caused variation within the total variation of a trait. It is accepted to distinguish heritability in broad and narrow sense of a word, understanding as last a share of additive variance in total phenotypic heritability. Heritability indexes in the broad sense of the word characterize independence of phenotypic display of a trait from a variation of environmental conditions. Heritability indexes in narrow sense characterize genetic heterogeneity of populations. Estimation of coefficient of heritability in the narrow sense of the word, characterizing the additive effects of genes transferred from parents to progenies is of great value [11]. Despite existing lacks, heritability coefficients are the simplest parameters allowing to judge roughly reliability of selection of the best genotypes by phenotypes in a certain situation.

Before revealing the phenomenon of redetermination of the genetic organization of a quantitative trait the conception of polygene inheritance by Kenneth Mather predominated in genetics of quantitative traits, created on the basis of Mendel's model [12]. The given concept assumed that environmental variability of a trait is based on change of activity of genes in a stable polygene system of a trait, and any fluctuations of environmental conditions generate ecological variability, changing only the activity of loci in the polygene system, constant by its gene set [13].

In 1984 a new model of the organization of a complex quantitative trait of plants [14] is published. Unlike methods of "*the genetic analysis*," the model of the ecological-genetic control studies not just the traits but six genetic-physiological systems by means of which the breeders can improve species on complex quantitative properties of productivity [13, 15]. The phenomenon of redetermination of the genetic formula of quantitative trait radically changes former representations about rigid unequivocal determination of complex traits. Two phenomena leading to loss of stability on a way "products of genes – a quantitative trait" are established: change in the spectrum of the loci determining the level and genetic variability of a quantitative trait at change of the limiting factor and change in the spectrum of modules [16, 17]. The module is an elementary unit of the description of the organization of a quantitative trait consists of three interrelated traits: one resultant and two componential. They are arranged in "pyramids" of modules. At change of external limits componential modules change their contributions into a final resultant trait [15]. At increase of level of a trait in hierarchical system of modules degree of its variability increases too.

Revealing and the account of the basic laws of display of correlation dependences between productivity traits in various ecological and cenotic conditions allows to raise efficiency of breeding of new varieties [13, 15, 18]. In process of reduction of values of ecological intravarietal correlation among modules degree of influence of abiotic factors on display of these correlations increases. Softening of pressure of limiting factors strengthens considered correlations [19].

The new model explains why and how parameters of genetic inventory will move from one environment by another [13], gives the chance to identify a genotype of a single organism on its phenotype without test of its progeny. However, to studying of genetic features of self-pollinated lines a method of diallel crossings [20–22] is applied more often. The diallel analysis allows to check up additive-dominant model – to reveal presence of non-allelic interactions and to estimate components of a genetic dispersion; gives the exact information for given exact dynamics of environmental limiting factors. Complexity of an establishment of laws of inheritance of quantitative trait consists that separate genes can possess different degree of dominance up to overdominance. Actions of separate polygenes can be not equal each other, and sum action of several genes can be not only simple additive but also become complicated with various forms of intergene interaction. In one cases some genes weaken action of others, and in others, on the contrary, strengthen it. All of these are reflected in segregation in F_2 generation [23].

Additive genetic effects are most important in plant breeding. Not additive genetic effects cannot be used in selection directly as a lot of time is required to achievement of full homozygozity of forms. Thus it is important to define presence of positively operating genes (increasing a trait) and to spend selection depending on dominant or recessive type of their inheritance [24]. In a number of works with oats the contribution of additive genetic variance into the total one is shown. In one cases prevalence of an additive part over not additive is shown for traits "grain mass per panicle," "number of grains in a panicle," "number of panicles per 1 m²," "panicle lengths," "number of spikelets per panicle," "number of branches per whorl," "number of whorls and branches in a panicle" [25], in others cases – for traits "1000 grains mass," "panicle length," "number of spikelets in panicle" and overdominance for trait "grain mass per plant" [26].

There are indications on presence of non-allelic interactions in inheritance of such traits as "plant weight," "diameter of a stem," "1000 grains mass," "number of grains per panicle," "grain mass per panicle," "yield per 1 m²" [27]. It is shown strong and constant non-allelic interaction for a trait "grain mass per plant," but for other traits (such as "length of a culm," "number of spikelets per plant," and "plant height") non-allelic interaction had very small value [28].

It is noticed that epistatic variance is more important component of genetic variance of yield of oat near-homozygous lines then additive variance [29]. At prevailing epistatic action of genes relation between display of trait at parents and progeny in late generations is weak or is absent. Populations F_2 and F_3 cannot contain enough number of stable epistatic forms to determined advantage of homozygous progeny in the subsequent generations. There is an opinion that screening of hybrid populations on productivity and elements of yield structure in early generations is impractical as effects of dominance prevail in the genetic control of these traits [30, 31].

Development and expressiveness of a quantitative trait appreciably depend on action of environmental factors, which can modify effect of polygenes. The quantitative trait is formed under interaction of various genetic systems and fluctuating environmental factors. As a result at identical value of a final trait componential traits have the diversified values as for the same variety in different conditions of growth and for different varieties in the same growth conditions [32, 33]. Thus, Diallel analysis gives the exact information only for given exact dynamics of limiting environmental factors; in other environment these characteristics can become others.

This character of display of quantitative trait complicates carrying out of selection after crossing as genetic variability will intertwine with the ecological variability in progeny. The success of selection depends both on effect of action of genes and from character and degree of inheritance of those trait on which selection is led [34].

At creation of valuable genotypes of oats in the conditions of Volga-Vyatka region of Russia the most effective was selection by traits "number of grains per panicle," "grain mass per panicle" and more rare by trait "1000 grains mass". Oats variety Kirovsky (Russia) and Ryhti (Finland) contain mainly dominant alleles for "grain mass per panicle," "grain mass per plant," and "1000 grains mass" but recessive alleles for grain number. Trait "grain mass per plant" in variety Cravache (France) is strongly influenced with recessive alleles but traits "number of grain in panicle," "1000 grains mass," "grain mass per panicle," and "length of panicle" – with dominant alleles. Variety Astor (the Netherlands) has approximately identical ratio of dominant and recessive alleles for the majority of traits [31].

The aim of study is to establish character of inheritance of the basic yield-forming traits and indexes of grain quality in spring oats (*Avena sativa* L.).

2.2 MATERIAL AND METHODOLOGY

The territory of the Kirov region is located in the northeast of the European part of the Russian Federation between 560 and 610 northern widths, in Predural, and occupies about 120.8 thousand km². A region climate is moderate-continental with long, multisnow cold winter and moderately warm summer. The growth period makes 155–170 days (from April, 20–29th till September, 26th October, 8th) that exceeds the period necessary for cultivation of spring grain crops for 40–60 days. The sum of active temperatures of air for a growth season makes 1550–2175°C. A soil cover is mosaic with 72% of poor sod-podzolic loamy soils; about 15% make sandy and sandy-loam soils of various mechanical structure. The average agrochemical parameters of soils make: humus 2–3%, pH_{KCl} = 5.0, the content of mobile phosphorus – 119 mg/kg, exchangeable potassium – 120 mg/kg. As a whole a climate and soils are favorable enough for oats cultivation for the seed-growing and food purposes.

Crossings are spent in the North-East Agricultural Institute (Kirov, Russia) on full diallel scheme with use of five covered varieties of oats (*Avena sativa* L.) – Argamak, Ulov, E-1643 (Russia), Petra, Freja (Sweden), and one variety of naked oats (*Avena nuda* L.) – Torch (Canada) (Table 2.1).

At first set of experiments F_1 hybrids and their parental forms were sowed in the field conditions in triple replications by rows in plots of 1 m width. Width of a row-spacing was 15 sm, distance between seeds in rows was 5 sm. Sowing of hybrids was spent under scheme $P_1 F_1 P_2$.

At second set of experiments parental forms as well as F_2 and F_3 hybrids were sowed on 1 м² and 2 м² plots in triple replication.

At sowing of F_2 and F_3 progenies grain of direct and reciprocal hybrids from crossing with naked variety Torch was divided on naked and covered and was sowed separately (covered and naked forms).

TABLE 2.1   The Scheme of Crossings

Maternal genotype	Paternal genotype					
	E-1643	Petra	Argamak	Ulov	Freja	Torch
E-1643		+	+	+	+	+
Petra	+		+	+	+	+
Argamak	+	+		+	+	+
Ulov	+	+	+		+	+
Freja	+	+	+	+		+
Torch	+	+	+	+	+	

After harvesting plants were analyzed in laboratory by following traits: "plants height," "panicle length," "number of spikelets per panicle," "number of grains per panicle," "mass of a panicle," "grain mass per panicle," and "1000 grains mass".

The Diallel analysis on elements of yield structure and on biochemical parameters were spent under incomplete scheme [20] with use of software AGROS 2.07. The average data on direct and reciprocal combinations of crossing were used for the analysis because these types of hybrids differed significantly. Statistical data processing was spent with use of software STATGRAPHICS plus for Windows 5.0.

2.3   RESULTS AND DISCUSSION

2.3.1   PLANT HEIGHT

An intraspecific variety of genus *Avena* includes high enough potential of variability on plant height. The height of oats plants except genetic determinacy is subject to considerable variability in dependence of cultivation conditions. At crossing of varieties differing on height transgressions are possible [35]. There are data that inheritance of the given trait has polygene character [36]. In our researches incomplete dominance took place in the control of a trait "plant height" in F_1 generation. Recessive genes prevail in a studied set of varieties. The majority of hybrids turned aside taller parent by the trait, for example, the genes increasing height of a plant were dominant. The high negative coefficient of correlation between average values of height of a stem of each variety and degree of dominance ($r = -0.85$) also testifies to this. The greatest number of dominant genes had varieties Freja and Torch. The great number of recessive genes causing a low stem was in varieties Argamak and Ulov (95% and 75% accordingly). In hybrid Freja x Ulov in F_1 generation heterosis on plant height is noted.

The hypothetical variety, which includes all dominant alleles, available for studied samples, will have height 92.83 sm. This parameter corresponds to naked variety Torch, which had made average expressiveness of the trait equal to 95.57 sm. Thus, the given variety possesses all dominant genes available for a studied set of varieties. In covered varieties Argamak and Ulov the trait was controlled mainly by recessive genes.

In the second generation of hybrids additive effects of genes prevailed in the control of a trait "plant height". It gives the basis to assume that the most part of genetic variability on the given trait between parental varieties should be highly inherited. Dominant and recessive alleles are distributed non-uniformly. Dominance was not unidirectional: correlation between expressiveness of a trait at parental forms and number of dominant genes was low negative ($r = -0.47$). In the majority of varieties dominant genes prevailed. In the control of a trait in variety Ulov having lower stem in comparison with other varieties the greatest number of recessive genes is noted (70%), in variety E-1643, in contrary, greatest number of dominant genes (90%). Hybrids with participation of varieties E-1643 and Ulov had higher stem in comparison with their initial forms, hence, dominance has been directed towards increase in trait value. In varieties Argamak, Freja, and Torch dominant genes were responsible for smaller value of the trait.

In F_3 generation additive effects of genes prevailed in the genetic control of the trait. Dominant genes operated towards reduction of stem height and effects of dominance were mainly monodirectional. Correlation between degree of dominance and average value of the trait was average positive ($r = 0.52$), hence, dominance is directed towards reduction of the trait value. In variety Petra the greatest number of dominant genes and the least height of a plant in comparison with other varieties is established. Varieties E-1643 and Torch with prevalence of recessive genes had a high stems. Value of the trait in variety Torch has made 101.0 sm that is 5.8 sm higher then value for hypothetical completely recessive parent (95.2 sm).

2.3.2 PANICLE LENGTH

The given trait in large part is related with plant height. According to the data obtained in All Russia Institute of Plant Industry (VIR, St. Petersburg), at the studied genofund of oats the length of a panicle of short-stem varieties varied in the conditions of the north-west of Russian Federation from 4.5 to 16.7 sm, at long-stem varieties from 20.1 to 29.1 sm. However, accurate laws of influence of height of plants on panicle length has not been revealed [35].

In inheritance of value of trait "panicle length" in the studied group of varieties overdominance prevailed at average degree of dominance in each locus. Dominant and recessive alleles defining the trait are distributed non-uniformly between parental forms, and dominance is directed towards increase of a studied trait. Correlation between expressiveness of the trait and dominance is average negative ($r = -0.47$). The greatest number of dominant alleles had parental form Petra. The panicle length in variety Torch was more than in hypothetical completely dominant parent.

Effects of overdominance prevailed in the genetic control of the trait in F_2 generation. The difference between average value of hybrids F_2 and total average mean of initial forms ($ml_1 - mL_0$) is equal 0.39 hence dominance is directed towards increase in panicle length. The negative coefficient of correlation between expressiveness of the trait and dominance ($r = -0.64$) testifies the same. Distribution of alleles with positive and negative action was non-uniform, dominant alleles prevailed a little. The greatest number of the dominant genes increasing length of a panicle had variety Freja; but recessive alleles prevailed in variety Ulov.

Additive effects prevailed in definition of value of the trait "panicle length" in F_3 generation. Average degree of dominance in each locus has made 0.38. Dominance has been directed towards decrease in length of a panicle (an index $mL_1 - mL_0 < 0$). The greatest number of recessive genes had variety Torch. The small number of dominant genes in variety Torch has caused presence of long panicle. A considerable number of dominant alleles and the shortest panicle are noted in hybrids with participation of covered variety Ulov as a parental form. Correlation coefficient between expressiveness of the trait and degree of dominance was high positive ($r = 0.87$) that specifies ability of recessive genes to increase the value of the trait.

2.3.3 NUMBER OF SPIKELETS PER PANICLE

Degree of dominance by trait "number of spikelets per panicle" has made 0.86 in F_1 generation that testifies to prevalence of additive effects in definition of value of the trait. In varieties E-1643, Freja, and Ulov recessive genes prevailed, but in varieties Argamak, Petra, and Torch – dominant genes. Overdominance was noted in control of the trait in F_2 generation $\left(\sqrt{H_1 / D} = 2.764\right)$, directed towards decrease in trait value

$(ml_1 - mL_0 = -0.90)$. Correlation coefficient between expressiveness of the trait and degree of dominance is moderate positive ($r = 0.60$) that specifies in influence of recessive alleles in trait express. Variety Freja with the greatest number of recessive alleles in comparison with other varieties had number spikelets per panicle at the level of hypothetical completely recessive parent – 27.4 pieces. As the given variety possesses extreme degree of expressiveness of a studied trait the probability of obtaining of positive transgressions on number of spikelets in the subsequent generations of hybrids with its participation is low.

In F_3 generation incomplete dominance is noted $\left(\sqrt{H_1 / D} = 0.49\right)$; recessive genes caused increase in value of the trait ($r = 0.64$). Variety Petra possessed the greatest number of recessive alleles; the number of spikelets per panicle was made 39.2 at 48.7 in completely recessive parent. Hence, it is possible to expect occurrence of a transgressive combination of polygenes of additive action that is display of stronger expression of the trait in hybrids with participation of variety Petra.

2.3.4 NUMBER OF GRAINS PER PANICLE

Earlier we established in covered oats that the number of grains is inherited as dominant character [36]. For interspecific hybrids the great number of new recombinants and positive transgressions [37, 38] has been noted.

In the first generation of the hybrids created with use of covered and naked parental varieties it is revealed mainly unidirectional incomplete dominance $\left(\sqrt{H_1 / D} = 0.53\right)$ on the given trait; coefficient of correlation between value of the trait and degree of dominance was low with a negative sign ($r = -0.44$). In varieties E-1643, Torch, and Ulov recessive genes prevailed; but in varieties Argamak and Petra – dominant genes. At the analysis of F_1 hybrids it is established that variety Petra had number of grains higher then hypothetical completely dominant parent – 74.3 (completely dominant parent – 63.7).

In F_2 generation in the genetic control of the trait "number of grains per panicle" overdominance took place $\left(\sqrt{H_1 / D} = 2.16\right)$ directed towards reduction of trait value ($ml_1 - mL_0 = -0.73$). The ratio of recessive and dominant alleles is not equal, but comes nearer to that. The correlation coefficient between a level of development of the trait and dominance degree has negative sign and testifies to high level of relation of an index and number of dominant alleles. The greatest number of dominant genes (approximately 80–90%) and the greatest expressiveness of the trait, as well as in F_1, are noted for varieties Argamak and Petra – 49 and 47 pieces at 51.2 for completely dominant hypothetical parent.

Negative values of the parameters estimating dominant effects have been received in F_3 germination. It specifies in the considerable contribution of additive effects to formation of the trait. As a whole dominant alleles prevailed ($F > 0$), in all varieties they operated towards reduction of number of grains, except for variety Petra. Plants of variety Petra having the greatest number of recessive genes had the greatest value of the studied trait – 63.4 (in completely recessive hypothetical parent – 66.7 pieces).

2.3.5 GRAIN MASS PER PANICLE

Studies have shown that additive effects of genes prevail in the genetic control of a trait in F_1 generation. Parameters H_1, H_2 and F had negative values that specifies in prevalence of recessive genes. Coefficient of correlation between average values of "grain mass per panicle" and of dominance is equal 0.87, hence, dominant genes defined increase in trait value. Variety Petra with maximum number of dominant alleles for the given set of varieties had the greatest "grain mass per panicle" (productivity of a panicle) – 2.99 g.

In F_2 generation in genetic definition of a trait the dominant effects which action is directed towards increase of value ($r = -0.72$) prevailed. Varieties Argamak and Petra had the greatest number of dominant alleles. The greatest number of recessive alleles in F_2 generation is noted in variety Ulov.

In F_3 generation incomplete dominance was the main in the control of a trait; degree of dominance in each locus $\left(\sqrt{H_1 / D}\right)$ has made 0.55. The small "grain mass per panicle" is caused by dominant genes ($r = 0.85$). Variety Petra was characterized with the greatest number of recessive alleles with positive action. Variety Freja at which ratio of dominant and recessive genes was equally 50:50 was the second by number of recessive genes.

2.3.6   1000 GRAIN MASS

For trait "1000 grains mass" there was established positive dominant alleles, which operate towards increase of trait value. There is data that high heritability and stability on years of the trait allows to conduct selection in early generations (beginning with F_2) [25].

In our study effect of a complete dominance presented in the genetic control of "1000 grains mass" in F_1 generation. Recessive alleles prevailed in the majority of loci ($F = -4.22$). Dominant and recessive alleles determining trait value are distributed non-uniformly between parental varieties. Relation between expressiveness of a trait and degree of dominance was absent ($r = -0.13$) that specifies in presence of dominant genes both for low and for high mass of 1000 grains. Variety Petra possessed the greatest number of dominant alleles. In varieties E-1643, Freja, Torch, and Ulov the trait controlled mainly by recessive genes. In inheritance of trait "1000 grains mass" in F_2 generation in investigated group of varieties effects of overdominance ($H_1/D = 5.32$) (Table 2.2) prevailed.

Distribution of alleles with positive and negative action was non-uniform ($H_2/4H_1 = 0.19$). Dominance is directed towards increase in a value of studied trait ($ml\ 1 - mL\ 0 = 0.97$). Coefficient of correlation between average values "1000 grains mass" and degree of dominance ($r = -0.50$) specifies in existence of relation of an average level between expressiveness of a trait in varieties and presence of dominant alleles in them. The greatest number of dominant alleles is noted in naked variety Torch, covered varieties Freja and Petra. Recessive genes prevailed in variety Ulov.

TABLE 2.2   Degree of Dominance (H_1/D) for Traits of Yield Structure in F_1–F_3 generations of oats

Trait	Degree of dominance (H_1/D)		
	F_1	F_2	F_3
Plant height	0.62	0.49	0.22
Length of a panicle	1.54	1.89	0.14
Number of spikelets per panicle	0.75	7.62	0.24
Number of grains per panicle	0.28	4.65	-0.13
Grain mass per panicle	-0.33	4.84	0.30
1000 grains mass	0.58	5.32	0.58

Variety Freja had "1000 grains mass" equal to 32.86 g and approximately 75% of dominant genes determining the given trait.

Additive effects of genes ($H_1 < D$) prevailed in the genetic control of a trait in F_3 generation. Average degree of dominance in each locus has made 0.79 that testifies to almost complete dominance. Dominant genes caused decrease in value of a trait ($r = 0.77$). Varieties with prevalence of recessive effects Freja and Petra had the greatest expression of a trait – 36.96 and 35.64 g at 45.78 g in hypothetical completely recessive parent. Varieties Argamak and E-1643 had 50% of dominant and 50% of recessive genes. The additive combination of dominant alleles at crossing of these varieties has caused positive transgression by trait "1000 grains mass".

Estimation of coefficients of heritability characterizing the contribution of an additive part of genotypic variance has shown that highest heritability had traits "plant height" (0.70–0.82), "panicle length" (0.41–0.74) and "1000 grains mass" (0.55–0.75) (Table 2.3).

Low values of coefficients of heritability for traits "number of spikelets per panicle," "number of grains per panicle" as well as for "grain mass per panicle" and the different additive effect varying on years specify in complex character of realization of a genotype of variety on traits of productivity under various environment conditions and difficulty of selection of the necessary genotypes.

The Diallel analysis with use of the studied varieties has shown that additive effects ($H_1 < D$) prevailed in the genetic control of all traits except

TABLE 2.3   Heritability Coefficients of Traits of Yield Structure in F_1–F_3 Generations of Oats

Trait	Heritability coefficient					
	In narrow sense h^2			In broad sense H^2		
	F_1	F_2	F_3	F_1	F_2	F_3
Plant height	0.78	0.82	0.70	0.82	0.86	0.72
Length of a panicle	0.41	0.69	0.74	0.54	0.87	0.75
Number of spikelets per panicle	0.28	0.14	0.45	0.36	0.75	0.45
Number of grains per panicle	0.10	0.28	0.28	0.22	0.79	0.19
Grain mass per panicle	0.30	0.17	0.66	0.30	0.73	0.70
1000 grains mass	0.67	0.55	0.75	0.71	0.79	0.79

"panicle length" in F_1 generation, but effects of overdominance – in F_2 generation with an exception of "plant height".

As overdominance took place in the control of considered traits in F_2 generation then selection on elements of yield structure except "plant height" will be inefficient in early generations of hybrids owing to high level of heterozygozity. Additive effects prevailed in the control of all traits in the third generation that testifies to possibility of carrying out of selection in F_3.

A close relation is established between number of dominant alleles in parental varieties and average expression of a trait in them at analysis of traits "plant height," "panicle length" and "grain mass per panicle" in F_1 and F_2 generations. Recessive genes influenced inheritance of traits "number of spikelets per panicle" and "number of grains per panicle". Dominant genes participated in determination of "1000 grains mass," both for high and for low value of a trait (correlation was weak negative).

Dominant genes reduced value of all studied traits in F_3, but coefficients of correlation were the average or high positive.

Distribution of positive and negative alleles was non-uniform as a whole, in F_2 came nearer to uniform for traits "panicle length," "number of spikelets per panicle," "number of grains per panicle," "grain mass per panicle," and "1000 grains mass".

Such traits as "plant height" and "panicle length" were controlled by recessive genes increasing degree of display of a trait in F_2 and F_3 generations in naked variety Torch.

At the analysis of F_1 and F_2 hybrids it has been established that varieties E-1643 and Freja had the higher plant height and length of a culm and the dominant genes defining these signs. Low-stem and shorter panicle in variety Argamak was controlled by recessive alleles. Distribution of genes in F_3 generation was opposite – high-stem and long panicle in variety E-1643 was determined by recessive genes, but small plant height and length of panicle in variety – by dominant genes. The same law is noted for other traits too.

As it appears from the analysis of hybrids of the first and second generations parental variety Petra possessed positive dominant genes for traits "panicle length," "number of spikelets per panicle," "number of grains per panicle," "grain mass per panicle" and "1000 grains mass". On number of grains and grain mass per panicle in F_1 the given grade had values above the hypothetical completely dominant parent. At analysis of F_2 generation variety Petra had a great number of dominant alleles determining the given traits. Thus, the probability of obtaining of positive transgressions on a number of quantitative traits is great at selection within hybrid populations made with participation of variety Petra. In the third generation of hybrids the given traits were controlled by recessive genes increasing value of a trait.

In a variety Argamak traits "number of grains per panicle," "grain mass per panicle" and "1000 grains mass" were controlled by dominant genes, but in the first and second generations they raised values of a trait, whereas in the third generation – reduced. For varieties E-1643 and Ulov analysis in F_1 and F_2 generations revealed that higher number of spikelets and grains in a panicle is determined by recessive genes, but in F_3 generation traits were controlled by dominant genes.

In variety Freja the raised number of spikelets was controlled by recessive genes. A level of development of traits "grain mass per panicle" and "1000 grains mass" was influenced by both dominant and recessive alleles. The variety had mass of 1000 grains at level of the hypothetical completely dominant parent, but in F_1 generation in the genetic control of a trait prevailed dominant and in F_2 – recessive alleles. In F_3 level of display of a trait was controlled by recessive genes.

Data have shown that among the studied set of hybrids in F_3 and the subsequent generations selection of genotypes with more favorable combination of traits than at parental forms is possible. In this respect the greatest value is represented by combinations with participation of varieties Argamak, Freja, and Petra.

2.3.7 STUDYING OF F_1–F_3 HYBRIDS ON ELEMENTS OF YIELD STRUCTURE

The greatest number of hybrids with high plants has been received in combinations of crossing with participation of tall naked variety Torch for which extreme expression of a trait is revealed by results of diallel analysis; in this relation in hybrids with its participation positive transgressions is not revealed. Only in F_1 hybrid from crossing of variety E-1643 with variety Torch values of a trait was significantly higher than in initial forms.

In most cases there were not observed significant distinctions on *plant height* between covered and naked forms of the same combination of crossing. The exception was made by combination Torch x Petra, which naked form had significantly lower height of plant in comparison with the covered form. Hybrids Torch x Petra (naked form), Petra x Torch (covered and naked forms) had plant height significantly lower than parental form Torch in F_2 and in F_3 generations. Hybrids Ulov x Torch (covered and naked forms) and Freja x Torch (covered form) with height of plants 84.3, 86.4 and 85.8 sm accordingly are obtained in F_3 at 98.4 sm in naked variety Torch. It is possible to explain the given phenomenon by action in F_3 generation of dominant genes reducing value of a trait which greatest number is noted in covered varieties Freja, Petra, and Ulov.

Panicle length was inherited in F_2 together with height of a plant in hybrids E-1643 x Argamak and Ulov x Petra. In F_3 generation significant excess on the given trait over covered parents are noted only for hybrids with using of variety Torch. No one hybrid combination had the forms significantly exceeding initial variety Torch on length of a panicle. In F_3 generation of naked combination Torch x Petra low-stem (77.6 sm) was combined with the raised length of a panicle (18.4 sm).

Heterosis effect was observed on *number of spikelets per panicle* in a number of F_1 hybrids:Argamak x Torch (41.8 pieces, at 21.8 in Argamak and 27.0 in Torch), Freja x Ulov (44.7, 29.8, and 29.9 pieces accordingly). In the subsequent generations the given effect was not shown. Both initial forms of covered hybrids E-1643 x Petra, E-1643 x Ulov, Ulov x Petra, Torch x Argamak, and Freja x Torch have significantly exceeded level of display of a trait in F_2 generation. There was not significantly greater number of spikelets in F_3 hybrids in comparison with initial forms. Values of trait at level of initial forms had naked hybrids Torch x Freja, Torch x Argamak and covered hybrids Petra x Argamak, Freja x Petra, E-1643 x Petra.

Significantly higher *number of grains per panicle* in comparison with parental forms in covered F_1 hybrids Freja x Torch and Freja x Ulov remained in F_2 generation. Besides, covered hybrids E-1643 x Petra, Ulov x Petra and naked hybrids Torch x Ulov were selected on the given trait in F_2

generation. In F_3 hybrids all deviations from parental forms on number of grains in a panicle were within limits of an error. The highest expressiveness of a trait was observed in naked hybrid Freja x Torch – 55.2 pieces (in parental forms 51.2 and 33.2 pieces accordingly) and in covered hybrids Freja x Ulov (60.4, 51.2 and 53.4 accordingly), E-1643 x Petra (51.1, 42.7 and 63.4 accordingly), Ulov x Petra (50.0, 53.4 and 63.4 accordingly). The greatest number of transgressions on the given trait was received in combinations with participation of varieties Freja, Petra, and Ulov.

Significant excess over initial forms by trait *grain mass per panicle* in F_1 and F_2 generations is received in hybrid Freja x Ulov but in F_3 the given hybrid significantly exceeded an initial variety Ulov only. In F_2 positive transgressions on the given trait were noted in covered hybrids E-1643 x Ulov, Freja x Torch, Ulov x Petra and naked hybrid Torch x Ulov. In the third generation these hybrids had mass of grain at level of parental forms.

Significantly high grain mass per panicle (productivity) concerning parental naked form Torch was observed in F_1-F_3 generations of naked hybrids Freja x Torch (2.64 g in F_1, 1.56 g in F_2 and 1.82 g in F_3), Torch x Freja (1.64 g in F_1, 1.44 g in F_2 and 1.65 g in F_3), Torch x Argamak (2.20 g in F_1, 1.35 g in F_2 and 1.54 g in F_3), at 1.26, 1.15 and 0.91 g in initial variety Torch.

1000 grain mass significantly exceeding values of initial forms was observed in hybrid E-1643 x Argamak: 43.2 g in F_1, 36.6 g in F_2 and 38.6 g in F_3. The greatest number of transgressions on the given trait was noted for hybrids with participation of the parental variety E-1643 having recessive genes of positive action. The raised 1000 grain mass in F_2 and F_3 generations was in covered hybrids Petra x E-1643, Petra x Freja, Freja x Petra, E-1643 x Ulov. Among naked hybrids the raised 1000 grain mass was noted for hybrid Freja x Torch – 29.52 g in F_2 and 32.60 g in F_3.

2.4 CONCLUSION

Thus, the method of diallel crossings has allowed to establish some laws of inheritance of quantitative traits in five covered varieties of oats (*Avena sativa* L.) – Argamak, Ulov, E-1643 (Russia), Petra, Freja (Sweden) and one naked oat (*Avena nuda* L.) variety Torch. In the genetic control of all studied traits of productivity, except for length of a panicle, additive effects ($H_1 < D$) prevailed in F_1 generation. In F_2 generation effects of over-dominance were predominating, the exception was made by a trait "plant height". In the third generation additive effects prevailed in the genetic control of all traits that testifies to possibility of carrying out of success-ful screening in F_3. The analysis of hybrids on elements of yield structure has allowed to select the best forms on a set of breeding valuable traits: covered hybrids Freja x Ulov, E-1643 x Ulov, and E-1643 x Argamak, Ulov x Argamak, naked hybrids Freja x Torch, Torch x Freja, and Torch x Argamak. The greatest number of transgressive forms has been selected in combinations with participation of varieties Freja, E-1643, Petra, and Ulov. The probability of obtaining of positive transgressions on a number of quantitative traits is great at selection from hybrid populations with participation of variety Petra.

KEYWORDS

- additive effects
- alleles
- correlation
- covered and naked oats
- dominance
- hybrids
- quantitative traits

REFERENCES

1. Vyushkov, A. A., Malchikov, P. N., Syukov, V. V., Shevchenko, S. N. Breeding-and-genetic improvement of spring wheat. 2-nd edition. Samara: Samara Scientific Center of RAS. 2012, 266 p. [in Russian].
2. Ginsburg, E. H., Nikoro, Z. S. Expansion of dispersion and problems of breeding. Novosibirsk: Nauka Publ., 1982, 168 p. [in Russian].
3. Merezhko, A. F. Problem of donors in plant breeding. St. Petersburg: VIR publishing office, 1994, 127 p. [in Russian].

4. Sokolov, I. D. Mendel's approach to description of inheritance of quantitative distinction. Lugansk: Elton-2. 2000, 179 p. [in Russian].

5. Ginsburg, E. H. Description of inheritance of quantitative traits. Novosibirsk: Nauka Publ., 1984, 249 p. [in Russian].

6. Tikhomirova, M. M. Genetic analysis. Leningrad: Leningrad State University Publ., 1990, 280 p. [in Russian].

7. Tompson, J. Quantitative variation and gene number. Nature. 1975, N. 5537, 665–668.

8. Merezhko, A. F. Use of Mendelian principles in computer analysis of inheritance of varying traits. Ecological genetics of cultural crops. Krasnodar: Rice research institute, 2005, 107–117. [in Russian].

9. Batalova, G. A. Oats in Volga-Vyatka region. Kirov: Orma Ltd., 2013, 288 p. [in Russian].

10. Lerner, M. Population genetics and animal improvement. Cambridge: Cambr. Univ. Press, 1950, 342 p.

11. Kuznetsov, V. M. Methods of pedigree estimation of animals with introduction to BLAP theory. Kirov: North-East Agricultural Research Institute, 2003, 358 p. [in Russian].

12. Mather, K., Jinks, J. L. Biometrical genetics. The study of continuous variation. London, Chapman and Hall, 1971, 382 p.

13. Dragavtsev, V. A. Ecological-and-genetic screening of genofund and methods of design of varieties of agricultural crops on productivity, resistance and quality. St. Petersburg: VIR publishing office, 1997, 49 p. [in Russian].

14. Dragavtsev, V. A., Litun, N. P., Shkel, I. M., Nechiporenko, N. N. Model of ecological-genetic control of plant quantitative traits. Report of Russian Academy of Sciences. 1984, Vol. 274. N. 3. 720–723. [in Russian].

15. Dragavtsev, V. A. About problem of genetic analysis of polygenic quantitative traits in plants. St. Petersburg: VIR publishing office, 2003, 35 p. [in Russian].

16. Dragavtsev, V. A., Averianova, A. F. Mechanism of genotype x environment interaction and homeostasis of plant quantitative traits. Russian Genetics. 1983, Vol. XIX. N. 11. 1806–1810. [in Russian].

17. Dragavtsev, V. A., Averianova, A. F. Re-estimation of genetic formulas of wheat quantitative traits under different environment conditions. Russian Genetics. 1983, Vol. XIX. N. 11. 1811–1817. [in Russian].

18. Nechiporenko, N. N., Dragavtsev, V. A. About possibility of prediction of levels and signs of coefficients of ecological correlations. Russian Genetics. 1986, Vol. XXII. N. 4. 616–623. [in Russian].

19. Sultanov, I. M., Dolotovsky, I. M. Intravarietal correlations in plants of soft spring wheat under influence of ecological and cenotic factors. Agricultural Biology. 2000, N. 1. 41–48. [in Russian].

20. Hayman, B. I. The theory and analysis of diallel crosses. Genetics. 1954, Vol. 39. N. 6. 789–809.

21. Griffing, B. A. A generalized treatment of diallel crosses in quantitative inheritance. Heredity. 1956, Vol. 10. N. 1. 31–50.

22. Batalova, G. A., Lisitsyn, E. M., Rusakova, I. I. Biology and genetics of oats. Kirov: North-East Agricultural Research Institute, 2008, 456 p. [in Russian].

23. Mather, K., Jinks, J. L. Biometrical genetics. Moscow: MIR Publ., 1985, 463 p. [in Russian].

24. Sampson, D. R. Choosing the best parents for a breeding program from among eight oat cultivars crossed in diallel. Can. J. Plant Sci. 1976, Vol. 56. N. 2. 263–274.

25. Youngs, V. L., Forsberg, R. A. Protein-oil relationship in oats. Crop Sci. 1979, Vol. 19. N. 6. – 798–802.

26. Kozlenko, L. V. Genetic principles in oat breeding. Herald of agricultural science. 1981, N. 9. 51–54. [in Russian].

27. Sampson, D. R., Tarumoto, I. Genetic variances in an eight parent half diallel of oats. Can. J. Genet. Cytol. 1976, Vol. 18. N. 3. 419–427.

28. Turbin, N. V., Khotyliova, L. V., Tarutina, L. A. Diallel analysis in plant breeding. Minsk: Science and Technic. 1974, 184 p. [in Russian].

29. Stuthman, D.D, Stucker, R. E. Combining ability analysis of near-homozygous lines derived from a 12-parent diallel cross in oats. Crop Sci. 1975, Vol. 15. N. 6. 411–414.

30. Kozlenko, L. V., Egorova, A. V. Breeding-and-genetic estimation of oat varieties. Bulletin Applied Botany, Genetics Plant Breeding. St. Petersburg: VIR publishing office, 1989, Vol. 129. 134–140. [in Russian].

31. Batalova, G. A. Oats: technology of cultivation and breeding. Kirov: North-East Agricultural Research Institute, 2000, 206 p. [in Russian].

32. Omarova, R. N. Character of inheritance and heritability of barley quantitative traits under different environmental conditions. Soil-protecting system of agriculture and grain production in Eurasian continent in XXI century. Novosibirsk, 1998, 79–81. [in Russian].

33. Dolotovsky, I. M. Phytocenogenetic aspects of formation of plant quantitative traits. Moscow: Agrarian Russia, 2002, 243 p. [in Russian].

34. Boroevitch, S. Principles and methods of plant breeding. Moscow: Kolos, 1984, 344 p. [in Russian].

35. Loskutov, I. G. Oats (*Avena, L.*). Distribution, systematic, evolution, and breeding value. St. Petersburg: VIR publishing office. 2007, 336 p. [in Russian].

36. Rodionova, N. A., Soldatov, V. N., Merezhko, A. F., Yarosh, N. P., Kobylansky, V. D. Cultivated flora. Oats. Moscow: Kolos, 1994, Vol. 2. Part 3. 367 p. [in Russian].

37. Soldatov, V. N., Batalova, G. A. Inheritance of panicle productivity traits in oats. Bulletin Applied Botany, Genetics Plant Breeding. St. Petersburg: VIR publishing office. 1989, Vol. 129. 129–133. [in Russian].

38. Helsel, D. B. Grain yield improvement through biomass selection in oats (*Avena sativa, L.*). Zeitschrift Pflanzenzuchtung. 1985, Vol. 94. N. 4, 298–306.

CHAPTER 3

ROLE OF SOMACLONAL VARIATION IN CEREAL BREEDING FOR ALUMINUM RESISTANCE

EUGENE M. LISITSYN and LYUDMILA N. SHIKHOVA

CONTENTS

ABSTRACT

The opportunity of obtaining of high-resistant oat samples by methods of traditional selection and in vitro was investigated. It was shown, that increased resistance of plant regenerants not always has a genetic nature. On the other hand, the transfer of a level of aluminum resistance from parents to hybrids at traditional crossings has no precise laws. The mechanisms of aluminum resistance acting at the cellular level provide an opportunity of obtaining resistant forms of plants by the cell culture methods. In practice, however, this approach has not become widely used because of difficulties of plant regeneration from callus culture. We consider the main causes of low efficiency of the approach. They include high intraspecific heterogeneity of cereal crops with regard to aluminum resistance; conventionality of the division of genotypes into resistant and sensitive one, accepted in practice; appearance of acid- and aluminum-resistant regenerants not only under stress but also under control (favorable) conditions without any action of the stress factor under study; lack of appropriate methods, connected to the specific behavior of aluminum ions in various growth media, when incorrectly chosen medium composition or pH conceal the action of aluminum per se. As a result of joint action of all the reasons mentioned in the chapter, the offered techniques of creation of high-resistant regenerants of cereal crops are behind traditional methods of intravarietal selection in duration, labor consumption, cost, and efficiency.

3.1   INTRODUCTION

Breeding of plants by any desirable trait is based on two general features of living organisms: heritability and variation. Variation per se may be detectable in existing populations of plants or created artificially with traditional crossing of different genotypes, treatment with mutagens, and with methods of cell selection. In any case the question of heritability of changed trait must be investigated separately because not all phenotypic variations are transferred through seed (sexual) generation. In last decades many research groups all over the world and, particularly, in Russia conduct breeding and genetic investigations on increasing of plant resistance against unfavorable abiotic environmental factors. Many new varieties and hybrids of cereal crops are created and assortment of initial material for such selection is enlarged considerably.

All mentioned are characteristic for oat breeding on aluminum resistance. Ions of trivalent aluminum are the main toxic agent of sod-podzolic acid soils covered 38% of arable area of Russia [1]. New knowledge about genetic control of plant resistance against aluminum action and features of its transferring at different breeding methods is need for both breeders and geneticists.

Selection of Al-resistant plant species and varieties for large-scale cultivation has been recommended as alternative for chemical amelioration in overcoming of aluminum toxicity of different acid soils all over the world. Physiological and biochemical mechanisms of aluminum resistance displaying at cellular level of plant organization gives reasons for some investigators to assume that it is possible to obtain plant form having resistance against the stressor with a method of selection in tissue culture [2, 3].

Somaclonal variation and cell selection in vitro is recognized as principally new tool for creation of plants with high potential of resistance against different abiotic environmental stressors [4–7].

Some authors [8, 9] suggest that this variation which nature is not clear yet together with high sensibility of isolated cells to stress impact makes it possible obtaining of aluminum-resistant somaclones of agricultural crops by a method of cell selection. However, in practice this approach does not find wide use. More over this fact is connected with difficulties of regeneration of plants in callus culture [10–12], with necessity to optimize regeneration conditions for any individual genotype [13]. With in vitro breeding program, selection must be followed by plant regeneration. The choice of the potential genotypes that could be improved depends mainly on their capacity to regenerate plant. The success of in vitro culture depends on the growth conditions of the source material [14, 15], medium composition and culture conditions [16] and on the genotypes of donor plants. Among those factors, the genotype appears to be important factor influencing the efficiency of in vitro culture. In *Triticum*, for the explants with same age and the same growth regulator combination, callus production and plant regeneration capacity depend essentially on genotype [17]. The same results were reported in *Oryza sativa* [18] and *Primula* ssp. [19]. Callus production ability in sugarcane is genotype dependent [20].

So the aims of a given article were: firstly, to characterize levels of potential aluminum resistance of newly selection material; secondly, to compare the effectiveness of different methods of increasing of variation of "aluminum resistance" trait for breeding purposes, and thirdly, to propose some reasons of low effectiveness of creation of aluminum resistant genotypes by cell selection. Next tasks must be solved for reaching the aims: to estimate potential level of aluminum resistance of genetically different sets of oats genotypes, and to investigate features of inheritance of this level in oat samples obtaining by different breeding methods.

3.2   MATERIAL AND METHODOLOGY

Three sets of oat genotypes were formed: (a) oat varieties of Russian and foreign breeding (62 varieties); (b) oat hybrids obtaining in different crossings of Russian and foreign varieties (64 hybrids); (c) regenerant genotypes created in North-East Agricultural Research Institute (Kirov, Russia) on different growth media with aluminum and osmotic stresses (32 genotypes). Seeds of all mentioned oats genotypes were given by laboratory of oat breeding of North-East Agricultural Research Institute (Kirov, Russia).

Estimation of potential aluminum resistance of oats genotypes was conducted in roll culture with use of distilled water (pH 6.0) as a control treatment and 1.0 mM aluminum solution (aluminum sulfate salt) at pH 4.3 (test treatment) by the method described earlier [21]. Level of aluminum resistance was scored as root tolerance index (RTI, %) – ratio of average root length of oat seedlings in test treatment to the average root length of seedlings in control treatment. The higher the RTI the higher a level of aluminum resistance.

Degree of similarity of seed generations of regenerant genotypes with initial genotypes on parameters of root growth in control and test treatments was estimated with χ^2 criteria by formula [22]

$$\chi^2 = \frac{N^2}{n_1 * n_2} * \left(\sum \frac{f_1^2}{S} - \frac{n_1^2}{N} \right) \geq \chi_{st}^2 (v = g - 1)$$

where, $N = n_1 + n_2$ – sum of sample volumes; n_1 and n_2 – volume of each compared samples; f_1, f_2 – frequency of the same group of seedlings in comparing samples; $S = f_1 + f_2$ – sum of frequencies on each group of seedlings; v – degree of freedom; g – total number of groups in both comparing samples.

Statistical processing of obtained data was conducted with use of Statistica 10 software.

3.3   RESULTS AND DISCUSSION

Distribution of seedlings on root length (within any genotype) in control and test treatments corresponded statistically to normal distribution so all statistical counts were done with use of parametric criteria. As a whole root length of oat seedlings was higher in varieties, some less – in regenerant genotypes, and the lowest – in genotypes of hybrid origin both in control and in test treatments. Hybrid genotypes had also biggest variability of root length (17–20%) whereas regenerant genotypes had lowest one (9%). Perhaps this fact is explained by genetic similarities of regenerants, which were obtained from limiting set of varieties and by high genetic variability of varieties including in hybrid crossings.

Oats hybrid genotypes had high levels of coefficients of pair correlations between parameters "root length in control treatment" and "root length in test treatment" as well as between parameters "root length in control treatment – RTI" ($r = 0.657$ and -0.680). For varieties these coefficients were twice lower ($r = 0.381$ and -0.342). Regenerant genotypes had levels of pair correlation between these two groups ($r = 0.461$ and -0.494). Variability of root length of control seedlings explains 46% of variability of potential aluminum-resistance level in hybrid genotypes, 24% – in regenerant genotypes, and only 11% – in oat varieties.

It is possible that all oats genotypes (or any other cereal) have some general mechanisms of root growth (general metabolic "base") and different additional mechanisms, which are specific for single genotype. Then in varietal level coefficient of determination in pair "root length in control treatment – RTI" will reflects degree of similarity of different varieties on genetic formula of trait "aluminum resistance" estimated by change of root growth. Highest coefficients of determination in regenerant genotypes may be explained by narrower genetic base of investigated set of genotypes, but in hybrid genotypes – by the fact that any hybrid has only half of parent genotype. Difference in allelic state of genes governing development of additional metabolic mechanisms of resistance leads to their incomplete displaying or lack of such displaying at all. In such case genes of general ("base") mechanisms will have much more effect.

Separately study of group of regenerant genotypes leads us to conclusion that the higher is the level of potential aluminum resistance of initial oat genotype the lower was level of overcoming of regenerant genotype on initial genotype by investigated trait. So initial genotype 10–05 has RTI = 77.0%. Average level of aluminum resistance of six regenerant genotypes was 67.0% and only one regenerant genotype (k-2246) has RTI higher then initial genotype (78.2%).

Initial genotype 9–05 has RTI = 71.4%; only one regenerant genotype (k-2231) has RTI = 77.0% (statistically higher then initial genotype), two other regenerant genotypes (k-2223, k-2228) have some lower level of RTI (70.3 and 70.7%), one more regenerant genotype is much more sensitive (RTI = 68.6%).

RTI level of initial genotype 282h00 was 61.4%. Ten regenerant genotypes were obtaining with its participation for analysis of aluminum resistance level. Average RTI level of these regenerants was 74.0%; all regenerant genotypes overcome initial form on RTI level, some of them – by 30% (regenerant genotypes k-2068, k-2070, and k-2084 had RTI about 78–80%).

Comparing of regenerant genotypes and initial genotype on parameters of root growth in control and test treatments within each sample for groups of sample of initial genotype 282h00 had shown the next levels of χ^2 criteria (Table 3.1).

The data of table allow to conclude that overcoming of regenerant genotypes on initial genotype by level of aluminum resistance is supplied basically by lower indexes of root growth of regenerants under control treatment. Under aluminum influence seedling root length of initial and regenerant genotypes is not differing statistically but for genotype k-2061. This regenerant genotype differs from initial genotype both in control and in test treatment. Perhaps it has genetic difference from initial genotype. All the other regenerant genotypes show general depression of root growth processes that is governing by some other factors of cultivation in vitro but not by direct influence of aluminum ions.

We must underline especially that we study not regenerants obtaining in culture in vitro per se but their seed generation. It has great importance for interpreting of the data. It is known that somaclonal variation being theoretical basis of regenerant obtaining is explaining by changes taking part on stage of indifferential growth in culture in vitro, which frequently (but not always) have genetic nature. If changes belong to genetic apparatus

TABLE 3.1 Character of Distribution of Oats Seedlings by Root Length Under Control and Test Treatments

Regenerant genotype	Groups by root length													χ^2 fact*	χ^2 theor*
	1	2	3	4	5	6	7	8	9	10	11	12	13		
282h00 (initial)	1	2	1		1	2	2		1	7	41	13	1		
Control treatment (distilled water, pH 6.0)															
k-2057			5		3			2	6	22	14			46.7	19.7
k-2061			2	2		3	4	3	10	37	1			47.3	21.1
k-2063	2	3	2	2	1	1		3	12	20	7			35.9	19.7
k-2068	1	3	2	2		1	1	3	16	19				28.4	21.1
k-2070	4	3		3	6		3	7	24	25				51.7	21.1
k-2073	2	2	2	2	2	2	1	3	11	45	4			95.5	21.1
k-2075	1	1	2	3	1		2	6	25	27	6			94.4	21.1
k-2078	1	1	2	1	1	1			6	16	21	7		22.0	19.7
k-2082	4	2	2			1	1	5	15	27	1			49.6	21.1
k-2084		1		1	1	3	1	3	23	28				58.4	21.1

TABLE 3.1 Continued

Regenerant genotype	Groups by root length													χ^2 fact*	χ^2 theor*
	1	2	3	4	5	6	7	8	9	10	11	12	13		
282h00 (initial)				3	8	40	21	2	2	1					
Test treatment (1.0 mM aluminum, pH 4.3)															
k-2057			1	1	3	32	19	4	2					6.92	14.1
k-2061				1	1	22	18	4	3					17.2	12.7
k-2063				1	7	36	24	2	1					22.9	12.7
k-2068				1	4	38	34	3	2	1				1.6	12.7
k-2070				1	5	41	17	1						12.3	12.7
k-2073				1	13	39	14	4						6.0	12.7
k-2075				1	5	35	25	7	3					0.0	12.7
k-2078					5	33	16	3	1					2.6	12.7
k-2082			2	3	3	30	33	7	2					6.7	14.1
k-2084					6	47	23	3	1					8.6	12.7

Note: * in comparison with initial genotype.

of cells then they will be hereditable in the next seed generations. But if there are modificational or epigenetical variations only then high level of aluminum resistance characteristic for the regenerant genotypes will not transfer by sexual generations.

The level of aluminum resistance in seed generation of regenerant genotypes obtaining with participation of initial genotype 10–05 as well as initial genotype 9–05 (with exception of genotype k-2231) does not exceed the level of initial genotype one may concludes that high level of regenerant genotypes' resistance has not genetic nature. In contrary, regenerant genotypes obtaining from initial genotype 282h00 display heritable level of aluminum resistance higher than in initial genotype but it may be explained by worse growth of regenerant genotypes in control treatment.

Comparison of data on root growth in oat plants both in control and test treatments (and correspond count of RTI) in two sets of genotypes – varieties and regenerants – suggests opinion [23] that amplitude of somaclonal variation rare exceeds limits of the given plant species.

Results of estimation of potential aluminum resistance of oats hybrids did not allow to determine exactly varieties – donors of high resistance at the modern state of investigation. Thus highly resistant oats varieties taking as parental or maternal components of crossing transfer their potential of resistance not always. Hybrids with participant of Swedish variety Freja which has RTI = 90.3% as a maternal component of crossing with variety Ulov (Russia), Torch (Canada), and Petra (Estonia) had RTI = 57.6–65.7%. When variety Freja plays role of paternal component then its hybrids with varieties IL-86–5698 (USA), Argamak, Ulov and E-1643 (Russia) had levels of resistance lower then each of initial varieties. But its hybrids with varieties Manu (Germany) and AC Lotta (Canada) were highly aluminum resistant (RTI is high then 90%).

In combination of crossing Freja (maternal component) and Torch (paternal component) hybrid genotypes (675h05, 683h04, and 397h04) had potential aluminum resistance levels 57.6–65.7%. But when variety Freja was used as parental component then RTI of hybrids was 51.1 (genotype 668h05), 69.5 (89h04) and 93.6% (64h04).

Two hybrids of the same combination (Ulov × Torch) have shown RTI level differing in 1.5 time: hybrid genotype 104h04 has RTI = 69.1%, and genotype 106h04–45.2% only. On the other hand crossing of varieties Dolphine (Australia) and I-2049 (Russia) leads to significant rise of RTI (61.8 and 78.8% for initial varieties and 90.8% for hybrid genotype).

Thus the data support possibility of obtaining of highly aluminum resistant oats hybrids at crossing of varieties having low level of potential resistance to the stressor and, contrary, significant lowering of this level in hybrids obtaining in crossing of highly resistant parental genotypes, which was pointed out by us earlier [24].

On our opinion, one of the main reasons of success of somaclonal selection and lack of regularities in transfer of resistance level at classical crossings is disturbance of basic principle of genetic analysis – principle of homozygosity of initial genotypes. Let consider briefly a process of creating of newly variety. For getting a maximum number of hybrid seeds crossings are conducted under non-stress conditions where varietal differences by stress resistance does not display. Then in F1 hybrid population there are dominant and recessive alleles of resistance gene simultaneously; as a following – in first generation of crossing one may get genetically different hybrids: in a case of monogenic governing of the trait there are genotypes aa, Aa, and AA.

In a line of self-pollinated generations number of heterozygotes reduces considerably and finally equilibrium of two homozygotes aa and AA will established. In a case of polygenic governing of the trait dominant and recessive alleles of different genes will combine at different ratio in different plants of the same "variety".

So, transferring material from big number of plants of any one variety or from hybrid plants of first generation into condition of in vitro culture we may select material having high level of stress-resistance initially. Then "regenerant"-plants obtaining in tissue or cell cultures indeed will represent result of simple selection which success depends in high degree on volume of sample taking into work from genetically heterogeneous initial plant population and un-controlled successful selection of those plants having maximum number of genes of stress-resistance in dominant state.

The same explanations may by apply for the case of classical crossings when different plants of the same variety carry different allelic variants of the same genes. In fact result of crossing will depends on genotype of two exact plants taking into crossing.

Let consider some reasons of low effectiveness of somaclonal selection, which are rare pointed out by much of researchers. Firstly we must say about theoretical disadvantages of the method because the character of reading information is determined by features of reader [25], and scientific explaining is depended not on logic demands but initial conceptual choice of the researcher.

1. **All plants of the same variety of self-pollinated crop are similar genetically; differences between them by the level of aluminum resistance are based on non-inheriting factors.**

   In practice, however, biotechnologists and breeders work with initial materials presented by varieties, selection lines and hybrids of different crops. Till beginning of XX century it is known that initial material for breeding must be heterogeneous genetically for selection of outstanding plants.

   But most of modern varieties of agricultural crops were created by crossing, for example, were populations in which different genotypes may be finding in different ratio. If for investigation in culture in vitro material was collected from some dozen or ever more plants of the same variety then probability of selection of genetically heterogeneous material was much more higher then probability of induction of mutations (which are proposed the main factor leading to obtaining of regenerant plants differing from initial material by level of stress resistance). Earlier we underlined high intravarietal heterogeneity of cereal crops on trait "aluminum resistance" [21, 26, 27].

2. **Differences between plant varieties on aluminum resistance have biological nature.**

   Firstly, we must pointed out that practical approach of dividing of genotypes on resistant and sensitive is rather agronomical (by degree of depression in development of any trait, most often – plant productivity, under stress conditions) but not biological. Biological resistance *per se* is determined by ability of plant to produce viable seeds [28], for example, variety decreasing productivity under stress conditions by 10–20% and variety decreasing its productivity by 80–90% are both resistant biologically because both of them produce viable seeds but in different amounts. More over till now no one gene is found that specifically governed exactly plant resistance to aluminum ions but total number of genes which activity is changed under aluminum influence reaches some thousand [29].

According to results of our investigations [30, 31] genetic control of aluminum resistance in different varieties of the same agricultural crop may be governed by different quantitative or qualitative sets of genes. More over sometimes the same genotype is determined as resistant in one publication and as sensitive in other publication of the same authors. For example barley genotype 999–93 was used by [3] as aluminum tolerant genotype but in next study [2] – as sensitive to aluminum toxicity. It is rather naturally if we assume the fact of relativity of level of genotype's aluminum resistance.

But, firstly, if we assume this point of view then conclusion of high reasonability of using in culture in vitro of varieties having low aluminum resistance level [32, 33] in comparison with resistant varieties will has not theoretical basis. Secondly, this fact makes it questionable the principle of obtaining of aluminum resistant genotypes both by method in vitro and by traditional breeding methods at all. It is necessary to point out that practically all authors underlie genotypic differences in morphogenic and regenerating abilities. Conditions that are optimum for regeneration of one genotype are not suitable for other genotypes, for example, obtaining of resistant genotypes in vitro are not always determined by their level of potential aluminum resistance.

3. **Resistance is development against the factor, which is inputted in culture in vitro as a stressor.**

However acid- and aluminum resistant regenerant genotypes are obtained in control condition as well [3, 34, 35] without action of any stressful agent. It is well known that effect of any mutation may be phenocoping, for example, induced without action of genetic changes and show the same morphogenetic nature [25].

According to theory of ecological-genetic organization of quantitative trait [36] there are not genetic systems constant on composition, which governs development of any quantitative trait under all possible growth conditions. It is established that aluminum taking at different concentration has different physiological mechanism of toxic action; any plant has different genetic systems determining reaction of the same genotype at different concentration of toxic agent [37]. It is assumed that different mechanisms take part in governing trait "aluminum resistance" in different plants obtaining in culture in vitro [6]. The methods of obtaining of regenerant plants in vitro include influence of large set of chemical compounds on biological material; many of these compounds may act as mutagens on cells and tissues. Highly important is ratio of phytohormones and its quantitative content in growth media. If we remember that disturbance of hormone balance in plants is one of a symptom of aluminum toxicity then artificial change of the ratio of these biologically active substances may change considerably the character of aluminum impact on plants – increases or decreases its toxicity.

Thus aluminum ions as selecting agent input in growth media at last stages of cultivation it is possible that regenerant plants are in pre-adapted state already and so they suffer aluminum influence easier, and aluminum ions lose their role of selecting agent. Then regenerant plants having higher level of general non-specific resistance will be scored as highly resistant to aluminum but not be genetically resistant to it.

4. **In callus culture the resistance against soil stresses is expressed.**
    In this relation it is interesting to underline that at initial stage of in vitro work calli have not roots at all, or these roots were removed artificially [3]. Thus researchers try to reach adaptive rearrangements in those plant organs, which must not principally and will not later suffer direct influence of stressor. It is necessary to point out that effect of this selection is estimated on degree of depression of development of aboveground parts of plants or of physiological reactions taking part in leaves – organs that never suffer direct impact of soil stresses but developing under conditions of finished adaptive rearrangement of plant which may suffers action of different environmental factors by this time. Unfortunately many authors did not take into account specificity of aluminum action

in chemically different media [38]; very often incorrectly chosen media or pH of growth media masks aluminum action.

It is well known that trivalent ions of aluminum has maximal toxicity at media pH = 4.3 [39]. Lowering of this value at constant concentration of aluminum significantly reduces its toxicity [40]. For example at pH = 3.7 transport of aluminum through cell membranes decreases by 10 times [41] that is explained by occupation of exchangeable sites of cell membranes by hydrogen ions [40]. More over at pH lower than 4.0 rapid polynuclear aggregations of trivalent ions of aluminum takes place into complexes designated as Al_{13} which are non-toxic at solid phase [42]. As a result researchers try to construct aluminum stress but indeed create acid stress or stress of lack of nutrients. Of course plants resistant to aluminum ions have resistance against high acidity but not contra verse [43, 44]. In worst case, stress conditions in culture in vitro will not have any relation to soil stresses.

Another important methodical mistake is related with correct choice of stressor concentration. For example, [45] take a task to increase rigidity of selective systems modeling action of stressor on plant at cellular level in vitro as one of the most important in obtaining of aluminum resistant regenerants in culture in vitro. Trying to increase rigidity of aluminum action in growth media (i.e., create rigid selection backgrounds) it is necessary to taking into account features of aluminum behavior in different chemical compositions. At using of sulfate salt of aluminum [2, 3, 32] increasing of aluminum concentration leads to significant decreasing of rigidity of stress influence. This is explained by the fact that under condition of Murashige-Skoog media [46] increase of concentration of ions Al^{3+} and SO_{4-} leads to significant producing of alunite and correspond decrease of activity of trivalent aluminum. It is well known that aluminum toxicity is determined by activity of Al^{3+} ions but not by its concentration [47]. So such "increasing" of rigidity of selection backgrounds indeed leads to its lowering. This conclusion may be supported by higher level of resistance of some regenerant plants obtaining in soft stress media in comparison with rigid stress media [3].

5. **Aluminum resistant regenerants are created in culture in vitro.**
   On our opinion it is not clear what authors are recognized as "created" of aluminum resistant regenerants – selection of most desirable genotypes from pre-existed variability or real creation of new
   genotypes. This question has principal meaning: if genotypes are
   created *de novo* then there are genetic rearrangements of genotype
   which may take place only as results of mutation.

   It is assumed that introducing of plant material into culture in vitro
   leads to change of genotype automatically by mutation process.
   More over [3] suggests that only point mutations which do not
   reduce sharply plant viability transferred into progeny of primary
   regenerants. The authors assume that existence of desirable mutation among somaclonal lines allows using somaclonal variation
   for creation of new initial material for breeding. On the authors'
   opinion at introducing of sensitive variety into culture in vitro the
   possibility of advantageous mutations is higher than at using of
   resistant varieties.

   We already mention the relativity of term "resistant" or "sensitive"
   genotype. According to synthetic theory of evolution new variation
   is supplied by mutations, which appeared un-directed and accidentally; their influence on living organism's adaptivity is independent on environmental conditions existing at given moment
   [48]. In other words mechanism of appearance of new variation
   proposes that variants adapted to the given growth media were
   pre-existed, for example, were within a population earlier before
   impact of the given stressor. Frequency of gene mutation in culture
   in vitro by estimation of [49] is about 10^{-5}–10^{-7} per cell generation. Of course, frequency of epigenetic mutation is higher (10^{-3}
   per cell generation) but in this case there are much of inversions
   and inheritance of these changes in sexual progeny is estimated by
   authors only as "possible". Within frame of genetic theory environment is thought as "appraiser" of inheritable variation of population only. By opinion of [48] frequency of appearance of acquired
   adaptive trait with stable transferring in generations is low (at the
   level of frequency of gene mutations) and so such cases is hard
   to find. M. Ivanov [35] studied populations of regenerant whose

initial explants were commercial varieties. He pointed out that factor limiting both genetic and morphologic variations is selective death of recombinant gametes and zygotes due to its imbalance so number of plants – carriers of lethal mutations (albinism, sterility of pollen and ear) that die at early stage of forming of populations of regenerant plants is rather high. As a result only little part of plants stays alive *de novo* mainly those having morphotype of initial genotype. In other words if mutant plants are appeared it is very high possibility that they cannot reach stage of mature plant.

Some researchers [4, 32] assume that fact of different resistance against stressors of regenerant plants obtaining from the same callus supports idea of appearance of somaclonal variation during cultivation in vitro. Other authors [50] propose genetic heterogeneity of embryo cells appeared as a result of double fertilization as a main reason of this variation. This point of view coincides with idea of pre-existing variability of initial material.

Some authors [35] explain different resistance of regenerants from the same callus by disadvantages of the method. On their opinion regeneration of callus cells takes place in the border medium – callus only and other cells are out of the process. So not all cells sensitive to stressor's action are died and may be initial cells for stress-sensitive regenerant plantlets. If we consider a simple selection of most adaptive variants from pre-existing variability of genotypes (due to genetic intravarietal heterogeneity) then we have long last, cost and highly inefficient alternative to simple selection in nutrient solution like one used by us [21] for estimation of potential level of aluminum resistance in different agricultural crops and creation of high-resistant populations of different cereal crops.

6. **As an evidence of effectiveness of selection of acid- or aluminum resistant regenerants researchers use results of field tests [2, 3, 34, 35].**

This approach has two main methodical mistakes. Firstly, development of aboveground parts is linked with features of root systems' development as objects of aluminum and hydrogen ions influence but this link is not simple; aboveground parts per se suffer influence of some other environmental factors. For example in our field tests during 1996–2005 years for such cereal crops as oats, barley, wheat and winter rye coefficients of pair correlation between level of aluminum resistance (counted by depression of root growth in presence of aluminum) and changes in development of single elements of yield structure or biochemical parameters of plant leaves (aluminum acid soil background in comparison with neutral soil background) only in some cases reach value $r = 0.5$ [51]. More often these correlations were insignificant statistically.

Secondly, all field tests of regenerants have significant methodical mistake, which destroys all reasons of such works – they ignore principle of "the only difference". All authors compare regenerants obtaining in selective conditions in vitro with initial variety or with standard variety of investigated crop. But the only correct solution in the given case is comparison of regenerants obtaining in presence of aluminum ions with the same regenerants obtaining under control treatment without stressor in selective media. Only at such design of test we may estimate effectiveness of method in vitro.

So till now theoretical problems connected with possibility of obtaining (creating) of aluminum resistant regenerants of cereal crops stay developed insufficiently. Such position leads to groundlessness of statements about possibility of increasing of aluminum resistance level by using selection of regenerants on selective backgrounds with aluminum in culture in vitro. Plants obtaining in such conditions may display higher level of resistance during further growth and development but, firstly, this result was not regular and repeated; secondly, obtaining of highly resistant regenerants in control treatment without any stressor indicates low effectiveness of offered methods. We may agree with the statement that somaclonal variation is one of the variants of using of intravarietal heterogeneity on aluminum resistance. At the same time the offered techniques of creation of high-resistant regenerants of cereal crop are behind traditional methods of intravarietal selection in duration, labor consumption, cost, and efficiency.

3.4 CONCLUSIONS

1. High level of aluminum resistance in regenerants obtaining in culture in vitro not always has genetic nature.
2. In a case when high level of aluminum resistance in regenerants may be inheritable (regenerants from genotype 282h00) it is explained by worst root growth in control treatment.
3. Till now we cannot select donor varieties of high level of aluminum resistance because highly resistant varieties not always transferred its potential of resistance to hybrids; crossing of varieties having low level of resistance may leads to significant increase of hybrid RTI.
4. Further investigation on large set of initial material are needed for statistically correct conclusions about advantages or disadvantages of different methods of creation of aluminum resistant genotypes of cereal crops.
5. Statements about wide use of biotechnological methods of obtaining of aluminum resistant material of cereal crops in breeding practice will be incorrect until problems connected with overcoming of above-mentioned theoretical and methodical disadvantages will be solved.

KEYWORDS

- aluminum
- callus
- cereal crops
- hybrids
- oat
- regenerant
- resistance
- selection
- tissue culture

REFERENCES

1. Agricultural biology (Ed. V. S. Shevelukha). 2-nd edition. Moscow: Higher School Publ., 2003, 421 p. [in Russian]. Shirokikh, I. G., Ogorodnikova, S.Yu., Dal'ke, I. V., Shupletsova, O. N. Physiological and biochemical parameters and productivity of barley plants regenerated from callus at selection systems. Reports of the Russian Academy of Agricultural Sciences N. 2011, N. 2. 6–9. [in Russian].

2. Shirokikh, I. G., Shupletsova, O. N., Shchennikova, I. G. Obtaining in vitro of barley forms resistant against toxic action of aluminum in acid soils. Biotechnology. 2009, N. 3. 40–48. [in Russian].

3. Bertine, P., Kinet, J. M., Bouharmont, J. Heritable chilling tolerance improvement in rice through somaclonal variation and cell line selection. Australian Journal of Botany. 1995, Vol. 44. 91–105.

4. Biswas, J., Chowdhury, B., Bhattacharya, A., Mandal, A. B. *In vitro* screening for increased drought tolerance in rice. *In vitro* Cell Development and Plant Biology. 2002, Vol. 38. 525–530.

5. Mandal, A. B., Basu, A. K., Roy, B., Sheeja, T. E., Roy, T. Genetic management foe improved aluminum and iron toxicity tolerance in rice – A review. Indian Journal of Biotechnology. 2004, Vol. 3. No. 3. 359–368.

6. Roy, B., Mandal, A. B. Towards development of Al-toxicity tolerant lines in *indica* rice by exploiting somaclonal variation. Euphytica. 2005, Vol. 145. 221–227.

7. Vnuchkova, V. A., Nettevich, E. D., Chebotareva, T. M., Khitrova, L. M., Molchanova, L. M. Use of in vitro methods in barley breeding for resistance to toxicity of acid soils. Reports of the Russian Academy of Agricultural Sciences. 1989, N. 7. 2–5. [in Russian].

8. Muyuan, Y. Z., Jianwei, P., Lilin, W. Qing, G., Chunyuan, H. Mutation induced enhancement of Al tolerance in barley cell lines. Plant Science. 2003, Vol. 164. 17–23.

9. Komatsuda, T., Enomoto, S., Nakajima, K. Genetics of callus proliferation and shoot differentiation in barley. Journal of Heredity. 1989, Vol. 80. No. 5. 345–350.

10. Litovkin, K. B., Ignatova, S. A., Bondar, G. P. Morphogenesis in culture of immature embryos of barley isogenic lines. Cytology and Genetics. 1999, Vol. 33. N. 5. 14–18. [in Russian].

11. Ovchinnikova, V. N., Varlamova, N. V., Melik-Sarkisov, O. S. Obtaining of regenerants of *Hordeum vulgare*, L. in culture in vitro. Reports of the Russian Academy of Agricultural Sciences. 2004, N. 3. 8–10. [in Russian].

12. Bakulina, A. V., Shirokikh, I. G. Genotypic reaction of barley on antibiotics canamycin and cephotacsim in culture in vitro. Proceedings of All Russian conference "Biological monitoring of natural-technogenic systems". Kirov, 29–30 November 2011, Kirov: Publishing House "Loban," 2011, Vol. 2. 69–72. [in Russian].

13. Caswell, K., Leung, N., Chibbar, R. N. An efficient method for in vitro regeneration from immature inflorescence explants of Canadian wheat cultivars. Plant Cell Tiss. Org. Cult. 2000, Vol. 60. 69–73.

14. Delporte, F., Mostade, O., Jacquemin, J. M. Plant regeneration through callus initiation from thin mature embryo fragments of wheat. Plant Cell Tiss. Org. Cult. 2001, Vol. 67. 73–80.

15. Saharan, V., Yadav, R. C., Yadav, R. N., Chapagain, B. P. High frequency plant regeneration from desiccated calli of *indica* rice (*Oryza sativa,* L.). African Journal of Biotechnology. 2004, Vol. 3. No. 5. 256–259.

16. Zale, J. M., Borchardt-Wier, H., Kidwell, K. K., Steber, C. M. Callus induction and plant regeneration from mature embryos of a diverse set of wheat genotypes. Plant Cell Tiss. Org. Cult. 2004, Vol. 76. 277–281.

17. Hoque, M. E., Mansfield, J. W. Effect of genotype and explant age on callus induction and subsequent plant regeneration from root-derived callus of *indica* rice genotypes. Plant Cell Tiss. Org. Cult. 2004, Vol. 78. No. 3, 217–223.

18. Schween, G., Schwenkel, H.-G. Effect of genotype on callus induction, shoot regeneration, and phenotypic stability of regenerated plants in greenhouse of *Primula ssp.* Plant Cell Tiss. Org. Cult. 2003, Vol. 72, 53–61.

19. Gandonou Ch., Abrini, J., Idaomar, M., Skali Senhaji, N. Response of sugarcane (*Saccharum sp.*) varieties to embryogenic callus induction and in vitro salt stress. African Journal of Biotechnology. 2005, Vol. 4. No. 4. 350–354.

20. Lisitsyn, E. M. Intravarietal level of aluminum resistance in cereal crops. Journal of Plant Nutrition. 2000, Vol. 23. No. 6. 793–804.

21. Plokhinsky, N. A., Markelova, I. O. Estimation of significance of segregation of two empirical distributions. Methods of modern biometry. Moscow: Moscow State University Publishing House, 1978, [in Russian].

22. Kunakh, V. A. Law of homological rows of, N. I. Vavilov in somaclonal variation of plants. Genetics in XXI century: modern state and perspectives of development. Moscow: Nauka, 2004, Vol. 1. 73. [in Russian].

23. Lisitsyn, E. M. Physiological basis of plant edaphic breeding at European North-East of Russia. Kirov: Publishing office of North-East Agricultural Research Institute, 2003, 196 p.

24. Shishkin, M. A. Individual development and lessons of Darwinism. Ontogenesis. 2006, Vol. 37. N. 3. 179–198. [in Russian].

25. Lisitsyn, E. M. Polymorphism of adaptive reactions of variety at a level of morphological and biochemical parameters. Bulletin of Applied Botany, Genetics and Breeding. 2006, Vol. 162. 150–154. [in Russian].

26. Tiunova, L. N., Lisitsyn, E. M. Aluminum resistance of oats genotypes created by traditional methods and methods of cell selection. An Agrarian Science of Euro-North-East. 2009, N. 4. 9–13. [in Russian].

27. Udovenko, G. V. Plant resistance to abiotic stresses. Physiological basis of plant breeding. Saint-Petersburg. 1995, Vol. 2. Part 2. 293–352. [in Russian].

28. Houde, M., Diallo, A. O. Identification of genes and pathways associated with aluminum stress and tolerance using transcriptome profiling of wheat near-isogenic lines. BioMed Central Genomics. 2008, doi: 10.1186/1471–2164-9-400.

29. Lisitsyn, E. M., Shchennikova, I. N., Shupletsova, O. N. Cultivation of barley on acid sod-podzolic soils of north-east of Europe. Barley: Production, Cultivation and Uses [Ed. S. B. Elfson]. New York: Nova Publ. 2011, 49–92.

30. Batalova, G. A., Lisitsyn, E. M., Rusakova, I. I. Biology and genetics of oats. Kirov: North-East Agricultural Institute, 2008, 456 p. [in Russian].
31. Van Sint, J. V., Costa de Macedo, C., Kinet, J., Bouharmont, J. Selection of Al-resistant plants from a sensitive rice cultivar, using somaclonal variation, in vitro and hydroponic cultures. Euphytica. 1997, Vol. 97. 303–310.
32. Roy, B., Mandal, A. B. Towards development of Al-toxicity tolerant lines in *indica* rice by exploiting somaclonal variation. Euphytica. 2005, Vol. 145. 221–227.
33. Zobova, N. V., Konysheva, E. N. Using of biotechnical methods for increasing of salt- and acid resistance in spring barley. Novosibirsk: Publishing office of Siberian Branch of Russian Academy of Agricultural Sciences, 2007, 124 p. [in Russian].
34. Ivanov, M. V. Biotechnological basis of creation of initial material of spring barley. Saint-Petersburg – Pushkin: Publishing office of All Russian Institute of Plant Industry, 2001, 205 p. [in Russian].
35. Dragavtsev, V. A., Litun, N. P., Shkel, I. M., Nechiporenko, N. N. Model of ecological-genetic control of quantitative traits of plants. Reports of the Academy of Sciences of the USSR. 1984, Vol. 274. No. 3. 720–723. [in Russian].
36. Kochian, L. V., Hoekenga, J. A., Pineros, M. A. How do crop plants tolerate acid soils? Mechanisms of aluminum tolerance and phosphorous efficiency. Annual Review of Plant Biology. 2004, Vol. 55. 459–493.
37. Lisitsyn, E. M. Chemistry of solutions and the estimation of the potential Al-resistance of plants (Methodological problems). Agrochemistry. 2007, No. 1. 81–91. [in Russian].
38. Taylor, G. J., McDonald-Stephens, J. L., Hunter, D. B. Direct measurement of aluminum uptake and distribution in single cell of *Chara coralline*. Plant Physiology. 2000, Vol. 123. 987–996.
39. Kinraide, T. B. Reconsidering the rhysotoxicity of hydroxyl, sulfate, and fluoride complex of aluminum. Journal of Experimental Botany. 1997, Vol. 48. 1115–1124.
40. Kinraide, T. B., Ryan, P. R., Kochian, L. V. Interactive effects of Al3+, H+ and other cations on root elongation considered in terms of cell-surface electrical potential. Plant Physiology. 1992, Vol. 99. 1461–1468.
41. Kinraide, T. B. Identity of the rhysotoxic aluminum species. Plant and Soil. 1991, Vol. 134. 167–178.
42. Voigt, P. W., Staley, T. E. Selection for aluminum and acid resistance in white clover. Crop Science. 2004, Vol. 44. 38–48.
43. Yang, J. L., Zheng, S. J., He, Y. F., Matsumoto, H. Aluminum resistance requires resistance to acid stress: a case study with spinach that exudes oxalate rapidly when exposed to Al stress. Journal of Experimental Botany. 2005, Vol. 414. 1197–1203.
44. Shupletsova, O. N., Shchennikova, I. N., Sheshegova, T. K. Study of barley genotypes having complex resistance created by the method of cell selection in the given book.
45. Ramgareeb, S., Watt, M. P., Marsh, C., Cooke, J. A. Assessment of Al^{3+} availability in callus culture media for screening tolerant genotypes of *Cynodon dactylon*. Plant Cell Tissue Org. Cult. 1999, Vol. 56. 65–68.
46. Kinraide, T. B., Parker, D. R. Cation amelioration of aluminum toxicity in wheat. Plant Physiology. 1987, Vol. 83. 546–551.

47. Zhivotovsky, L. A. Inheritance of acquired traits: Lamarck was true. Chemistry and Life. 2003, No. 4. 22–26. [in Russian].
48. Butenko, R. G. Biology of cells of higher plants in vitro and biotechnologies on their basis. Moscow: Publishing House FBK-Press, 1999, 160 p. [in Russian].
49. Chowdhury, B., Mandal, A. B. Microspore embryogenesis and fertile plantlet regeneration in salt susceptible – salt tolerant rice hybrid. Plant Cell Tissue Org. Cult. 2001, Vol. 65. 141–147.
50. Lisitsyn, E. M. Potential aluminum resistance of agricultural crops and its realization under conditions of European North-East of Russia. Doctoral Thesis. Moscow. 2005, 361 p. [in Russian].

CHAPTER 4

ECOLOGICAL STABILITY OF SPRING BARLEY VARIETIES

IRINA N. SHCHENNIKOVA and EUGENE M. LISITSYN

CONTENTS

ABSTRACT

Results of long-term studying of varieties of spring barley (*Hordeum vulgare* L.) in competitive field varietal test of the North-East Agricultural Research Institute and on State variety test stations of the Kirov region are presented in the article. Varietal specificity in their reactions to change of soil-environmental conditions is revealed. A share of influence of environmental conditions on formation of productivity and effect of "genotype – environment" interaction in a studied set of varieties is established. Ranging of barley varieties on productivity in favorable and stressful conditions of growth is presented; the variation of productivity on years is defined. The various methods based on an estimation of productivity of varieties in years differing on meteorological conditions and sites of study were used for obtaining the varieties characterized by high ecological stability. The botanical and economic-biological characteristic of new high-yielding varieties of barley different by high ecological stability is given. The obtained data allows to recommend a set of new ecologically stable barley varieties adapted for conditions of cultivation for wider using in agricultural industry.

4.1   INTRODUCTION

Barley is the reliable culture for Volga-Vyatka region capable to use biological potential for formation of stable yield as much as possible. Economic value of barley's grain is of great importance: it is used in animal industries and for manufacture of groats, a flour, beer, and coffee drinks [1]. The basic direction of use of barley in Volga-Vyatka region is for grain-forage. More than 60% of grain is used on preparation of mixed fodders and on the fodder purposes directly. The successful decision of a problem of production of fodder grain in the volumes necessary for satisfaction of requirements of region is possible at the complex decision of some other problems. On the one hand, it is increasing of productivity at the expense of expansion of the sowing areas under barley, observance of optimum technologies of its cultivation. On the other hand – it is purposeful breeding, for example, creation of the high-yielding varieties adapted for local conditions of cultivation. Breeding as well as any other science solves problems of a human society on exact period of its development. New problems put the new purposes in the face of breeding. Ecological direction of plant breeding is objective requirement of the breeding theory and practice in relation with change of the basic priorities of an agricultural production: high productivity and resistance of agrocenoses against abiotic and biotic stresses [2, 3].

The species and variety have important environment-forming value, defining level of anthropogenous loading on environment. It is caused by what features and all elements of technology of cultivation with plant species and variety are linked: doses; terms and types of fertilizers and pesticides; ways and frequency rate of soil processing; degree of its compaction and development of erosion processes; mass of stubbly rests for restoration of soil fertility; necessity of application of an irrigation. If the variety is not adapted genetically for a wide spectrum is soil-environmental conditions, that is does not possess corresponding norm of reaction it cannot resist to action of various stresses. The adaptive variety is the variety adapted not only to optimum conditions, but also to a minimum and a maximum of external factors of environment. Creation of such agro-ecologically address varieties is the major problem of plant breeding [2].

At any direction of breeding of spring barley the yield per area unit in a combination with precocity and resistance against adverse factors remains the main criterion of an estimation of a new variety [4]. It is proved that with increasing of potential productivity of varieties their resistance to adverse factors of environment decreases that influences actual productivity of these varieties – it decreases too [5]. As a result breeders are faced now with a problem not only to increase productivity of plants but also to combine it with resistance to cultivation conditions.

The genotype and environment interaction linked with various norm of reaction of genotypes and change of their ranks in various conditions of environment is a biological basis of environmental problems of plant breeding [2]. It is possible to define norm of reaction of a variety in case of its binding to exact limiting factors of environment on time and on place. Existing methods of an estimation of ecological stability of varieties are based on different criteria of an estimation of a studied material and are widely presented in the modern agricultural and genetics literature [5–10].

With a view of reduction of ecological dependence of variety it is necessary to spend purposeful selection on adaptability to contrast weather conditions and, first of all, to the extreme one. It is important because the crop shortage in adverse years brings more essential economic losses than the income of a high yield in favorable years [9]. Ability of varieties to keep high productivity in various ecological conditions is highly appreciated by breeders and agriculturists. State variety testing is the most extensive set of environments for an estimation of the genotypes allowing to receive the objective information on their adaptive possibilities.

The aim of study – to define ecological stability of the area-specific and perspective varieties of spring barley under conditions of environment differ on time (competitive variety testing) and on place (state variety testing) with use of various methods of an estimation.

4.2   MATERIALS AND METHODOLOGY

Experimental work was carried out on fields of North-East Agricultural Research Institute (Kirov, Russia). The data of productivity of varieties on State variety test stations of Kirov region was used also. There are stations near towns Malmyzh, Podosinovets, Slobodskoy, Sovetsk, Yaransk, and Zuevka.

Objects of researches were area-specific for Kirov region and new perspective varieties of spring barley.

Observations, estimations and yield accounts were spent according to the State commission technique on variety testing of agricultural crops [11].

In competitive variety tests (2007–2011) the following varieties bred in the North-East Agricultural Research Institute were studied: Dina, Fermer, Kupets, Lel, Novichok, Pamiaty Rodinoy, Rodnik Prikamia, and Tandem. Variety Bios 1 was used as the standard. Studying was spent on plots with the registration area of 10 m^2, with four replications. Sowings were settled down in a breeding crop rotation. Soil of experiment site is sod-podzolic, middle loam, well cultivated. The experiments were sown in optimum early terms; mineral fertilizers were input in dose $N_{60}P_{60}K_{60}$.

As material for an estimation on State variety test station of Kirov region next varieties was served: Bios 1, Moscowsky 2 (Moscow Agricultural Research Institute), Zazersky 85 (Belarus), Abava (Latvia), Dina, Dzhin, Ecolog, Novichok, Lel, Tandem (North-East Agricultural Research Institute), and Hlynovsky (Vyatka State agricultural academy, Kirov).

Reliability of the received results of researches was estimated by a method of the two-way dispersion analysis (first factor – a variety; second factor – year or location of variety test station) according to B.A. Dospekhov technique [12]. According to Goncharenko [9] the difference between the minimum and maximum productivity of a variety for years of studying ($Y_{min} - Y_{max}$) reflects level of resistance of a variety to stressful factors: the low is the index the higher is level of stress-resistance of variety. Average productivity of a variety in years contrast on growth conditions $[(Y_{min} + Y_{max})/2]$ characterizes genetic flexibility of a variety, for example, the higher is conformity between a genotype and environment factors the higher this indicator is. An average index of environment was defined as average value of productivity of all varieties in competitive variety tests for years of estimation, in state variety tests – on the average mean on different State variety test stations. The estimation of effect of interaction of genotype and environment, adaptive ability and stability of varieties is spent according to A.V. Kilchevsky's technique [3]. At selection on ecological stability determination of the general adaptive ability of a genotype (GAA) was used which characterizes average value of a trait in various environmental conditions as well as specific adaptive ability (SAA) – a deviation from GAA in the exact environment. An average index of environment was determined as average value for all sites. A line of methods of estimation of genotype – environment interaction is based on regression analysis [7, 9]. Two parameters for an estimation of fitness of variety are used for this purpose: average value of variety in all environments and linear regression of productivity on average yield of all varieties for each year. The linear regression coefficient (b_i) thus serves as a measure of phenotypic plasticity. As it follows from model [6], the high-yielding variety should have b_i close to 1. Change of factor towards increase ($b_i > 1.0$) indicates responsiveness of a variety on change of conditions of cultivation, for example, characterizes a variety as intensive. Coefficient of regression of productivity of varieties on environmental conditions counted on Eberhart, Russell technique [13]. The technique offered by S.P. Martynov [14] ranges genotypes on their ability to combine high potential productivity in favorable conditions with its minimum decrease in adverse conditions of cultivation.

4.3   RESULTS AND DISCUSSION

The estimation of barley varieties in competitive variety tests in years differ on environmental conditions has revealed significant distinctions between them on productivity. Results of the dispersion analysis have shown that the variation of productivity of varieties has been defined basically by ecological factors. The share of influence of weather conditions made 77.4%, effect of interaction "genotype – environment" – 9.6%. Influence of a variety on a productivity variation was much less and made only 5.4%. It confirms conclusions [3, 9, 10] that in varieties the share of a variation of productivity is caused basically by ecological factors, and the share of influence of a genotype on this variability much more low that accordingly reduces level of ecological reliability of new barley varieties. Accordingly the urgency increases of researches on creation of varieties characterized by the lowered reaction to change of growth conditions or selection of them among existing.

Average productivity for years of researches (an average index of environment) made 5.6 t/ha (LSD_{05} = 0.4 t/ha). The characteristic of years on hydrothermal condition types has shown that optimum growth conditions for formation of high productivity have developed in 2008 and 2009; in 2011 – at level of average value for years of estimation. Significant decrease in productivity by 0.9 t/ha in compare with average index of environment was marked in 2007 as a result of combination of the long period of abnormal heats with practical absence of precipitations during the period of grain filling and maturing; the lowest productivity has been fixed in 2010 (Figure 4.1).

Various reactions of varieties on change of growth conditions were revealed in studies. The maximum productivity under stressful conditions of 2010 had variety Dina, significant excess over a standard variety

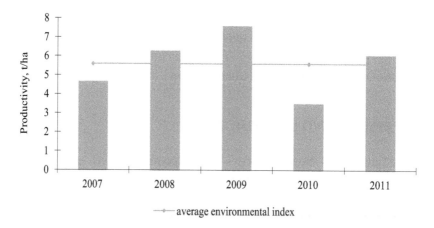

FIGURE 4.1 Influence of conditions of cultivation on productivity of barley varieties, 2007–2011.

Bios 1 (LSD_{05} = 0.9 t/ha) is noted for varieties Lel, Pamiaty Rodinoy, and Rodnik Prikamia. Varieties Kupets and Tandem have proved as sensitive to stress which has lowered productivity more than twice (on 62.9 and 58.4% accordingly).

Under favorable growth conditions (2009) ranging of varieties has changed: the maximum productivity had variety Lel; some less – varieties Kupets, Pamiaty Rodinoy, and Tandem. Undoubtedly, such reaction of varieties is caused by their biological features and distinctions at genotype – environment interaction.

According to A.A. Goncharenko's technique [9] the highest average productivity in contrast growth conditions is noted in variety Lel. Rather high level of productivity had varieties Dina, Kupets, Pamiaty Rodinoy, and Tandem (Table 4.1).

Average productivity of varieties for all years of study has revealed high potential possibilities of varieties of Lel, Pamiaty Rodinoy, and Tandem which have significantly exceeded all other varieties including a standard variety Bios 1. But use of regression analysis for estimation has not allowed to reveal essential distinctions between varieties on studied trait; in experiment the index b_1 in all varieties was close to 1. The exception was made by varieties Novichok and Dina. In variety Novichok value b_1 less than 1 is explained by low average productivity in study; but the early ripening variety Dina because of short growth period in 2010 has

TABLE 4.1 Adaptive Potential of Studied Barley Varieties on Productivity, 2007–2011 (calculated according to Ref. [9])

Variety	Productivity, t/ha		$Y_{min} - Y_{max}$	$(Y_{min} + Y_{max})/2$
	Y_{min}	Y_{max}		
Bios 1, standard	2.9	7.2	–4.3	5.0
Dina	4.4	6.7	–2.4	5.6
Kupets	3.0	8.2	–5.1	5.6
Lel	3.9	8.5	–4.6	6.2
Novichok	3.1	6.7	–3.6	4.9
Pamiaty Rodinoy	3.8	8.1	–4.3	5.9
Rodnik Prikamia	3.3	7.5	–4.1	5.4
Tandem	3.5	8.3	–4.9	5.9
Fermer	2.8	6.8	–4.0	4.8

generated yield before a drought and as a result has lowered productivity less than other varieties that was reflected most likely in value of coefficient of regression (Table 4.2).

The obtained data confirms Kilchevsky and Khotyleva's opinion [7] that the coefficient of regression depends on average value of a trait, for example, varieties with high value of a trait will have higher coefficient of regression. Hence, use a method of regression analysis only does not provide enough full information on ecological stability of all set of varieties at this exact growth conditions. However, as Kilchevsky and Khotyleva [7], the coefficient of regression at all specified disadvantages characterizes responsiveness of a genotype on environment and can be used for a rough estimate of ecological stability. Proceeding from the received data, varieties Dina and Novichok can be characterized as plastic; but varieties Bios 1, Lel, Pamiaty Rodinoy, and Rodnik Prikamia – as less sensitive to change of growth conditions. At the same time in agree with [9] increase in varietal plasticity leads to reduction of their fitness and stability at the expense of increase of sensitivity of a variety not only to favorable, but also to adverse conditions of growth.

Parameters of ecological stability, being a quantitative measure of fitness of genotypes, do not give the information on the general (GAA) and specific adaptation ability (SAA) to exact growth conditions. The method of the genetic analysis developed by A.V. Kilchevsky [3] based on test of

TABLE 4.2   Parameters of Phenotypic Plasticity of Barley Varieties, 2007–2011

Variety	Average productivity, t/ha (Y_{aver})	b_i	$V, \%$
Bios 1, standard	5.0	1.0	32.6
Dina	5.6	0.6	16.6
Kupets	5.3	1.1	35.5
Lel	6.0	1.1	28.7
Novichok	5.3	0.9	28.9
Pamiaty Rodinoy	6.2	1.1	27.9
Rodnik Prikamia	5.5	1.0	31.9
Tandem	5.9	1.1	29.4
Fermer	5.4	1.1	33.8
LSD_{05}	0.4	—	—

genotypes in various environmental conditions allows to reveal *GAA* and *SAA* of genotypes and their ecological stability. Estimation of *GAA* allows to select genotypes providing the maximum average yield in all set of environments. At estimation on *GAA* in our study varieties Lel, Pamiaty Rodinoy, and Tandem were selected that confirms the previous data about a combination in the given varieties of high productivity and ecological stability. For establish of a deviation from *GAA* in exact environmental conditions we used variance of *SAA* ($\sigma^2 SAA_1$). The least variation of productivity on years is noted in ascending order in varieties Dina, Novichok, Bios 1, Lel, and Pamiaty Rodinoy (Table 4.3). However, in varieties Bios 1 and Novichok it is explained by low productivity at all years of study; on productivity of variety Dina the drought of 2010 has affected to a lesser degree as it already was marked earlier.

The value of index of "relative stability of genotypes" (S_{gi}) to which according to Kilchevsky [3] it is necessary to prefer at determination of stability of varieties confirms high ecological stability of varieties Lel, Pamiaty Rodinoy, and Tandem. Similar results are received at simultaneous selection of genotypes on productivity and stability; for this purpose the author [3] suggests to determine "breeding value of a genotype" (BVG_i).

For specification of the received data results of competitive variety tests have been analyzed according to S.P. Martynov's technique [14]. Results have also confirmed the previous conclusions about high ecological stability of varieties Dina, Lel, Pamiaty Rodinoy, and Tandem (Table 4.4).

TABLE 4.3 Parameters of Adaptive Ability and Stability of Barley Varieties, 2007–2011 (Calculated According to Ref. [3])

Variety	Productivity, t/ha	GAA_i	$\sigma^2 SAA_i$	S_{gi}	BVG_i
Bios 1	5.0	−0.56	2.62	32.2	2.52
Dina	5.6	0.05	0.91	16.0	2.82
Kupets	5.3	−0.31	3.44	35.1	2.64
Lel	6.0	0.42	2.90	28.3	3.01
Novichok	5.3	−0.30	2.37	29.1	2.65
Pamiaty Rodinoy	6.2	0.60	2.92	27.6	3.10
Rodnik Prikamia	5.5	−0.10	2.99	31.5	2.75
Tandem	6.0	0.40	3.04	29.1	3.00
Fermer	5.4	−0.18	3.28	33.5	2.71

TABLE 4.4 Stability Indexes of Productivity of Barley Varieties (Calculated According to Ref. [14])

Variety	Stability index	
Bios 1	−5.54	below average
Dina	0.88	above average
Kupets	−2.88	below average
Lel	2.71	above average
Novichok	−0.59	below average
Pamiaty Rodinoy	3.58	above average
Rodnik Prikamia	−0.74	below average
Tandem	2.80	above average
Fermer	−0.24	average
Average mean	0.00	—
LSD_{05}	0.32	—

As a result of carrying out of the analysis of experimental data of competitive variety tests it is established that ranging of varieties on their ecological stability remained at use of different techniques of estimations (Table 4.5).

Presence of the electronic version of S.P. Martynov's technique "AGROS" [15] simplifies considerably determination of ecological ability of varieties.

The data received at the analysis of competitive variety tests are co-ordinated with results of an estimation of barley varieties on State variety test stations of Kirov region. Considerable influence of growth conditions on productivity was established by method of dispersion analysis too. The share of influence of year made 46.3%, influence of a genotype of variety and a growth place on a productivity variation was much less and made only 4.7 and 14.4%, accordingly. Effect of interaction "variety – variety test stations" was 4.4%. An average index of environment in the given study was 4.8 t/ha (LSD_{05} = 0.2 t/ha). The estimation of variety test stations by type of growth conditions has shown that Slobodskoy and Sovetsk were characterized by higher productivity of barley varieties (6.0 and 5.9 t/ha accordingly) (Figure 4.2).

It is established [7, 16] that under industrial conditions the genotypes were selected more adapted for low-yielding environments, but under conditions of variety test stations – to high-yielding environments. In this

TABLE 4.5 Ranging of Barley Varieties of Competitive Variety Testing Nursery on Ecological Stability by Different Techniques of An Estimation

Estimation technique			
A.A. Goncharenko [9]	S.A. Eberhart, W.A. Russell [13]	A.V. Kilchevsky, L.V. Khotyleva [7]	S.P. Martynov, [14]
Lel	Pamiaty Rodinoy	Dina	Pamiaty Rodinoy
Pamiaty Rodinoy	Lel	Pamiaty Rodinoy	Tandem
Tandem	Tandem	Lel	Lel
Kupets	Fermer	Tandem	Dina
Dina	Kupets	Novichok	Fermer
Rodnik Prikamia	Rodnik Prikamia	Rodnik Prikamia	Novichok
Bios 1	Bios 1	Bios 1	Rodnik Prikamia
Novichok	Novichok	Fermer	Kupets
Fermer	Dina	Kupets	Bios 1

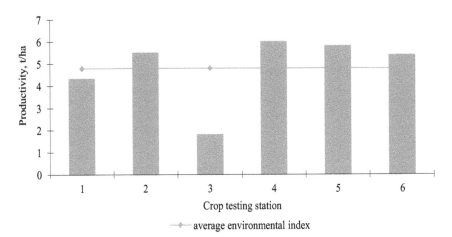

FIGURE 4.2 Productivity of barley on different Crop testing stations of Kirov region: 1 – Zuevka, 2 – Malmyzh, 3-Podosinovets, 4 Slobodskoy, 5 – Sovetsk, 6 – Yaransk.

relation there is a necessity at test at different background to select geno-types adapted for an average background typical for industrial conditions along with varieties of intensive type. In region it is necessary to grow up group of complementary varieties with various types of fitness which use various ecological and agrotechnical conditions as much as possible and successfully resist to limiting factors [17]. Variety Novichok can serve as an example of such variety created especially for sod-podzolic soils with high acidity caused by the high contents of ions of hydrogen and aluminum. On variety test stations with high level of fertility a variety Novichok did not exceed the standard, but at the same time at Podosinovets station the significant increase to the standard Bios 1 is received. Productivity of the given variety on the average for three years on different variety test stations varied from 2.2 t/ha (Podosinovets station) to 5.5 t/ha (Slobodskoy station). The maximum productivity is noted also at Podosinovets station – 2.8 t/ha and at Slobodskoy station – 6.5 t/ha.

Among area-specific barley varieties by potential productivity for all years of studying have caused a variety Ecolog; its productivity of grain in 2003 at Slobodskoy station made 7.5 t/ha. Variety Dzhin characterized by stability of high yields; coefficient of variation for 20 years of researches has made only 15.6%. (Table 4.6).

Multirow barley varieties Lel and Tandem were characterized by equally high productivity of 4.6 and 4.6 t/ha accordingly on the average

TABLE 4.6   Productivity of Barley Varieties on Crop Testing Stations of Kirov Region

Variety	Number of years of investigation	Productivity, t/ha		V, %
		Average for 2001–2011	maximum year, Crop test station	
Dzhin	20	4.4 ± 1.9	6.9 2003, Slobodskoy	15.6
Dina	25	4.1 ± 0.2	6.9 2006, Yaransk	17.6
Kupets	2	4.5	6.5 2011, Zuevska	—
Lel	10	4.8 ± 0.3	6.9 2011, Malmyzh	18.2
Novichok	11	4.3 ± 0.2	6.5 2003, Slobodskoy	19.7
Rodnik Prikamia	4	4.5 ± 0.4	6.3 2011, Malmyzh	19.2
Tandem	8	4.6 ± 0.4	6.9 2004, Sovetsk	21.9
Ecolog	22	4.4 ± 0.2	7.5 2003, Slobodskoy	17.3

for all years of studying. It confirms conclusions of long-term experiences in laboratory of selection and primary seed-growing of barley of the North-East Agricultural Research Institute (Kirov, Russia) proving doubtless advantage of varieties of multi-row barley over two-row barley under conditions of Kirov region [18].

The estimation of varieties by results of ecological test on variety test stations of Kirov region has shown that all varieties were divided into three conditionally accepted groups by reaction to environmental conditions. Varieties Ecolog and Moscowsky 2 were segregated in group with minimum level of stability (Table 4.7) which had an index of stability of less than one (-4.10 and −8.57 accordingly). This data confirms our previous conclusions [19] that variety Ecolog is recommended for economies with a high standard of farming.

Varieties Dina and Dzhin were segregated to group of varieties with average level of stability; their productivity laid in interval 4.6±0.3 t/ha and index of stability – within a confidential interval (–0.70–0.13). High level of stability (5.22–2.17) was characteristic for varieties Bios 1, Lel, Novichok, and Tandem.

Over the last 10 years varieties of barley bred in North-East Agricultural Research Institute, (Kirov, Russia) characterized by stably high productivity Lel, Pamiaty Rodinoy, and Tandem were regionalized in Volga-Vyatka region of Russia.

TABLE 4.7 Stability Indexes of Spring Barley Varieties (Calculated According to Ref. [14])

Variety	Productivity, t/ha (average for Crop testing stations of Kirov region)	Stability index
Bios 1	4.8	2.17 (above average)
Dzhin	4.6	0.13
Dina	4.6	−0.70
Lel	5.1	5.22 (above average)
Moskovsky 2	3.9	−8.57 (below average)
Novichok	4.7	2.23 (above average)
Tandem	5.0	3.62 (above average)
Ecolog	4.5	−4.10 (below average)
Average mean	4.6	0.00
LSD_{05}	0.3	0.97

Variety Lel (multi-row barley) is included in the State register of the breeding achievements admitted to use in Volga-Vyatka region since 2005. A subspecies – *pallidum*. Plants of variety Lel have an upright bush, wide (1.2–1.5 sm) leaves with light green coloring. Coloring of stem knots and crescent ears the light. A culm is long and strong. An ear is yellow, cylindrical, average length and the raised density, weakly dropped. Awns are long, jagged, and yellow. Grain of average size, of semi-extended form, 1000 grain mass – 36.9 g.

Variety Lel is middle maturing. It ripens on the average for 80 days, is characterized by faster passage of the period from heading till maturing in comparison with the standard. It is resistant against lodging, despite tall. High-yielding, average productivity for years of competitive test made 6.8 t/ha that is 1.0 t/ha higher the standard, maximum productivity – 9.2 t/ha. In all years of studying variety Lel exceeded significantly the standard on productivity at the expense of higher productive tilling capacity, the best grain content in the ear and grain mass per plant. Calculation of a biological yield has shown that variety Lel can surpass a standard variety Bios 1 by 29–33%. High productivity of a new variety proves to be true on State variety test stations of Kirov region. In 2011 on Malmyzh station its productivity reached 6.9 t/ha, excess over the standard made 0.7 t/ha, in 2010 on Slobodskoy station – 0.6 t/ha. In droughty 2010 on seed-growing sowings of laboratory of selection and primary seed-growing of barley of the North-East Agricultural Research Institute productivity of a variety reached 6.7 t/ha.

In the State register of the selection achievements admitted to use in territory of the Russian Federation, multi-row barley is presented basically with winter forms. Among the varieties of spring barley included in the Register only 9.7% consists of multi-row barley. However, as it was marked earlier [16] multi-row barley varieties possess higher potential of productivity in comparison with the double-row barley in Volga-Vyatka region. So, for example, the maximum productivity of multi-row *variety Tandem* in our sowings is made about 10.3 t/ha.

Variety Tandem is regionalized since 2008. Subspecies – *pallidum*. Plants of variety Tandem have a semi-upright bush, green coloring of leaves. Coloring of stem knots and crescent ears is light. A culm is long, strong. An ear is yellow, cylindrical. Awns are long, jagged, and yellow.

1000 grains mass – 39.5 g. The variety is middle-maturing, ripens on the average for 80 days. Variety Tandem is high-yielding, average productivity in competitive variety tests in the North-East Agricultural Research Institute made 7.5 t/ha that is 1.3 t/ha higher the standard. The superiority of the variety over the standard was provided by higher productive tillering ability and grain mass per ear. The variety is resistant against lodging (4.5–5.0 points), spring frosts and drought; differs among other varieties with good teachability of grains.

According to data of ecological test on experimental fields of Mari Agricultural Research Institute (Mari El, Russia) (2008–2009) high responsiveness of variety Tandem on input of fertilizers is established. Advantage of a variety over the standard at background without application of fertilizers made 0.8 t/ha, at background of entering of a full dose of mineral fertilizers ($N_{60}P_{60}K_{60}$) – 1.1 t/ha at productivity of the standard – 4.3 t/ha.

Variety Tandem provided significantly high increase in yield in comparison with standard at Yaransk, Malmyzh, and Zuevka State variety test stations of Kirov region. Productivity of the variety at Malmyzh station in 2011 made 6.9 t/ha, excess over the standard – 0.7 t/ha.

Economic efficiency calculation has shown that wide using of new varieties Lel and Tandem in agricultural industry will allow to increase manufacture of barley grain counting on sowing area of Kirov region by 90 thousand tons, profitability will increase by 15.7 and 25.7% accordingly.

New barley *variety Pamiaty Rodinoy* is regionalized in Volga-Vyatka region since 2014. Subspecies – *nutans*. Plants of the variety have semi-upright bush. Antocyan coloring of ears and stem knots is absent or very weak. Occurrence of plants with the inclined flag leaf is very low. An ear is double-row cylindrical form with weak wax touch, average length and density, straw-yellow colored. Awns are long, jagged on all length, straw-yellow colored.

In competitive variety tests in the North-East Agricultural Research Institute the variety formed productivity of 3.9–7.9 t/ha. For years of competitive variety tests the variety Pamiaty Rodinoy significantly exceeded the standard on productivity by 0.7–1.4 t/ha. In droughty 2010 productivity of the variety made 3.9 t/ha that is 0.9 t/ha higher the standard. The maximum productivity of the variety is received in 2009–7.9 t/ha.

The variety is middle-maturing (73–80 days), ripens, as a rule, for 3 days earlier standard variety at the expense of faster passage of the period from the beginning of tillering to leaf-tube formation stage. The variety is characterized by ability to form high productive plant stand per unit of area at the expense of good tillering ability of plants. The new variety possesses a strong culm of the average length resistant against lodging. Distinctive feature of a grade is the grain having high 1000 grains mass (47.0–51.7 g) and good technological properties. It gives the chance to recommend the variety for use as for food-processing industry on groat and for brewing purposes.

The variety passed ecological test in two scientific research institutes of region. In Perm Agricultural Research Institute productivity of the variety in 2011 made 3.4 t/ha, excess over the standard was 0.6 t/ha (LSD_{05}=0.2 t/hectare). Ecological variety test (at experiment fields of Mari Agricultural Research Institute, 2008–2010) on two backgrounds of mineral nutrition (without fertilizers; and input of $N_{60}P_{60}K_{60}$) has shown high economic efficiency of cultivation of variety Pamiaty Rodinoy both without input of fertilizers, and at pre-sowing entering of mineral fertilizers. Productivity of a new variety for years of tests was 0.7–0.9 t/ha higher than in the standard. The increase of productivity has made solving impact on drop in the cost price of grain (by 28.4–31.7%) in comparison with the standard at cultivation on both backgrounds.

At the background without input of fertilizers the cost price of manufacture of a new variety is 22.1% lower than in the standard, thus expenses for 1 hectare remained invariable. The net profit of use of a new variety per 1 hectare is almost 60% higher than in the standard variety. Level of profitability of cultivation of the variety Pamiaty Rodinoy on both backgrounds was 57.6 and 108.5% higher the standard.

4.4 CONCLUSION

Thus, on the basis of the long-term data on productivity in various on time and growth place conditions ecological stability of spring barley varieties regionalized in Kirov region is determined. The varieties differ by high ecological stability are selected in competitive variety tests at experiment fields of the North-East Agricultural Research Institute – varieties Lel,

Pamiaty Rodinoy, and Tandem; on State variety test stations of Kirov region – varieties Lel and Tandem.

Efficiency of use of S.P. Martynov's technique for determination of ecological stability of barley varieties is established.

The obtained data allows to recommend wider introduction into agricultural industry of ecologically stable barley varieties Lel, Pamiaty Rodinoy, and Tandem adapted for conditions of cultivation.

KEYWORDS

- adaptive ability
- competitive variety test
- genotype
- index of environment
- linear regression coefficient
- state variety test
- variance

REFERENCES

1. Gryaznov, A. A. Karabalycsky barley (fodder, groat, beer). Kostanay: Kostanajskij Pechatnyj Dvor, 1996, 448 p. [in Russian].
2. Zhuchenko, A. A. Adaptive plant production (ecological and genetic backgrounds) Theory and practice. Volume II. Biologization and ecologization of intensification processes as a basis for transition to adaptive development of agroindustrial sector. Fundamentals of adaptive utilization of natural, biological and technogenic resources. Moscow: Publishing House Agrorus, Ltd. 2009, 1104 p. [in Russian].
3. Kilchevsky, A. V. Ecological organization of breeding process. Ecological genetics of cultivated plants. Krasnodar: All Russian Rice Research Institute. 2005, 40–55. [in Russian].
4. Plishchenko, V. M., Golub, A. S. Structure of yield of spring barley in dependence on growth conditions during pass of organogenesis stages. Agro XXI. 2009, N. 1–3. 30. [in Russian].
5. Kosyanenko, L. P. Coarse grain crops in East Siberia. Krasnoyarsk: Krasnoyarsk State Agrarian University Publ, 2008, 300 p. [in Russian].

6. Pakudin, V. Z., Lopatina, L. M. Estimation of ecological plasticity and stability of varieties of agricultural crops. Agricultural biology. 1984, N. 4. 109–103. [in Russian].

7. Kilchevsky, A. V., Khotylyova, L. V. Ecological plant breeding. Minsk: Tehnalogiya. 1997, 372 p. [in Russian].

8. Golovochenko, A. P. Peculiarities of adaptive breeding of soft spring wheat in forest-steppe zone of Middle Povolzhye. Kinel: Povolzhsky Institute of Breeding and Seed growing, 2001, 380 p. [in Russian].

9. Goncharenko, A. A. About ecological plasticity and stability of productivity of varieties of cereal crops. Ways of increasing of resistance of agricultural production in modern conditions. Orel: Orel State Agrarian University Publ., 2005, 46–56 [in Russian].

10. Kurkova, I. V., Rukosyev, R. V. Estimation of parameters of stability in spring barley varieties of Far East breeding. Bulletin of Altai State Agricultural University. 2013, Vol. 99. N. 1. 13–14. [in Russian].

11. Methods of State Commission on testing of agricultural crops. Moscow, 1983, 269 p. [in Russian].

12. Dospekhov, B. A. Technique of field experiment (with basis of statistical treatment of results of investigations). Moscow: Agropromizdat, 1985, 351 p. [in Russian].

13. Eberhart, S. A., Russell, W. A. Stability parameters for comparing varieties. Crop Science. 1966, Vol. 6. N. 1, 36–40.

14. Martynov, S. P. Estimation of ecological plasticity of agricultural crops. Agricultural Biology. 1989, N. 3. 124–128. [in Russian].

15. Package of breeding-directed programs AGROS version 2.07. Tver. 1993–1997. [in Russian].

16. Boroevitch, S. Principles and methods of plant breeding. Moscow: Kolos, 1984, 343 p. [in Russian].

17. Rodina, N. A. Barley breeding in North-East of Non-Chernozem Zone. Kirov: North-East Agricultural Research Institute, 2006, 488 p. [in Russian].

18. Rodina, N. A., Shchennikova, I. N., Kokina, L. P. Reaction of new barley varieties on different technological procedures. Achievements of science and technique of Agrarian-Industrial Complex. 2009, N. 8, 14–16. [in Russian].

19. Shchennikova, I. N., Kuts, S. A., Abdushaeva Ya.M. Estimation of varieties and hybrids of barley in conditions of North-East of Non-Chernozem Zone. Advances in Current Natural Sciences. 2007, N. 4, 21–24. [in Russian].

CHAPTER 5

BARLEY GENOTYPES (*HORDEUM VULGARE* L.) CREATED BY THE METHOD OF CELL SELECTION

OLGA N. SHUPLETSOVA, IRINA N. SHCHENNIKOVA, and TATYANA K. SHESHEGOVA

CONTENTS

ABSTRACT

Genotypes of barley resistant against high soil acidity, toxicity of aluminum and osmotic stress are created by method of cell selection. Influence of various selective systems on survival rate of callus and on ability of callus tissue to form regenerant plants at a stage of morphogenesis is studied. Optimum selective systems are developed for selection of callus cultures. Monitoring of level of stress resistance of initial varieties and their regenerant forms in laboratory conditions has shown that cultivation in the conditions of tissue culture and subsequent selection of regenerant plants has varietal-specific mode. As a result of field researches of resistance to osmotic stress the decisive superiority of regenerant plants over the standard variety on degree of decrease in efficiency of plants and productivity of genotypes under drought conditions has been revealed. Immunological estimation of regenerant lines of barley at natural and artificial *Helminthosporium* backgrounds has not revealed immune genotypes. Wide enough intravarietal differentiation on susceptibility is found out. The material perspective for creation of new high-yielding adaptive varieties of barley is created and is at various stages of selection process.

5.1   INTRODUCTION

On the acid sod-podzolic soils occupying more than a half of all arable land of the North-East of the Non-Chernozem zone of Russia the basic negative factors influencing plants are high content of exchangeable ions of aluminum and hydrogen along with low natural fertility that reduces productivity of cultivated grain crops by 20–50% [1]. At the same time meteorological conditions of region are characterized with non-uniform drop of precipitations during the growth season. Even in areas of sufficient humidifying where the Non-Chernozem zone enters essential change of an amount of precipitation during growth season is observed. Deviations from mid-annual norm reach 3–5 folds [2]. Last years in the Kirov region there was a change of intensity of precipitations during the period of grain filling which is necessary, as a rule, for July. Since 1977 till 1995 the amount of precipitation in July was marked close or above norm. And since 1996 for 17 years of observations in 41.2% cases (7 years) precipitations dropped out less than 60% [3], in 1997, 2001 and 2010 the amount of precipitation made 16, 22 and 11% of mid-annual norm accordingly that characterizes these years as droughty [4]. The raised temperatures rendered negative influence on growth and development of plants during the given period as well.

Toxicity of aluminum in soil retards growth and degree of penetration of roots in depth that in addition reduces resistibility of plants to a drought especially in the conditions of deficiency of precipitations; as consequence there is insufficient use of nutrients in subsoil that as a result increases losses of a crop by 85% [2, 5]. In the light of predicted climate change the special urgency is got by purposeful creation of varieties with the adaptive reactions providing complex resistance to high soil acidity, toxicity of aluminum and to deficiency of moisture (osmotic stress).

Now, despite an intensification of the gene-engineering works, traditional selection is still the unique real decision of a question of creation of genotypes of the plants capable to maintain toxic action of metals without decrease in productivity. As a way of increase of adaptive potential of cultivated plants use of cell selection is probably. Realization of mechanisms of resistance to ionic toxicity and to dehydration at cellular level [1, 2, 6] gives the grounds to use cell technologies of selection of resistant genotypes. Owing to somaclonal variability arising in cultures of cells and tissues in vitro, obtaining is possible of genotypes with the changed properties which selection in selective systems and the subsequent regeneration of plants allow to create a perspective material for adaptive selection.

Now the facts are known about occurrence in callus cultures treated to selection on resistance to one defined stressor resistance to stressor of other nature or even for several stressors simultaneously [7, 8]. This phenomenon which has received the name of cross-adaptation becomes possible owing to activation at cellular and molecular level of some mechanisms participating in formation of the general response of a plant on stressful influences [8]. However, similar phenomena under conditions in vitro do not give in to forecasting; the continuity of relations of signs of the conjugated resistance in vitro with respect to conditions in vivo is studied insufficiently. Therefore, for obtaining of plants with complex resistance to abiotic stresses in callus cultures selection of resistant genotypes in the selective systems including simultaneously or consequently all complex of stressful factors should consider most reliable.

It is necessary to develop selective systems for successful carrying out of cell selection with various combinations and "rigidity" of selective factors which would simulate in vitro the stress adequate on force of influence on plants in vivo.

One of limiting factors of reception of high yield of qualitative grain of barley in the main zones of cultivation including the Kirov region are helminthosporium blotch as well: net blotch (activator *Drechslera teres,* teleomorpha *Pyrenophora teres* Shotm.) and stripe blotch (activator *Drechslera graminea*, teleomorpha *Pyrenophora graminea* Jto et Kurib.). Their display is to some extent observed here annually. Defeat of plants with blotch leads to a crop shortage which can reach 10–30% and in epiphytotics years – up to 60% [9]. Owing to a wide circulation and injuriousness of these diseases there is a necessity of creation of the technologies focused on cultivation of agricultural crops, adaptive to influence of harmful organisms. Application of chemical methods of protection often demands large material expenses and can cause environmental contamination. Besides, application of the same pesticides for a long time leads to resistance of pathogens to acting substance of a preparation [10].

Therefore the most effective, ecologically safe, and power- and resource-saving method of protection of plants from phyto-pathogens is the breeding-genetic method, for example, creation of resistant varieties. The directed selection of agricultural crops on long resistance to illnesses can be provided by use of genetically various sources and donors of a sign. Search and creation of immunologically valuable genotypes is the beginning of any selection work.

At any direction of selection the yield per area unit in a combination to precocity and resistance to adverse factors remains the main criterion of an estimation of a new variety [11]. Selection is focused on creation of varieties capable to give stable high yield on a high agro-background and not to reduce it in the presence of stressful factors. Aluminum tolerant varieties, as a rule, possess low efficiency in comparison with the standard variety on soils not subject to stress [12]. As a result, breeders must now not only to raise productivity of plants but also to combine it with resistance to stressful conditions of cultivation.

The present work has the purpose to working out methodology of obtaining and estimation of barley genotypes with complex resistance to stressful factors.

5.2   MATERIAL AND METHODOLOGY

Six selection lines of barley (*Hordeum vulgare* L.) were used in the study: NN 370–05, 472–06, 552–98, 999–93, 917–01, 781–04 and regenerant lines obtained from them in generations R_4 and R_5. An induction of callusogenesis, regeneration of plants and obtaining seed progenies from them carried out according to earlier described methods [13].

Callus tissue was induced on Murashige – Skoog medium (MC) without introduction of selective agents. At stages of callus proliferation and morphogenesis selective conditions were created with addition in acid nutrient medium of aluminum sulfate and/or polyethyleneglycol as osmotic with molecular mass 6000 (PEG). In the control callus was passed on MC medium without selective agents with the reaction close to neutral (pH 6.0). In experiments applied schemes with consecutive and simultaneous introduction of selective agents (H^+, Al^{3+}, and PEG) in nutrient mediums at stages of proliferation and morphogenesis. Reaction of callus on selective agents was compared in two-factorial experiment, where the factor A is concentration of stressor in selective system (by three gradation for ions Al^{3+} and PEG), the factor B is sequence of introduction of the agent at various stages of callus development (direct sequence is Al – PEG, and reverse sequence is PEG – Al) (Table 5.1).

TABLE 5.1 The Experiment Scheme on Revealing of Optimum Concentration of Selective Agents for Consecutive Introduction in Nutrient Mediums

Variant number	Direct sequence		Reverse sequence	
	Input in stages			
	proliferation Al^{3+}, mg/L	morphogenesis PEG, %	proliferation PEG, %	morphogenesis Al^{3+}, mg/L
Control	0	0	0	0
1	20	10	10	20
2	25	10	10	25
3	30	10	10	30
4	20	12.5	12.5	20
5	25	12.5	12.5	25
6	30	12.5	12.5	30
7	20	15	15	20
8	25	15	15	25
9	30	15	15	30

An acceptability of this or that variant of selective systems was judged by quantity of the survived cultures (%) and frequency of regeneration of plants (%). The analysis of yield structure was carried out in regenerants of the R_0-generations grown up in chambers of an artificial climate.

Estimation of resistance of regenerants and initial varieties to aluminum-acid and to osmotic stresses was spent in laboratory conditions according to methodical instructions [14]. For division of a studied set of varieties into groups on resistance degree to aluminum-acid stress additional pair comparison has been spent on root length in control and in stressful backgrounds with use of criterion of Student (t_{05}).

In competitive variety tests researches were spent per 2008–2012 according to a technique of the State commission on variety test of agricultural crops [15]. As a material for studying regenerants of barley of initial selection lines 441–05, 530–98, 780–04, 774–04, 770–04, 917–01, and 781–04 created by selection of somaclonal forms in callus culture on acid (pH 3.8–4.0) selective media with aluminum (20–40 mg/L Al^{3+}) and the subsequent regeneration of plants were used. Variety Bios 1 recommended by the State commission on variety test is used as the standard.

Field drought resistance (*FDR*) counted under the formula [16]:

$$FDR = (Y_{fav} - Y_{dr})/Y_{fav} \times 100\%$$

where Y_{fav} is yield in favorable conditions of growth; Y_{dr} is yield in drought conditions.

At creation of an artificial infectious background on the root rot a mix of pathogenic strains of fungi *Bipolaris sorokiniana* + *Fusarium. oxysporum* + *F. culmorum* in the ratio 70:20:10 was used. An infection in the form of infected grain mix in a dose 200–250 g/m² was input simultaneously with sowing of seeds observing direct contact to them [17, 18]. An artificial infectious background on stripe and net blotch was created by infection of flowers in a phase of green anthers [19]. As inoculum water-spores suspension of local populations of *Drechlera graminea* and *Drechlera teres* in concentration of spores 5×10^5 conidia/mL was used.

In 2008, 2009, 2011 and 2012 heat and moisture combination was favorable for growth and development of barley plants. In 2010, productivity of genotypes developed of indicators the majority from which was formed under favorable conditions of growth during the period "seedlings – ear formation". At development of kernel and it swelling abnormal hot dry weather is prevailed on 5.2°C above climatic norm, precipitations has dropped out only 9 mm at norm of 79 mm.

Statistical data processing is executed with use of a package of software AGROS version 2.07 [20].

5.3 RESULTS AND DISCUSSION

Working out of selective conditions for selection of genotypes of barley with complex resistance to toxicity of aluminum and water deficiency was began with study of reaction of callus tissue of various genotypes on simultaneous input in composition of a nutrient medium of 20 mg/L Al^{3+} and 10% PEG. Reaction of medium lead up to pH 4.0. The choice of concentration of selective agents has been caused by results of earlier spent researches on their separate input in nutrient mediums [13]. At joint input of toxic ions and osmotic at a stage of proliferation callus tissue perished for 7–10 day after a passage or stopped in the development. To overcome the arisen difficulty we tried change of reaction of medium from 4.0 to 4.5 and 5.0 units of pH at which mobility of aluminum in selective system remains else (Table 5.2).

As a result of increase of medium pH (by 0.5–1.0 pH units) the survival rate of callus has risen practically in all investigated genotypes and has averaged more than 60% of initial value. Frequency of regeneration in callus cultures, unlike survival rate which on the average for all genotypes

TABLE 5.2 Reaction of Callus Tissue on Joint Input of 20 mg/L Al^{3+} and 10% PEG Depending on Acidity of Media

pH of medium	Initial selection line								Average mean	
	999–93		370–05		472–06		552–98			
	1	2	1	2	1	2	1	2	1	2
4.0	0	0	28	0	0	0	0	0	7	0
4.5	88	5.0	72	5.6	36	11.1	o.d.	o.d.	65.3	7.2
5.0	72	16.7	100	25.0	32	37.5	48	33.3	63	28.1

Note: 1 – survival rate of callus, %; 2 – frequency of regeneration, %; o.d. – out of data.

did not differ significantly at pH values 4.5 and 5.0, was at pH 5 considerably higher (28.1%), than at pH 4.5 (7.2%). Irrespective of acidity of media and of initial genotype joint input of aluminum ions and osmotic was accompanied by occurrence of a significant amount of albinos among plant-regenerants that testified to active mutational processes and further it could appear an obstacle for carrying out of cell selection.

The alternative approach to selection of resistant genotypes in vitro is separate input in nutrient mediums of selective agents on each of development stages of callus tissue. Studying of influence of various selective systems on four genotypes of barley resulted in Table 5.1 has allowed to come to conclusion that the survival rate of callus depended on concentration of selective agents and sequence of their input on media insignificantly changing by variants of experiment in limits from 36.4 to 50.6%. Both investigated factors (concentration of selective agents and sequence of their input in the selective media) have rendered significant influence on ability of callus tissue to form plant-regenerants at a stage of morphogenesis. It is revealed that the reverse order of input of selective agents (PEG at a stage of proliferation and Al^{3+} at a stage of morphogenesis) depressed regeneration ability of callus cultures of barley than direct sequence ($Al^{3+} \rightarrow$ PEG) (Figures 5.1 and 5.2) more considerably.

Besides, influence of osmotic concentration on regeneration ability in callus cultures exceeded influence of aluminum ions. At step increase in PEG concentration (10.0, 12.5, and 15.0%) frequency of regeneration was on the average 2, 3 and 5 times lower in comparison with the control

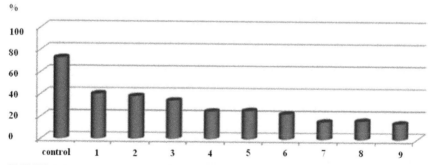

FIGURE 5.1 Frequency of regeneration in callus cultures of barley (an average for four genotypes) as a result of separate input of selective agents in direct sequences (numbers of variants, see Table 5.1).

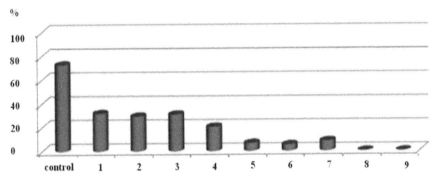

FIGURE 5.2   Frequency of regeneration in callus cultures of barley (an average for four genotypes) as a result of separate input of selective agents in reverse sequences (numbers of variants see Table 5.1).

accordingly and did not depend on concentration of aluminum. Frequency of regeneration of plants comprehensible for practical work took place at introduction on medium of 20–30 mg/L Al^{3+} and 10% PEG both in direct (34.3–40.2%) and in reverse (29.3–31.7%) sequences. The increase in PEG concentration to 12.5 and 15.0% irrespective of concentration of aluminum reduced ability of callus to form regenerants to 24.7 and 14.4% accordingly that specifies in low efficiency of these variants of selective systems.

Along with selective system, reaction of callus lines of barley on modeled in vitro stress was defined by genotypic features of an initial plant. Despite lacking of direct dependence between survival rate of callus lines and ability to regeneration (both in the control and in selective conditions) the increase in survival rate and regeneration ability of a genotype is revealed at its repeated introduction in culture in vitro [21].

Advantage of regenerants has proved to be true in complex selective systems too. Callus tissue initiated by genotypes of regenerative origin RA917–01 and RA781–04 passed earlier selection in callus cultures was characterized by higher survival rate (50–87%) in rigid selective conditions than callus received from an initial genotype 999–93 which in most cases is perished (Table 5.3).

By working out of selective systems for selection of callus lines resistant against a complex stressful edaphic factors it is necessary to consider possibility of lower efficiency genotypes of selected in vitro in comparison with initial varieties of plants. For the purpose of revealing of possible

TABLE 5.3 Reaction of Callus Tissue of Regenerants and Its Initial Genotype in Rigid Selective Conditions

Genotype	Without selection		Selection scheme					
			30 mg/L Al^{3+} – 15% PEG Step by step		15 PEG % – 30 mg/L Al^{3+} step by step		15 PEG % + 30 mg/L Al^{3+} simultaneously	
	1	2	1	2	1	2	1	2
999–93	88.0	18.2	54.5	0	25	0	0	0
RA 917–01	96.0	12.5	55.0	27.2	87.0	0	0	0
RA 781–04	96.0	16.7	50.0	13.6	81.8	50.0	86.9	30.0

Note: 1 – survival rate, %; 2 – frequency of regeneration, %.

negative influence of selective systems the analysis of structure of productive signs of regenerants (R_0) grown up in the conditions of an artificial climate was carried out. In all plant-regenerants received in selective systems independence of the selection scheme signs "plant height" and "number of spikelets in an ear" were lower in comparison with the control plants which have been not subjected to selection on selective media. At the same time duration of their growth season was reduced in comparison with control plants by 4–7 days. Such signs as "productive tillering ability," "grain mass per ear and per plant" at the majority of the genotypes received on selective media with consecutive entering of selective agents did not differ significantly from control plants (Table 5.4).

On the contrary, selection on media with simultaneous input of both toxic ions and PEG led to decrease in productive signs at the majority of plants in comparison with the control. The exception was made by selection line 917–01 representing "secondary" regenerant initiated by selection line 999–93 at which "productive tillering capacity" and "grain mass per plant" significantly exceeded values of signs in the control.

Seed progeny of regenerants received by selection in selective systems on resistance to a complex of stressful factors was studied further in field conditions on the standard scheme of selection process accepted for self-pollinated crops.

At studying of resistance of genotypes to the high content of ions of hydrogen and aluminum it is revealed that the variation of value "root tolerance index" (RTI) changed from 0.61 to 1.20. The received results

TABLE 5.4   Influence of Conditions of Selection In Vitro on Productive Signs of Regenerants (R_0 Generation)

Regenerant	Control (without of selection)	Input of selective agent		
		Step by step Al^+(pH 4.0) – PEG	Step by step PEG – Al^+ (pH 4.0)	Simultaneously Al^+ (pH 5.0) + PEG
Productive tillering capacity, numbers				
RA370–05	11.4±2.9	12.3±4.2	14.8±3.0	12.5±2.1
RA999–93	11.3±3.7	20.0±2.0*	Out of data	9.4±2.5
RA917–01	14.3±1.5	24.6±2.1*	Out of data	23.9±1.9*
RA552–98	16.8±2.2	16.3±3.2	15.8±1.2	14.5±2.0
Grain mass per main ear, g				
RA370–05	1.2±0.1	1.2±0.5	1.6±0.2*	1.1±0.3
RA999–93	1.1±0.1	0.9±0.2	Out of data	1.0±0.4
RA917–01	1.4±0.1	1.6±0.2	Out of data	1.3±0.4
RA552–98	1.8±0.1	1.7±0.2	1.6±0.2	1.5±0.3*
Grain mass per plant, g				
RA370–05	10.4±0.4	12.8±0.8*	15.7±2.9*	12.8±1.9*
RA999–93	10.3±2.1	11.5±2.5	Out of data	11.7±1.4
RA917–01	9.1±0.7	13.3±1.3*	Out of data	14.0±3.8*
RA552–98	11.7±0.7	10.5±2.0	12.7±1.0	9.2±1.7*

Note: * distinction from control variant is significant at $p \leq 0.05$.

indicate that as initial genotypes and their regenerant forms which have been purposefully selected on selective media in vitro on the basis of resistance possessed high level of resistance to aluminum-acid stress. It is established that significant decrease in root length on a stressful background was marked in the standard Bios 1 and regenerants from selection lines 440–05 and 441–05. Simultaneously stimulating action of aluminum on genotype Novichok and selection lines 530–98, 780–04 is revealed (Table 5.5). It is noted significant decrease in root length at influence of stress at regenerant from selection line 496–98.

Comparison of stress resistance of selection lines and their regenerant forms has also shown that cultivation in the conditions of tissue culture and the subsequent selection has genotype-specific character. Selection from the initial selection line 999–93 (RTI = 0.86) promoted increase in

TABLE 5.5　Comparative Characteristic of Regenerants of Barley

Regenerant	Origin	Change of root length at stress treatment in comparison with control treatment, cm	RTI	t_f	t_t
Bios 1	standard	−0.9*	0.89	3.09	2.0
Novichok	—	+1.3*	1.20	2.86	2.1
440–05	RA Novichok	−1.5*	0.82	4.88	2.0
441–05	RA Novichok	−1.3*	0.83	2.63	2.0
999–93	1200–85 x 2867–80	−1.0	0.86	1.43	2.1
917–01	RA 999–93	−0.4	0.94	1.39	2.0
781–04	RA 999–93	−0.5	0.91	0.62	2.1
780–04	RA 539–98	+0.9*	1.11	2.23	2.0
530–98	RA 173–85	+1.3*	1.17	2.8	2.1

Note: t_f Student's coefficient in fact; t_t – Student's coefficient theoretical; * distinction between variants is significant at $p \leq 0.05$.

level of aluminum resistance in regenerant forms 781–04 (RTI = 0.91), 917–01(RTI = 0.94), and 780–04 (RTI = 1.11). But in genotype Novichok selection at cellular level has not caused essential increase in RTI value of the selected forms in comparison with an initial genotype. This feature of genotype Novichok was marked in our previous work [22].

The laboratory estimation of drought resistance of regenerants has shown distinctions between genotypes on ability to germinate in the conditions of osmotic stress. The majority of the studied regenerants by results of a laboratory trial have appeared sensitive to stress (resistance degree is lower than 33.3%). It is not revealed regenerants with high resistance to stress (more than 66.6%). From all set of regenerants it is selected regenerant 496–98 differ with average degree of resistance to osmotic stress (45.2%).

Results of field researches on resistance to osmotic stress have shown a real advantage of regenerants over the standard on degree of decrease in efficiency of plants and productivity of genotypes in drought conditions of 2010 in comparison with favorable conditions of growth in 2009.

The objective sign characterizing ability of a genotype not to react to a drought is "grain mass per ear" [23] in which formation sign "1000 grains mass" plays an important role; the sign "1000 grains mass" influences plant productivity as well [24]. In the present study significant dependence of efficiency of an ear and a plant from grain size is also revealed; the correlation coefficient made $r = 0.73$ and $r = 0.57$ accordingly.

Varietal distinctions on reaction to stressful conditions during the period of grain swelling are noted. In the standard Bios 1 grain mass per ear has decreased by 0.57 g (FDR = 32.2%) whereas in regenerants 917–01, 770–04, 441–04 (Table 5.6) it is not revealed significant change of efficiency of an ear under the influence of a drought. But in regenerants 770–04 and 441–04 some increase of mass of grain per ear is fixed. This is explained with the fact that heats and the limited amount of precipitation have not affected on grain swelling and as result 1000 grains mass in these genotypes has increased by 1.7 (FDR = −4.0%) and 0.5 g (FDR = −1.1%) accordingly. But in standard genotype Bios 1–1000 grains mass has decreased by 6.7 g (FDR = 11.9%).

Ability of plants to form a considerable number of productive stems as a rule is defining in formation of grain mass per plant ($r = 0.51$–0.65). By our researches it is established that in favorable 2009 efficiency of plants almost directly depended upon productive tillering capacity ($r = 0.70$) whereas in a drought conditions of 2010 the given dependence was not detected ($r = -0.11$).

In 2009 the maximum efficiency is noted in the standard Bios 1–3.45 g per plant whereas in regenerants it changed from 2.60 (regenerant 530–98) to 2.13 g (regenerant 781–04). In 2010 regenerant 780–04 had the biggest grain mass per plant 2.60 g at 2.00 g in standard. Regenerants 917–01,

TABLE 5.6   Field Drought Resistance in Barley Regenerants, %

Regenerant	Grain mass per ear	Grain mass per plant	Productivity
Bios 1, standard	32.2	42.2	35.3
530–98	46.2	46.2	0.4
917–01	1.1	−16.8	14.1
781–04	6.98	24.9	21.1
780–04	12.3	−1.98	−4.08
770–04	−20.5	0.4	4.76
774–04	15.1	22.4	22.5
441–05	−22.2	−4.35	6.59

441–05, and 770–04 were characterized with high efficiency of plants too. The same regenerants were had lower indexes FDR which changed from -16.8% (regenerant 917–01) to 0.4% (regenerant 770–04) at 42.2% in the standard Bios 1. All regenerants in the conditions of stress have lowered productivity much less in comparison with the standard Bios 1 with an exception of regenerant 774–01 which field drought resistance was at the level of the standard. Index FDR of genotypes made from 0.4% (regenerant 530–98) to 21.1% (regenerant 781–04). Regenerant 780–04 exceeded productivity in droughty year by 0.2 t/ha at FDR $= -4.08\%$.

In our researches all studied regenerants have shown productivity up to the standard – variety Bios 1. The regenerant 917–01 with productivity of 5.11 t/ha that is by 0.24 t/ha (4.9%) higher the standard and by 0.34 t/ha (7.1%) higher than its initial form 999–93 was the best. The regenerant 917–01 is characterized by high combination ability. With its participation new hybrid combinations are created which are at all stages of selection process and are characterized by high productivity. In control nursery regenerant lines 382–10, 368–10, 378–10, and 372–10 and in preliminary nursery-regenerant lines 138–09, 458–09, and 459–09 are obtained with productivity more than 5.0 t/ha (the increase to the standard made more than 1.0 t/ha). All noted regenerant lines are perspective for creation of the new high-yielding varieties adapted for conditions of cultivation in Kirov region.

The immunological estimation of regenerant lines of barley against natural and artificial infectious helminthosporious backgrounds has not revealed immune genotypes. However, wide enough intravarietal differentiation on susceptibility is found out. In reaction to root rot regenerant lines are characterized as moderately resistant with an exception of regenerant line 781–04 which in 2012 was defeated in weak degree (Table 5.7).

At increase of infectious loading all regenerants pass in susceptible group. It is possible to note only regenerant lines 917–01 and 781–04 with the least development of illness. In relation to helminthosporious blotch of leaves regenerant lines show high resistance at background without infection. At artificial infection regenerant lines 917–01 and 530–98 are characterized by average level of resistance to strip and net blotch. The best immunological state has regenerant line 917–01. It is defeated with helminthosporious diseases to a lesser degree than standards and other regenerant lines on both backgrounds of developments of activators.

TABLE 5.7 Immunological State of Perspective Regenerant Lines of Barley, 2009–2012

Regenerant line	Degree of defeat, %			
	by root rot	by stripe blotch	by net blotch	by spot blotch
Bios 1 – standard 1	23.3/43.5*	16.3	21.3	7.0
Nur – standard 2	32.2/43.0	4.0	19.0	8.0
917–01	16.0/30.5	0.5/15.5	5.0/20.0	2.0
530–98	21.1/33.2	7.5/23.5	8.8/20.0	2.5
496–07	18.0/35.0	8.0	10.0	4.7
781–04	12.1/23.4	2.0	6.7	3.3

Note: * – in a denominator – defeat degree on an infectious background, in a numerator – on natural infectious background.

As a result of our researches the regenerant lines exceeding standard varieties at cultivation on a favorable agro-background are obtained. That fact is especially remarkable that all obtained regenerant lines except for 781–04 in droughty 2010 have significantly exceeded the standard on productivity (Table 5.8).

Hence the drought has not affected process of grain swelling in regenerants. The maximum productivity is noted in regenerant line 496–07 (7.50 t/ha) excess over the standard on the average for years of studying has made 1.72 t/ha.

Regenerant line 496–07 is received on selective medium with consecutive creation of osmotic and aluminum-acid stresses (10% PEG; 20 mg/L Al^{3+}, pH 4.0). High productivity of a new regenerant line is provided with good productive tillering capacity, high number of grains per ear, and more compact ear in comparison with the standard. The regenerant line

TABLE 5.8 Results of Competitive Test of Regenerant Lines in 2010–2012

Regenerant line	Productivity, t/ha				Relative to standard	
	2010	2011	2012	average	t/ha	%
496–07	7.50*	5.77	6.15*	6.47	+1.72*	136.2
917–01	5.30*	5.39	3.60	4.76	+0.49	111.5
781–04	4.90	5.17*	4.91	4.99	+0.01	100.6
530–98	5.40*	5.64	4.38	5.14	+0.57	112.4

Note: * statistically significant excess over the standard at $p \leq 0.05$.

differs with high resistance to lodging; on duration of the growth period it is characterized as mid-ripening. It confirms possibility of use of regenerant lines received on rigid selective media in creation of spring barley varieties resistant to stressful factors.

5.4 CONCLUSIONS

As a result of researches efficiency of use of a method of tissue culture in vitro for reception of forms of spring barley resistant to stressful factors is confirmed. By a method of cell selection new barley regenerant lines are created.

As a result of an estimation regenerant lines being sources of resistance to aluminum-acid stress are obtained: 530–98, 917–01, 780–04, 774–04, and 770–04; to a drought during the period of grain swelling: 780–04, 770–04, 441–05, and 917–01; regenerant lines 530–98, 917–01, 781–04, and 496–98 have smaller defeat by helminthosporious diseases in comparison with standards.

The combined resistance to aluminum-acid stress and a drought is characteristic to regenerant lines 917–01, 780–04, and 770–04; to aluminum-acid stress and helminthosporious diseases – 917–01, 781–04, and 530–98. Regenerant line 917–01 possessed a combination of high productivity and resistance to a complex of abiotic and biotic stressful factors at high combination ability. A material perspective for creation of new high-yielding adaptive varieties of barley is created with its participation and exists at various stages of breeding process.

High drought resistance of regenerant lines received on selective media with aluminum in conditions in vitro is an example of the non-specific resistance which mechanisms leads to creation of forms with complex resistance to stresses of the various nature.

KEYWORDS

- aluminum
- callus

- **helminthosporiosis**
- **osmotic stress**
- **productivity**
- **regenerants**
- **selective systems**
- **sod-podzolic soils**
- **soil acidity**

REFERENCES

1. Poukhalskaya, N. V. Problem questions of aluminum toxicity (review). Agrochemistry. 2005, N. 8. 70–82. [in Russian].
2. Nettevich, E. L. Selected works. Breeding and seed-growing of spring grain crops. Moscow, Nemchinovka: Agricultural Research Institute of Central Regions of Non-Chernozem Zone, 2008, 348 p. [in Russian].
3. Shchennikova, I. N. Influence of weather conditions on growth and development of plants of barley in the Kirov region. An Agrarian Science of Euro-North-East. 2014, N. 4. 9–12. [in Russian].
4. Agroclimatic bulletin of Kirov region 1977–2012. Kirov: Committee on Hydrometeorology, 2013, 320 p. [in Russian].
5. Klimashevsky, E. L. Genetic aspect of a mineral nutrition of plants. Moscow: Agropromizdat, 1991, 415 p. [in Russian].
6. Chernyadiev, I. I. Influence of water stress on the photosynthetic apparatus of plants and a protective role of cytokinins (review). Applied biochemistry and microbiology. 2005, Vol. 41. N. 2. 133–147. [in Russian].
7. Kuznetsov, V. V., Shevyakova, N. I. Stress responses of tobacco cells to high temperature and salinity. Proline accumulation and phosphorylation of polypeptides. Physiologia Plantarum. 1997, Vol. 100. N. 2. 320–326.
8. Gladkov, E. A. Biotechnological methods for isolation the plants possessing complex resistance to heavy metals and salinization. Agricultural biology. 2009, N. 6. 85–88. [in Russian].
9. System of conducting of agro-industrial manufacture of the Kirov region for the period till 2005, Kirov: North-East Agricultural Research Institute, 2000, 267 p. [in Russian].
10. Manzhelesova, N. E., Poljakova, N. V., Shchukanov, V. P., Korytko, L. A., Melnikova, E. V. Microbiological immunization of barley. Modern problems of immunity of plants to harmful organisms. Saint Petersburg: All Russian Institute of Plant Protection, 2008, 261–263. [in Russian].
11. Plishchenko, V. M., Golub, A. S. Yield structure of spring barley depending on growth conditions during passage of stages of organogenesis. Agro XXI. 2009, N. 1–3. 40–42. [in Russian].

12. Zhuchenko, A. A. Adaptive system of selection of plants (Ecologo-genetic bases). Moscow: Peoples' friendship university of Russia Publishing house, 2001, Vol. 1. 780 p. [in Russian].

13. Shirokikh, I. G., Shchupletsova, O. N, Shchennikova, I. N. In vitro obtaining of barley tolerant to Al in acid soil. Biotechnology. 2009, N. 3. 40–48. [in Russian].

14. Klimashevsky, E. L. Estimation of acid resistance in plants. Diagnostics of resistance of plants to stressful influences (a methodical guide). Leningrad, 1988, 97–100. [in Russian].

15. The technique of State commission on test of agricultural crops. [Ed. Fedin, M. A] Moscow. Publishing house of the Ministry of Agriculture of the USSR. 1985, 285 p. [in Russian].

16. Golovochenko, A. P. Feature of adaptive selection of spring soft wheat in a forest-steppe zone of the Middle Volga region (monograph). Kinel: Povolzhsky Research Institute of Breeding and Seed-Growing, 2001, 380 p. [in Russian].

17. Grigoriev, M. F. Methodical instructions on studying of resistance of grain crops to root rots. Leningrad: VASHNIL, VIR, 1976, 59 p. [in Russian].

18. Sheshegova, T. K., Kedrova, L. I. Methodical recommendation on creation of artificial infectious backgrounds and breeding of winter rye on resistance to fusarium diseases. Kirov. North-East Agricultural Research Institute, 2003, 30 p. [in Russian].

19. Rodina, N. A., Efremova, Z. G. Methodical recommendation on barley breeding on resistance to diseases and their application in North-East Agricultural Research Institute. Kirov: North-East Agricultural Research Institute, 1986, 78 p. [in Russian].

20. A package of breeding-focused software AGROS version 2.07. Tver. 1993–1997 [in Russian].

21. Shirokikh, I. G., Shupletsova, O. N. Cell selection of barley and its practical results. Resistance of plants to adverse factors of an environment. Irkutsk: Siberian Institute of Plant Physiology and Biochemistry, 2007, 330–333. [in Russian].

22. Schennikova, I. N., Shupletsova, O. N., Butakova, O. I. Evaluation of acidity (Al^+) tolerance in spring barley cultivars. Bulletin Applied Botany, genetics and Plant Breeding. 2009, Vol. 165, 179–182 [in Russian].

23. Kumakov, V. A., Evdokimova, O. A., Zaharchenko, N. A., Pozdeev, A. I., Sher, K. N. Productivity and drought resistance of wheat in South-East. Problems and ways of overcoming of a drought at the Volga region. Saratov: Agricultural Research Institute for South-East Region, 2000, Vol. 1, 275–285. [in Russian].

24. Shchennikova, I. N., Kokina, L. P., Butakova, O. I. Estimation of world barley gene-pool on grain size under condition of Volga-Vyatka district. An Agrarian science of Euro-North-East. 2011, N. 1, 12–16. [in Russian].

CHAPTER 6

BASIC ELEMENTS OF MINERAL NUTRITION AT PLANTS OF GRAIN CROPS UNDER CONDITIONS OF ACID STRESS

LYUDMILA N. SHIKHOVA, EUGENE M. LISITSYN, and GALINA A. BATALOVA

CONTENTS

ABSTRACT

Influence of aluminum ions on requirements of plants of grain crops (wheat *Triticum aestivum* L. and oats *Avena sativa* L.) in macronutrients is studied. In the first series of experiments it is revealed that doubling of a dose of phosphorus in the acid media has caused doubling of its inclusion in metabolic processes in aluminum-resistant wheat Irgina on early stages of development already, but in aluminum-sensitive wheat

Priokskaya – only on a flowering stage. Development of root systems in wheat varieties differ in their Al-resistance level in the conditions of full supply with nitrogen and potassium does not depends on presence of ions of hydrogen, aluminum or phosphorus in soil. Aluminum-sensitive wheat variety is characterized by considerable dependence of photosynthesis on conditions of a mineral nutrition whereas intensity of photosynthetic processes of aluminum-resistant variety is influenced basically by a development stage. In the second series of experiments plants of aluminum-resistant oats variety Krechet were capable to support metabolic processes with participation of nitrogen, phosphorus and potassium in the conditions of action of the stressful factor at the same level as in the neutral growth media. Changes of relative requirements in macronutrients specifies in considerable reorganizations of metabolic reactions in plants of Al-sensitive oats variety Argamak under stressor action. Change of the used form of nitrogen fertilizer (nitric, ammoniacal, or mixed) leads to considerable changes of relative requirements of oats plants in macronutrients. The biochemical processes related with action of the photosynthetic apparatus of leaves in stressful conditions have a priority in their supply with macronutrients, that is specified with smaller variability of the triple ratio N:P:K necessary for the maximum development of the leaves of oats in comparison with root systems at aluminum influence.

6.1 INTRODUCTION

The stresses limiting plant growth on acid soils consist of proton rhizotoxicity (low pH), nutrient deficiency (primarily phosphorus but also potassium, calcium, and other minerals), and metal toxicity (aluminum and manganese) [1]. Among these constraints, toxicity of exchangeable Al^{3+} ions is considered to be the major limiting factor at cultivation of plants on acid soils [2], a major factor reducing efficiency of plants on 67% of all acid soils [3].

Adequate entering of nutrients is necessary for efficient crop production. As the majority of acid sod-podzolic soils in natural state are deficient in primary nutrients particularly nitrogen and phosphorus [4], and the part of these elements become gradually depleted by crops removal, it is necessary to fill this shortage at the expense of external sources. However, this strategy has become economically less feasible with increase of production cost, especially in the soils demanding phosphoric fertilizers. Therefore, efforts of scientists are directed toward breeding of plants which are capable to receive the maximum nutrient elements from the soil and/or make this process more efficient.

Aluminum interferes with absorption, translocation and utilization of many essential nutrients necessary for a plant, such, as nitrogen, calcium, potassium, magnesium, and phosphorus [5, 6].

Nitrogen fertilizers are widely used for increase productivity of grain crops and the protein content of grain in cereal crops. However, it is necessary to optimize their use in order to decrease the risks of environmental contamination and production costs [7]. For that reason, efficiency of use of nitrogen fertilizers by plants becomes a very important trait in studying and breeding of plants, including cereal [8, 9]. Increase of accumulation of nitrogen in plants not at the expense of increase of entering of nitrogen fertilizers, but at the expense of creating genotypes with the better ability of their root systems to absorb nitrogen from soil becomes the core task in the decision of the problem. On the other hand, in order to get higher values of grain yield, that process should be followed by an increased intensity of photosynthesis. Otherwise high concentration of nitrogen in grain and straw will be reached with lowering in efficiency of nitrogen utilization [10]. As percent of acid soils is high throughout the world, so there is plenty of references dealing with parameters of metabolism of nitrogen on such soils [11] and considerable efforts is directed to establishment of genotypic specificity of parameters of a nitrogen metabolism [7, 12].

Aluminum toxicity and phosphorus deficiency are two common constraints limiting crop production in acid soils [13]. Understanding the mechanisms underlying aluminum and phosphorus interactions will help to develop management principles to sustain production of agricultural plants in acid soils. Effects of phosphorus-aluminum interaction on adaptation of plants to toxicity of acid soils were studied in many researches [14–17]. Fukuda et al. [18] have assumed that a common metabolic system is responsive to both deficiency of phosphorus and to toxicity of aluminum in rice.

Increasing phosphorus supply substantially decreased extractable Al in bulk soil [19]. This decrease in aluminum ions extractability in soil is likely to have resulted from the chemical sedimentation of aluminum with the added phosphorus that leads to lowering the activity of trivalent aluminum ions in the soil solution [20, 21].

The increasing entering of phosphorus can ameliorate the toxic effect of aluminum ions on root growth, but degree of the amelioration is dependent on the severity of the stress. High doses of phosphorus were more effective on influence on development of shoots than on the root systems. Very strong positive effect of entering of raising doses of phosphorus on shoot growth was found out at high level of stressful influence (200 µM aluminum) [19]. The key moment was that moving of aluminum ions from the roots to the shoot was markedly reduced at entering of 80 mg of phosphorus per kg of soil. Irrespectively of the level of added aluminum wheat seedlings were able to absorb more phosphorus from the soil, translocated more phosphorus to the shoots and utilize phosphorus more effectively for shoot growth and development with increase of level of entering phosphorus although total absorption of phosphorus decreased with the aluminum influence. Similar findings have been received previously in the conditions of nutrient solution cultures [20, 22].

It has been suggested that immobilization of aluminum by phosphorus in cell wall of roots is a potential mechanism for Al-tolerance in buckwheat [17], barley [23], and maize [22]. Under Al stress in hydroponics phosphorus application was shown to stimulate malate exudation (an indicator of aluminum resistance) from the tap root tip of the P-efficient soybean (*Glycine max*) genotypes compared with the P-inefficient genotypes [11]. In field experiments with soybean [22] it is shown that phosphorus addition to acid soils stimulates aluminum resistance, especially for the genotypes capable to absorb this macroelement effectively. Subsequent studies in hydroponic culture conditions have shown that solution pH, levels of aluminum and phosphorus coordinately changed growth of roots of soybean and exudation of malate anions (as a basic mechanism of plant Al-resistance).

Iqbal et al. [17] assume at least four different ways in which phosphorus can lower toxicity of aluminum. First, phosphorus directly reacts with aluminum in soil forming Al-P precipitates, and thus reduces activity of Al^{3+} ions in a soil solution. Second, phosphorus reduces the amount of apoplastic aluminum, that was bound to the root cell walls and this binding was around 37% of the total Al uptake by the root. Third, high doses of entering of phosphorus reduce the total absorption of aluminum by plant (to 50%), thus simultaneously its concentration in roots decreases to a lesser degree (on 12%), than in shoots (on 88%) with high degree of stress. And, at last, fourth, phosphorus reduces moving of aluminum from roots to shoots by up to 90% in high doses of phosphorus and aluminum.

The aluminum can reduce the absorption of potassium by competitive inhibition [23]. It has been found that the potassium deficiency, induced by aluminum, affects nitrogen metabolism by stimulating the accumulation of putrescine [24]. At *Stylosanthes* aluminum also increases the adsorption of nitrogen in tolerant species [25], leading to increases of nitrate, free amino acids and proteins contents in tissues [26]. This increase in nitrogen metabolism in tolerant species subjected to aluminum is related to the synthesis of specific proteins [27, 28] that provides differential tolerance of plants to aluminum. Some researchers [26] have found that the toxic effects of aluminum are shown only when nitrogen is presented in nitric form. However, results of study [29] indicate that aluminum does not statistically influenced the growth rates of two *Stylosanthes* species when supplied with nitrate form of nitrogen fertilizer although under these conditions the tendency of decrease in relative growth rate in specie *S. guianensis* is observed. In the presence of nitric nitrogen, aluminum increased potassium concentrations in *S. macrocephala* plants, but not in *S. guianensis*. When the nitrogen source was supplied by ammoniacal form, aluminum did not influence the adsorption of potassium in both species. In the presence of the nitric source, the aluminum increases the potassium concentrations only in the tolerant specie, the *S. macrocephala*.

Thus, although mechanisms of interaction of aluminum and elements of a mineral nutrition of plants has been taken into consideration in a few studies on plant adaptation to acid soils, this subject is still quite poorly understood

Agricultural use of fertilizers must be as optimal as possible for ensuring increase of plant productivity with the same or higher quality of products, and minimizing environment contamination by their surplus. Therefore, it is necessary to optimize the ratio of all three basic macronutrients (N, P, and K) in fertilizer simultaneously. However, this problem is connected with a considerable amount of variants under investigation. In a case of acid soils presence of ions of exchangeable aluminum increases twice amount of variants of the experience necessary for finding-out of the put question.

Plant physiologists offered principally new approach in decision of the question – estimation of total N+P+K doze in fertilizer and an optimum of N:P:K ratio within this doze. This approach gives an exact approximation for optimum parity of nutrients in fertilizer and demands a little number of variants (10–15). Complexity of such research can be lowered even more, if a method of systematic variants [30] is used. The main advantage of a method is that only three variants are necessary for estimation the optimum N:P:K ratio. At studying influence of the stressful factor three more variants are added and, thus, total number of variants for each investigated variety of plants makes six.

Process of a mineral nutrition of plants is closely related with photosynthesis. A mineral nutrition provides the growing photosynthetic apparatus with building elements. Three main components of plant productivity of cereal crops – number of productive stems, number of grains per ear (panicle) and 1000 grain mass – positively influence increase in photosynthesis intensity at stages of formation of these components. The positive interrelation between intensity of CO_2 digestion and the specific leaf area (SLA) is an important regularity of varietal variability of photosynthesis. The heightened content of chlorophyll in plant leaves is also connected with the heightened plant productivity or quality of product [31]. These characters are stable enough for a genotype, but under influence of edaphic factors they change in different degree. Change of the ratio of mineral elements or the nitrogen form (ammoniacal NH_4^+, nitric NO_3^- or mixed NH_4NO_3) in fertilizer can change considerably reaction of the photosynthetic apparatus to stressful influence of aluminum. Moreover, differences in aluminum tolerance level between plants are often related with their ability to use this or that form of N-fertilizer.

Thus, the main tasks of a given study were: (a) to establish aluminum influence in acid soil on modification of requirement of plants of grain crops in the basic macronutrients, and (b) to estimate influence of different forms of nitrogen fertilizer (ammoniacal, nitric, and mixed) on parameters under investigation.

6.2 MATERIALS AND METHODOLOGY

1. Influence of soil acidity on the content of phosphoric complexes in wheat plants.

In the conditions of acid reaction of growth medium pot experiment with two spring wheat varieties (Irgina – acid-resistant variety, and Priokskaya – acid-sensitive variety) is put. The natural sod-podzolic soil (pH$_{KCl}$ 3.97, hydrolytic acidity = 2.48 mg-equivalent/100 g of soils) containing 16 mg of exchangeable aluminum per 100 g of soils and 1.7 mg of phosphorus per 100 g of soils (at extraction with 0.2 H HCl) was used. Plants grew up in pots with 4 kg of soil in triple replications on 4 nutrition backgrounds.

Optimum doses of nitrogen and potassium were calculated according to [32]: nitrogen – 120 mg/L, potassium – 150 mg/L of a nutrient solution. Phosphorus was brought in quantity equal to the aluminum content in soil (variant NP$_1$K), at the double by aluminum content in soil (variant NP$_2$K). For calculation of quantity of the phosphorus precipitated with aluminum, tables of recalculation were used [33]. Considering that 4 kg of soil correspond to 10 l of a solution, following quantities of nutritious salts (g/pot) were taken (Table 6.1).

Fertilizers were input in the form of chemically pure salts one week prior to sowing. Sowing was carried out with 15 dry seeds per pot and 10 most vigorous seedlings have been left after germination. Each variant had 12 pots in total. Samples for estimation of a chlorophyll content and fractional structure of phosphates consisted of the mixed plant sample (3 pots with 10 plants each per each variant of study). Estimation of the basic fractions of phosphorus was spent according to [34] after wet combustion. Selection of plants for the analysis was carried out at three growth stages: tillering, leaf-tube formation, and flowering. The content of a chlorophyll defined in acetone extract with "SHIMADZU UVmini-1240" spectrophotometer by a Ref. [35] technique.

TABLE 6.1 Entering of Nutritious Salts for Creation of Nutrient Backgrounds, g/pot

Nutrient background	NH$_4$NO$_3$	KCl	KH$_2$PO$_4$
The control (natural soil)	—	—	—
The control + NK	3.43	1.43	—
The control + NP$_1$K	3.43	—	2.59
The control + NP$_2$K	3.43	2.87	5.17

2. Influence of aluminum on modification of requirement of oats plants in the basic macronutrients.

Plants were grown up in the conditions of sand culture. 15 dry seeds were sown in pots with 4 kg of sand each and 8 most vigorous seedlings have been left after germination. In this study 2 oat varieties were investigated – Krechet (Al-resistant) and Argamak (Al-sensitive). Duration of the study has made 30 days according to the literary data [36, 37].

The content of micronutrients was estimated in mg-atoms, instead of weight quantities of "acting matter" (N, P_2O_5, K_2O) as it is accepted in agronomical practice. Content of real atoms (ions NO^{3-} NH^{4+}, PO_4^{3-} K^+, or atoms N, P, K) absorbed by roots is more important for plants than the contents of conditionally "acting matters". One kg of each of "acting matter" will contain the different quantity of atoms. Application of step schemes of change of quantity of input fertilizers (e.g., 30:30:30, 60:60:60 and so on) leads to disturbed of the requirement of "only distinction" between variants as it changes not only quantity of input substances, but also ratio in number of input atoms of each element. Besides, "acting matters" include different amount of oxygen atoms, which is not taken into account. Therefore, Vakhmistrov and Vorontsov [38] suggest to count a ratio of elements in mg-atoms, and in all variants of experiment the total content of elements should be identical and correspond to their sum in standard Hogland-Arnon-1 medium (22 mg-atom per 1 liter of a solution).

Thus research consisted of three variants at neutral reaction of growth medium (pH = 6.5), and the same three variants at acid reaction (pH = 4.3 + 1 mM aluminum in the form of sulfate): N:P:K = 70:15:15 atomic %, N:P:K = 15:70:15 atomic %, and N:P:K = 15:15:70 atomic %. Total N+P+K content are equal to 22 mg-atom per 1 kg of substrate. Each variant of experiment is put in four replications. Upon termination of experiments dry weight of roots and shoots, the total and specific leaf area, and the chlorophyll content were estimated.

3. Influence of the form of nitrogen fertilizer on modification of requirement of oats plants in the basic macronutrients under influence of aluminum.

The general conditions of study are described above, but each variant of research (control treatment pH 6.0, and stress treatment 1 mM Al, pH 4.3) consisted of three forms of nitrogen fertilizer ($NaNO_3$, $(NH_4)_2SO_4$, and NH_4NO_3) in three replications. After 30 days of growth following parameters have been estimated for each replication: specific leaf area (SLA), leaf area ratio (LAR), leaf weight ratio (LWR), and chlorophyll content in leaves. Each experiment is repeated twice within two years. Thus, the data resulted in tables, represents average value from 12–16 replications (3–4 biological replications × 2 replications in each year × 2 years). Statistical calculations and an estimation of an optimum ratio of macronutrients in growth media are spent according to [39]. Accuracy of experiment made 2.6–3.2% depending on the year.

6.3 RESULTS AND DISCUSSION

1. Influence of phosphorus on plant resistances to high soil acidity
Results of estimation of content of various fractions of phosphoric complexes in leaves of spring wheat are presented in Table 6.2.

The data about a ratio of organic and inorganic forms of phosphoric complexes which serves as an indicator of intensity of inclusion of phosphorus in exchange processes of an organism is most interesting to researchers and breeders. It is possible to indicate that input of nitrogen, phosphorus and potassium into acid soil at tillering stage of growth has lowered intensity of phosphorus metabolization though for Al-sensitive variety it has occurred only at the doubled dose of phosphorus. At the following stages of development, in contrary, improvement of a mineral nutrition promotes the strengthened inclusion of phosphorus in a metabolism, and it is manifested much more strongly it Al-resistant variety.

Doubling of a dose of phosphorus has caused doubling of its inclusion in exchange processes in a resistant variety at early growth stages, but in sensitive variety – only on a flowering stage. This finding indicates higher level of a metabolism in Al-resistant variety at early stages of growth. At first two stages of development input of nitrogen and potassium into soil was more effective for phosphorus metabolization than input of phosphorus at nitrogen and potassium background.

Synthesis of organic acid-soluble phosphorus (in % of the total phosphorus content) in the course of growth raises constantly, and at tillering stage it is more strongly shown in aluminum-sensitive variety, but at the

TABLE 6.2 Influence of Nutrient Backgrounds on the Content of Phosphorus Forms in Leaves of Spring Wheat (g/g of Fresh Mass)

Nutrient background	Tillering stage		Leaf-tube formation stage		Flowering stage	
	1*	2	1	2	1	2
Mineral phosphorus						
Control	0.04	0.04	0.07	0.03	0.02	0.04
Control + NK	0.04	0.02	0.03	0.05	0.03	0.04
Control + NP$_1$K	0.05	0.04	0.06	0.04	0.04	0.03
Control + NP$_2$K	0.02	0.05	0.03	0.05	0.02	0.01
Organic acid-soluble phosphorus						
Control	0.55	0.35	0.20	0.24	0.24	0.28
Control + NK	0.34	0.22	0.50	0.41	0.27	0.28
Control + NP$_1$K	0.32	0.43	0.39	0.33	0.52	0.43
Control + NP$_2$K	0.20	0.40	0.34	0.44	0.34	0.26
Organic acid-non-soluble phosphorus						
Control	0.14	0.11	0.11	0.05	0.08	0.07
Control + NK	0.11	0.08	0.11	0.11	0.09	0.06
Control + NP$_1$K	0.12	0.09	0.10	0.09	0.03	0.03
Control + NP$_2$K	0.12	0.09	0.10	0.07	0.02	0.02
Total phosphorus						
Control	0.73	0.50	0.37	0.33	0.34	0.39
Control + NK	0.49	0.32	0.64	0.57	0.36	0.38
Control + NP$_1$K	0.49	0.56	0.55	0.45	0.59	0.49
Control + NP$_2$K	0.34	0.53	0.47	0.56	0.37	0.29

Note: * 1 – variety Irgina, 2 – variety Priokskaya.

following stages of growth the resistant variety catches up sensitive variety and even overtakes it a little on the given parameter.

If to compare the content of total phosphorus in shoots of contrast varieties in absolute (g) instead of relative (g/g of dry matter) values, that is to consider higher shoot mass of Al-resistant variety, is possible to suggest that resistant variety takes significantly more amount of phosphorus from soil, than sensitive variety.

At a tillering stage there are no distinctions between varieties on biomass accumulation on all nutrient backgrounds, thus improvement of nutrition leads to double increase of a biomass of plants (Table 6.3).

TABLE 6.3 Influence of Nutrition Backgrounds on Accumulation of the Total Biomass by Spring Wheat

Nutrition background	Tillering stage		Leaf-tube formation stage		Flowering stage	
	1*	2	1	2	1	2
Control	0.095	0.112	0.327	0.271	0.527	0.438
Control + NK	0.185	0.206	0.391	0.399	0.691	0.479
Control + NP$_1$K	0.172	0.202	0.439	0.468	0.710	0.562
Control + NP$_2$K	0.202	0.199	0.608	0.624	0.906	0.655

Note: * – 1 – variety Irgina, 2 – variety Priokskaya.

At a leaf-tube formation stage differences between varieties were not revealed also, but differences on nutrition backgrounds were considerably showed. The doubled dose of phosphorus has led 1.5–2 fold increase in plant productivity. However, plants of a control background have strongly grown up by this stage also that was especially showed in aluminum-resistant variety. Varietal differences were especially showed at flowering stage on all nutrition backgrounds; resistant variety has appeared much more productive than sensitive one.

The particular interest represents the fact that entering of phosphoric fertilizers both into one- and in double dose has led to almost identical strengthening of accumulation of dry matter by both varieties (both NPK backgrounds in comparison with NK background). In other words, differences in resistance to aluminum have not affected action of phosphoric fertilizers. The same conclusion arises at the analysis of accumulation of a biomass of an underground part of plants that is development of root systems of plants contrast on Al-resistance level in the conditions of supply with nitrogen-and-potassium nutrition depends a little on presence of ions of hydrogen, aluminum or phosphorus in soil. The analysis of growth of plants on natural acid soil (control background) has allowed to show differences in schemes of development of the investigated varieties.

Though both varieties as a whole have shown great advance of a gain of roots from tillering stage to leaf-tube formation stage, and in a gain of shoot – from tillering stage to flowering phase, absolute value of a gain both roots and shoots is much more for aluminum-resistant variety. In other words, in acid soil resistant variety increases root mass more intensively at the beginning of growth that, possibly, allows it to increase shoot mass more intensively too.

Summarizing the above-stated it is possible to point out that selection of plants on acid soils can be conducted on development of root system at leaf-tube formation stage already, but on development of shoot mass – not earlier than a flowering stage.

It is known that autotrophic growth of organisms is provided, on the one hand, at the expense of root nutrition and, on the other hand, at the expense of assimilation of carbon of air in the course of photosynthesis. These two processes in an organism are interrelated and interdependent; synthesis of elements of the photosynthetic apparatus depends on absorption of necessary mineral substances from soil.

At the analysis of wheat varieties contrast on Al-resistance level it is revealed that in the control background (acid soil) resistant variety synthesized higher quantity of a chlorophyll (chlorophyll *a, b* and their sums), than sensitive variety at all investigated stages of development (Table 6.4).

Obviously given parameter (content of a chlorophyll per gram of dry mass) as index of resistance to lowered pH of soil solution requires more detailed research on more number of varieties in field conditions.

Other interesting fact which is necessary to noting in respect of influence of nutrition backgrounds: at input of phosphoric fertilizers into soil Al-sensitive variety synthesizes much more quantity of all forms of chlorophyll than resistant variety at all stages of development. In other words, at improvement of conditions of mineral nutrition compensation mechanisms of plants limit excessive power consumption and plastic substances on construction of the photosynthetic apparatus as receipt of substances and energy raises at the expense of root system.

TABLE 6.4   Influence of Nutrition Backgrounds on the Total Content of a Chlorophyll in Leaves of Spring Wheat (mg/g of Dry Mass)

Nutrition background	Tillering stage		Leaf-tube formation stage		Flowering stage	
	1*	2	1	2	1	2
Control	7.51	6.80	6.30	5.59	3.63	3.24
Control + NK	6.99	8.30	3.59	3.85	3.31	2.87
Control + NP$_1$K	6.85	9.40	3.61	4.78	2.01	2.89
Control + NP$_2$K	6.98	7.37	3.03	4.11	2.09	2.50

Note: *1 – variety Irgina, 2 – variety Priokskaya.

Thus if the resistant variety reduces the content of chlorophyll already at tillering stage then in sensitive variety decrease begins only at leaf-tube formation stage, besides decrease in synthesis of chlorophyll in resistant variety occurs much more sharply. It can testify higher plasticity of a metabolism of aluminum-resistant variety – it reacts on change of environment conditions much faster.

The parameter of specific leaf area (SLA) indirectly characterizes a thickness of leaf and a share of dry matter in it. Influence of input of phosphorus on the given parameter in our experiments was differing for the investigated varieties. Strengthening of a phosphoric nutrition has led to increase of an average SLA of all leaves of plants of both varieties at tillering stage (a background with a double dose of phosphorus), to its increase in resistant variety at leaf-tube formation stage and to drop at flowering phase whereas in sensitive variety, on the contrary, strengthening of a phosphoric nutrition has lowered this parameter at leaf-tube formation stage and has strengthened at flowering stage (in comparison with natural acid background and background with one dose of phosphorus) (Table 6.5).

This data indicates considerable differences in a metabolism of the studied varieties: at flowering stage resistant variety has strengthening of outflow of plastic substances from leaves into generative organs whereas a sensitive variety has not complete formation of the leaf apparatus by this time and continued to increase it to the detriment of generative organs.

Results of the two-way ANOVA have allowed to calculate shares of influence of growth stage and nutrition backgrounds on change of SLA.

TABLE 6.5   Influence of Nutrition Backgrounds on Specific Leaf Area of Spring Wheat Varieties, Contrast on Al-Resistance (mg/sm^2)

Nutrition background	Tillering stage		Leaf-tube formation stage		Flowering stage	
	1*	2	1	2	1	2
Control	2.51	2.67	3.47	3.49	4.18	2.41
Control + NK	2.33	2.37	4.21	4.83	4.19	4.54
Control + NP$_1$K	2.43	2.47	3.76	3.20	4.27	2.91
Control + NP$_2$K	3.28	3.06	3.94	3.17	3.06	4.45

Note: * 1 – variety Irgina (Al-resistant), 2 – variety Priokskaya (Al-sensitive).

Nutrition backgrounds have rendered twice a greater influence on SLA variability in sensitive variety (influence of factors "nutrition background" + "nutrition background x growth stage" = 61/9% against 27.5% in resistant variety). Thus, sensitive variety is characterized by considerable dependence of photosynthesis on conditions of mineral nutrition whereas intensity of photosynthetic processes in resistant variety is influenced basically by growth stage.

2. Influence of aluminum on modification of oat plants' requirement in the basic macronutrients

The study of mechanisms and genetic basis of aluminum resistance of agricultural plants and creating of varieties with high level of this type of resistance becomes more actual in relation with a drop in volumes of soil liming and, accordingly, increase in the areas of acid soils all over the world. However, presence of exchangeable Al^{3+} ions in growth medium determined low pH of these soils is not the sole stressful edaphic factor. Not the smaller role is played by shortage of some elements of a mineral nutrition and absence of a proper ratio between them. The sand culture allows to simulate soil key parameters, abstracting from complexity of this natural medium of plant habitat linked with features of mineral, salt structure, presence of organic structures and microbiological activity. The estimation of relative level of aluminum resistance in laboratory and greenhouse experiments was carried out by different researchers which used various nutrient solutions. The content and ratio of elements of a mineral nutrition in used media is so differ sometimes that can mask exact action of the aluminum. So, the particular ratio of nitrogen, phosphorus and potassium (in mg-atoms per liter of solution) makes the following values: [40] – 54:0:46; [41] – 77:5:18; [42] – 64.5:0.5:35; [43] (Steinberg's solution) – 87:1:12. Standard Hogland-Arnon-1 medium [44] contains the specified elements in the ratio 68:5:27. Probably, such considerable distinctions in the ratio of the basic macronutrients can affect a comparative estimation of Al-resistance level of plant varieties.

Using of Homes' method [30] for an estimation of optimum N:P:K ratio in a condition of sand culture for oat varieties differing in their reaction to aluminum, has shown that resistance can be explained by a higher degree of resistance of metabolic processes (maintenance of balance between synthetic and catabolic reactions) which was testified by smaller changes of relative need in macronutrients under stress (Table 6.6).

Al stress had little influence on the ratio of metabolic/catabolic processes in the roots of the resistant oats variety Krechet. Of course, the total level of reactions of synthesis and destruction can change (increase or decrease), but in equal degrees. Metabolic activity of roots of a sensitive variety Argamak has changed considerably: the requirement for phosphorus has increased by 2.5 times, and requirement in potassium has decreased by 1.5 times. These changes can mean that some biochemical processes have been switched from the basic metabolic pathway to some other.

Development of aboveground mass under stressful influence has also shown distinction between the varieties. If the resistant variety demanded increase in a relative share of phosphorus (at the expense of nitrogen) at constant requirement in potassium the sensitive variety has changed relative shares of all three elements (requirement for nitrogen and phosphorus have increased by 1.3 and 2.2 times, accordingly, at the expense of requirement reduction in potassium by 1.7 times). The forming of the maximum leaf area of a resistant variety has demanded relative increase in a share of phosphorus and reduction in share of potassium in the growth medium at the constant content of nitrogen, but for a sensitive variety the requirement for nitrogen has increased, and the requirement for other elements has decreased. However, modification of requirements of the leaf apparatus in macronutrients have not affected processes of synthesis of a chlorophyll neither at resistant, nor at a sensitive variety. Thus the interrelation between N, P and K at which the process of synthesis of photosynthetic pigments was maximum at a resistant variety actually coincides with the ratio of the macronutrients necessary for the maximum development of root system both in neutral, and in the acid growth condition. It can indicate the greater level of coordination of processes of a mineral nutrition and photosynthesis at aluminum-resistant oat variety than at a sensitive one.

However, from our point of view, more interesting is the fact that mathematically it is possible to calculate that ratio between macronutrients at which plants will not suffer depression of growth under the influence of the stressful factor. If we take into account the relation of dry weights of different parts of a plant (i.e., root-to-shoot ratio) or all plant in the control

TABLE 6.6 Optimum Ratio of Nitrogen, Phosphorus and Potassium for Some Indicators of Development of Plants of Two Oats Varieties, Contrast on Aluminum-Resistance

pH of growth media	Root dry mass	Shoot dry mass	Total dry mass	Specific leaf area	Total leaf area	Total content of chlorophyll
Variety Krechet (Al-resistant)						
pH 6.5	46:25:29	45:18:37	45:19:36	30:43:28	46:15:39	49:20:31
pH 4.3	43:25:32	29:34:37	33:31:36	24:41:35	45:32:23	47:17:36
Variety Argamak (Al-sensitive)						
pH 6.5	46:9:45	36:10:54	39:9:52	33:35:32	23:30:47	51:22:27
pH 4.3	45:25:30	46:22:32	48:21:31	29:44:27	47:18:35	54:23:23

and in the test treatment thus their maximum ratio (100%) will correspond to optimum N:P:K ratio.

In our case the following optimum N:P:K ratio are received for oat varieties Krechet and Argamak: for absence of depression of root growth – 30:34:36 and 36:10:54, accordingly; for absence of depression of shoot growth – 18:52:30 and 31:11:58, accordingly; for a total mass of plants – 21:49:30 and 32:11:57, accordingly.

3. Influence of the form of nitrogen fertilizer on modification of requirement of oat plants in basic macronutrients under influence of aluminum

Scientific literature suggests that nitrogen metabolism is involved in differential aluminum resistance among plant genotypes. Nutrition with NH_4 instead of NO_3 can strengthen Al-induced deficiency of magnesium, but reduce damages of roots [45]. Under field conditions ameliorating effect of NH_4 on damages of roots induced by aluminum, shown in the conditions of nutritious solutions, can be compensated by its negative effects when NH_4 absorption leads to decrease in rhizosphere pH and to the subsequent increase of aluminum toxicity [46]. The form of inorganic nitrogen in the growth medium can considerably affects growth and development of many plant species. Use of the oxidized form of nitrogen by plants and resultant acidification of growth medium can underline the effects of nutrient supply imbalance and toxicity of aluminum. If the nutritious solution contained up to 15% of nitrogen in the form of NH_4N, Al-resistant plants absorbed nutrients more effectively, than Al-sensitive plants [47]. The amelioration of aluminum toxicity by NH_4 ions has been explained by a cations competition for binding sites of apoplast [37].

The data of Tables 6.7 show that the nitrogen form could change essentially relative requirement of plants for these or those macronutrients. It can specify that presence of the certain form of nitrogen fertilizer in growth medium leads to different physiological modifications. Distinctions between varieties in their reaction to aluminum can be related not with their ability to uptake of this or that form of nitrogen, but that relative requirement for nitrogen changes and different mechanisms of maintenance of a homeostasis (different pathways of using of nutrient matters etc.) were switched up.

As it is seen from data presented in Tables 6.7 and 6.8 used forms of nitrogen fertilizer influenced development of the investigated parameters both in the control, and in the stress conditions. Soil acidity did not render

TABLE 6.7 Influence of the Form of Nitrogen Fertilizer and Aluminum on Optimum N:P:K Ratio for the Maximum Development of Oats Plants

Treatment	Part of plant	Form of nitrogen fertilizer		
		NO_3NH_4	NO_3	NH_4
Variety Argamak				
Control	Roots	55:15:30	22:60:18	13:56:31
(pH 6.0 without	Leaves	54:35:11	35:51:14	17:68:15
aluminum)	Stems	50:39:11	34:54:12	24:32:44
	Total shoots	52:37:11	35:52:13	18:60:22
Aluminum	Roots	32:47:21	26:52:22	16:34:50
(1 mM, pH 4.3)	Leaves	54:31:14	37:45:18	37:25:38
	Stems	58:30:12	32:50:18	43:27:30
	Total shoots	57:31:13	35:47:18	40:26:34
Variety Krechet				
Control	Roots	14:54:32	21:59:20	15:51:34
(pH 6.0 without	Leaves	38:41:21	36:43:21	41:33:26
aluminum)	Stems	19:61:20	25:57:18	22:51:27
	Total shoots	27:53:20	32:48:20	32:42:26
Aluminum	Roots	28:55:17	19:54:27	11:54:35
(1 mM, pH 4.3)	Leaves	44:22:24	20:59:21	48:27:25
	Stems	57:26:17	25:61:14	15:45:40
	Total shoots	55:24:21	22:60:18	31:37:32

influence on the maximum development of parameters SLA and LAR at the mixed type of a nitrogen nutrition in Al-sensitive variety Argamak whereas in case of Al-resistant variety Krechet these two parameters have demanded opposite modification of growth medium for their maximum development: SLA has demanded increase of potassium share at the expense of nitrogen and phosphorus; LAR – decrease share of potassium and strengthening of share of nitrogen, but their mutual compensation has led to stability of parameter in acid medium.

Maintenance of N, P and K ratio at which the maximum quantity of chlorophyll has been synthesized in leaves of both varieties, in stressful conditions of growth medium can specify relative independence of processes of pigments synthesis on presence of the stressful agent in rooting medium. Possibly, processes and reactions dealing with synthesis of photosynthetic

TABLE 6.8 Influence of Aluminum and the Form of a Nitrogen Fertilizer on N:P:K Ratio Necessary for the Maximum Development of Parameters of the Leaf Apparatus of Oats

Treatment	Form of nitrogen fertilizer	Parameters of leaf apparatus			
		Content of chlorophyll	SLA	LAR	LWR
Variety Argamak					
Control	NO_3	51:29:19	34:41:25	44:21:35	41:28:31
(pH 6.0 without	NH_4	48:30:22	29:46:25	40:19:41	44:30:26
aluminum)	$NO_3.NH_4$	30:25:45	31:35:34	50:25:25	40:20:40
Aluminum	NO_3	51:27:22	38:32:29	39:37:28	39:29:32
(1 mM, pH 4.3)	NH_4	35:24:41	36:38:26	43:22:35	43:28:30
	$NO_3.NH_4$	23:40:37	28:34:38	50:28:22	47:28:35
Variety Krechet					
Control	NO_3	44:18:38	31:38:31	55:17:28	55:19:26
(pH 6.0 without	NH_4	29:36:35	38:39:23	42:21:37	50:24:26
aluminum)	$NO_3.NH_4$	29:29:42	35:42:23	46:18:36	54:23:23
Aluminum	NO_3	51:17:32	32:30:38	41:28:31	42:27:31
(1 mM, pH 4.3)	NH_4	44:35:21	26:50:24	41:21:38	40:32:28
	$NO_3.NH_4$	32:33:34	25:25:50	65:15:20	58:17:25

pigments have a priority in their supply with macronutrients in a correct ratio even when the plant is under stressful conditions. In the same way it is possible to analyze action of soil acidity and various forms of nitrogen fertilizer on modification of requirements of plants in macronutrients for maximum development of the photosynthetic apparatus.

6.4 CONCLUSIONS

Doubling of a phosphorus dose in the growth medium has caused doubling of its inclusion in metabolic processes in Al-resistant wheat variety at early growth stages already, but in sensitive variety phosphorus inclusion amplified only at flowering stage. At first two stages of development addition of nitrogen and potassium into soil is more effective for phosphorus metabolization than input of phosphorus salt at nitrogen and potassium background.

Development of root systems in plants of wheat varieties contrast on aluminum resistance level in the conditions of supply with nitrogen and potassium nutrition depends a little on presence of ions of hydrogen, aluminum or phosphorus in soil.

Aluminum-sensitive variety of wheat is characterized by considerable dependence of photosynthesis on conditions of mineral nutrition whereas intensity of photosynthetic processes in aluminum-resistant variety is influenced basically by a growth stage.

Al-resistant oat variety maintained relative levels of N, K and P metabolism (increased or decreased them by equal extent) in roots under stress condition. Possibly, it occurs at the expense of modification of nitrogen and phosphorus metabolism in shoots. The Al-sensitive variety of oat under conditions of aluminum influence keeps N-metabolism level in roots, but levels of phosphorus and potassium metabolism are exposed to significant modifications. In shoots there is a reorganization of the metabolic processes occurring to participation of all three elements.

Changing the form of nitrogen fertilizer we not only change conditions of a nitrogen nutrition but also essentially influence on requirement of plants for phosphorus and potassium that, in turn, is reflected in start of these or those mechanisms of resistance to aluminum impact.

Choosing a certain ratio of elements of a mineral nutrition it is possible to reach a situation when aforementioned parameters will not expose negative action of aluminum on their development.

KEYWORDS

- aluminum
- nitrogen
- oats
- phosphorus
- potassium
- resistance
- spring wheat

REFERENCES

1. Kochian, L. V., Hoekenga, O. A., Pineros, M. A. How do crop plants tolerate acid soils? Mechanisms of aluminum tolerance and phosphorous efficiency. Annu. Rev. Plant Biol. 2004, Vol. 55. 459–493.
2. Jayasundara, H. P. S., Thomson, B. D., Tang, C. Responses of cool season grain legumes to soil abiotic stresses. Adv. Agron. 1998, Vol. 63. 77–151.
3. Eswaran, H., Reich, P., Beinroth, F. Global distribution of soils with acidity. Braz. Soil Sci. Soc. 1997, 159–164.
4. Lisitsyn, E. M., Shchennikova, I. N., Shupletsova, O. N. Cultivation of barley on acid sod-podzolic soils of north-east of Europe. Barley: Production, Cultivation and Uses. [Ed. S. B. Elfson] New York: Nova Publ. 2011, 49–92.
5. Guo, T. R., Zhang, G. P., Zhou, M. X., Wu, F. B., Chen, J. X. Influence of aluminum and cadmium stresses on mineral nutrition and root exudates in two barley cultivars. Pedosphere. 2007, Vol. 17. 505–512.
6. Olivares, E., Pena, E., Marcano, E., Mostacero, J., Aguiar, G., Benitez, M., Rengifo, E. Aluminum accumulation and its relationship with mineral plant nutrients in 12 pteridophytes from Venezuela. Env. Exp. Bot. 2009, Vol. 65. 132–141.
7. Le Gouis, J., Fontaine J-X., Laperche, A., Heumez, E., Devienne-Barret, F., Brancort-Hulmel, M., Dubois, F., Hirel, B. Genetic analysis of wheat nitrogen use efficiency: coincidence between QTL for agronomical and physiological traits. Proceedings of the 11th International Wheat Genetics Symposium. 2008 (URL: http://hdl.handle.net/2123/3217).
8. Hirel, B., Le Gouis, J., Ney, B., Gallais, A. The challenge of improving nitrogen use efficiency in crop plants: towards a more central role for genetic variability and quantitative genetics within integrated approaches. J. Exp. Bot. 2007, Vol. 58. 2369–2387.
9. Deletić, N., Stojković, S., Djurić, V., Biberdžić, M., Gudžić, S. Genotypic specificity of winter wheat nitrogen accumulation on an acid soil. Research Journal of Agricultural Science. 2010, Vol. 42, N. 1. 71–75.
10. Stojković, S. Genotipska variranja nekih pokazatelja akumulacije i iskorišćavanja suve materije i azota kod ozime pšenice. Magistarska teza, Poljoprivredni fakultet, Univerzitet u Prištini. 2001, [in Serbian].
11. Bednarek, W., Reszka, R. The influence of liming and mineral fertilization on the utilization of nitrogen by spring barley. Annales Universitatis Mariae Curie–Skłodowska, Lublin – Polonia. 2009, Vol. 64. N. 3. 11–20.
12. Habash, D. Z., Bernard, S., Schondelmaier, J., Weyen, J., Quarrie, S. A. The genetics of nitrogen use in hexaploid wheat: N utilization, development and yield. Theor. Appl. Genet. 2007, Vol. 114. 403–419.
13. Liao, H., Wan, H., Shaff, J., Wang, X., Yan, X., Kochian, L. V. Phosphorus and aluminum interactions in soybean in relation to aluminum tolerance: exudation of specific organic acids from different regions of the intact root system. Plant Physiol. 2006, Vol. 141. 674–684.
14. Dong, D., Peng, X., Yan, X. Organic acid exudation induced by phosphorus deficiency and/or aluminum toxicity in two contrasting soybean genotypes. Physiol. Plant. 2004, Vol. 122. 190–199.

15. Jemo, M., Abaidoo, R. C., Nolte, C., Horst, W. J. Aluminum resistance of cowpea as affected by phosphorus-deficiency stress. J. Plant Physiol. 2007, Vol. 164. 442–451.

16. Sun, Q. B., Shen, R. F., Zhao, X. Q., Chen, R. F., Dong, X. Y. Phosphorus enhances Al resistance in Al-resistant *Lespedeza bicolor* but not in Al-sensitive, *L. cuneata* under relatively high Al stress. Ann. Bot. (Lond). 2008, Vol. 102. 795–804.

17. Zheng, S. J., Yang, J. L., He, Y. F., Yu, X. H., Zhang, L., You, J. F., Shen, R. F., Matsumoto, H. Immobilization of aluminum with phosphorus in roots is associated with high aluminum resistance in buckwheat. Plant Physiol. 2005, Vol. 138. 297–303.

18. Fukuda, T., Saito, A., Wasaki, J., Shinano, T., Osaki, M. Metabolic alterations proposed by proteome in rice roots grown under low P and high Al concentration under low pH. Plant Sci. 2007, Vol. 172, 1157–1165.

19. Iqbal, T., Sale, P., Tang, C. Phosphorus ameliorates aluminum toxicity of Al-sensitive wheat seedlings. 19th World Congress of Soil Science, Soil Solutions for a Changing World 1–6 August 2010, Brisbane, Australia. 2010, 92–95.

20. Nakagawa, T., Mori, S., Yoshimura, E. Amelioration of aluminum toxicity by pretreatment with phosphate in aluminum-tolerant rice cultivar. J. Plant Nutrit. 2003, Vol. 26. 619–628.

21. Silva, I. R., Smyth, T. J., Israel, D. W., Rufty, T. W. Altered aluminum inhibition of soybean root elongation in the presence of magnesium. Plant Soil. 2001, Vol. 230. 223–230.

22. Gaume, A., Machler, F., Frossard, E. Aluminum resistance in two cultivars of *Zea mays* L: root exudation of organic acid and influence of phosphorus nutrition. Plant Soil. 2001, Vol. 234. 73–81.

23. McCormick, L. H., Borden, F. Y. Phosphate fixation by aluminum in plant roots. Soil Sci. Soc. Am. J. 1972, Vol. 36. 779–802.

24. Liang, C., Piñeros, M. A., Tian, J., Yao, Z., Sun, L., Liu, J., Shaff, J., Coluccio, A., Kochian, L. V., Liao, H. Low pH, aluminum, and phosphorus coordinately regulate malate exudation through *GmALMT1* to improve soybean adaptation to acid soils. Plant Physiol. 2013, Vol. 161. 1347–1361.

25. Malavolta, E., Vitti, G. C., Oliveira, S. A. Avaliação do estado nutricional das plantas: princípios e aplicações. 2.ed. Potafos, Piracicaba. 1997, 319 p. [in Portuguese].

26. Basso, L. H. M., Lima, G. P. P., Gonçalves, A. N., Vilhena, S. M. C., Padilha, C. do, C. F. Efeito do alumínio no conteúdo de poliaminas livres e atividade da fosfatase ácida durante o crescimento de brotações de *Eucalyptus grandis* x, *E. urophylla* cultivadas in vitro. Sciencia Forestalis. 2007, Vol. 75. 9–18. [in Portuguese].

27. Cordeiro, A. T. Efeito de níveis de nitrato, amônio e alumínio sobre o crescimento e sobre a absorção de fósforo e de nitrogênio em *Stylosanthes guianensis* e *Stylosanthes macrocephala*. Viçosa: 1981, 53 f. Dissertação (Mestrado em Fisiologia Vegetal) – Universidade Federal de Viçosa. [in Portuguese].

28. Amaral, J. A. T. do, Cordeiro, A. T., Rena, A. B. Efeitos do alumínio, nitrato e amônio sobre a composição de metabólitos nitrogenados e de carboidratos em *Stylosanthes guianensis* e, *S. macrocephala*. Pesquisa Agropecuária Brasileira, 2000, Vol.35. N. 2. 313–320. [in Portuguese].

29. Iuchi, S., Koyama, H., Iuchi, A., Kobayashi, Y., Kitabayashi, S., Kobayashi, Y., Ikka, T., Hirayama, T., Shinozaki, K., Kobayashi, M. Zinc finger protein STOP1 is critical

for proton tolerance in Arabidopsis and coregulates a key gene in aluminum tolerance. Proc. Natl Acad. Sci. USA. 2007, Vol. 104. N. 23. 9900–9905.

30. Kobayashi, Y., Hoekenga, O. A., Itoh, H., Nakashima, M., Saito, S., Shaff, J. E., Maron, L. G., Pineros, M. A., Kochian, L. V., Koyama, H. Characterization of *AtALMT1* expression in aluminum-inducible malate release and its role for rhizotoxic stress tolerance in Arabidopsis. Plant Physiol. 2007, Vol. 145. N. 3. 843–852.

31. Amaral, J. A. T. do, Rena, A. B., Cordeiro, A. T., Schmildt, E. R. Effects of aluminum, nitrate and ammonium on the growth, potassium content and composition of amino acids in *Stylosanthes*. IDESIA (Chile). 2013, Vol. 31, N. 2. 61–68.

32. Homes, M. V. L. Alimentation minerale equilibree des vegetaux. Wetteren: Universa, 1961, Vol. 1. 55 p. [in French].

33. Zelensky, M. I. Photosynthetic characteristics of major agricultural crops and prospects of their breeding application. Physiological principles of plant breeding. St. Petersburg, 1995, 466–554.

34. Rinkis, G.Ya., Nollendorf, V. F. Balanced nutrition of plants with macro- and microelements. Riga: Apgāds "ZINĀTNE," 1982, 301 p. [in Russian].

35. Arinushkina, E. V. Guide on chemical analysis of soil. Moscow: Moscow University Press, 1970, 491 p.

36. Pronina, N. B., Ladonin, V. F. (Eds.) Physiological-and-biochemical methods of studying of action of chemical means complex on plants (Methodical recommendations). Moscow: All Russian Research Institute of Agrochemistry, 1988, 68 p. [in Russian].

37. Lichtenthaler, H. K., Buschmann, C. Chlorophylls and carotenoids – Measurement and characterization by UV-VIS. Current Protocols in Food Analytical Chemistry. John Wiley & Sons, Madison, 2001, P. F4.3.1-F4.3.8. [Nr. 107].

38. Foy, C. D. Tolerance of barley cultivars to an acid, aluminum-toxic subsoils related to mineral element concentrations in their shoots. J. Plant Nutrit. 1996, Vol. 19. 1361–1380.

39. Blamey, F. P. C., Edmeades, D. C., Wheeler, D. M. Empirical models to approximate calcium and magnesium ameliorative effects and genetic differences in aluminum tolerance in wheat. Plant Soil. 1992, Vol. 144. 281–287.

40. Vakhmistrov, D. B., Vorontsov, V. A. Selective ability of plants is not directed on providing of their maximum growth. Russian Plant Physiol. 1997, Vol. 44. N. 3. 404–412. [in Russian].

41. Vakhmistrov, D. B. Separate estimation of optimal doze N+P+K and N:P:K ratio in fertilizer. 1. Foundation of problem. Agrochemistry. 1982, N. 4. 3–12. [in Russian].

42. Somers, D. J., Gustafson, J. P. The expression of aluminum stress induced polypeptides in a population segregating for aluminum tolerance in wheat (*Triticum aestivum*, L.). Genome. 1995, Vol. 38. 1213–1220.

43. Wagatsuma, T., Kawashima, T., Tamaraya, K. Comparative stainability of plant root cells with basic dye (methylene blue) in association with aluminum tolerance. Commun. Soil Sci. Plant Anal. 1988, Vol. 19. 1207–1215.

44. Grauer, U. E., Horst, W. J. Effect of pH and N source on aluminum tolerance of rye (*Secale cereale*, L.) and yellow lupin (*Lupinus luteus*, L.). Plant Soil. 1990, Vol. 127. 13–21.

45. Fleming, A. L., Foy, C. D. Root structure reflects differential aluminum tolerance in wheat varieties. Agron. J. 1968, Vol. 60. 172–176.

46. Hoagland, D. R., Arnon, D. I. The water-culture method for growing plants without soil. Circ. 347, Calif. Agric. Exp. Stn., Berkeley, CA. 1950.

47. Tan, K., Keltjens, W. G., Findenegg, G. R. Calcium-induced modification of aluminum toxicity in sorghum genotypes. J. Plant Nutrit. 1992, Vol. 15. 1395–1405.

48. Tan, K., Keltjens, W. G., Findenegg, G. R. Effect of N form on aluminum toxicity in sorghum genotypes. J. Plant Nutrit. 1992, Vol. 15. 1383–1394.

49. Foy, C. D. Effects of aluminum on plant growth. The plant root and its environment. Univ. Press of Virginia, Charlottesville. 1974, 601–642.

CHAPTER 7

EDAPHIC STRESS AS THE MODIFIER OF CORRELATION OF YIELD STRUCTURE'S ELEMENTS IN CEREAL CROPS

EUGENE M. LISITSYN and LYUDMILA N. SHIKHOVA

CONTENTS

ABSTRACT

Results of microplot trails have shown that under the influence of ions of heavy metals there is a considerable reorganization of system of interrelations of elements of plant productivity in oats (*Avena sativa* L.) and barley (*Hordeum vulgare* L.). In the majority of variants of treatment of plants with heavy metals the structure of correlations between elements of yield structure of plants characteristic for normal conditions is collapsed. Influence of intravarietal distinctions on aluminum resistance level of plants of cereal crops on system of correlations of elements of yield structure is found out. As a whole for all studied metals resistant varieties of cereal crops kept coordination of growth processes under stressful conditions in a greater degree than varieties sensitive to stressful influence. It is shown that fluctuations of level of correlations between elements of yield structure in oats and barley can indicate intravarietal heterogeneity on separate parameters. In two series of study it was showed specie-specificity in change of structure of correlations of the investigated parameters in barley and oats.

7.1   INTRODUCTION

At the present time exudation of anions of organic acids is considered the basic physiological mechanism of resistance of plants to aluminum influence [1, 2]. The genetic control of this mechanism [3, 4] is established. However, it is known that in reaction to stressful influence of aluminum or of ions of heavy metals the plant will involve the numerous physiological mechanisms accompanied by change of activity of a great number of genes (genetic systems) [5]. The genes involved in reactions of a plant organism on similar influences are genes of the general reaction of plants on external stimulus [6, 7]. Plant species can differ with time necessary for start of responses – from several minutes till several hours and days [8, 9]. In the second case change of activity of some nuclear genes takes place [10] whereas in the first case use of already synthesized organic complexes amplifies. Therefore, for studying of genetic features of plants' aluminum resistance in our opinion it is more logical to apply stressful influence within several days.

Integrated indicator of fitness of plants of cereal crops to growth conditions is formation of accurately working system of interrelations of separate organs and plant parts. Therefore, as total reaction to stressful factors it is possible to consider infringement of this system or, on the contrary, stability of separate interrelations under conditions of stressor impact. Thus consider [11, 12] that strengthening of stressful influence (i.e., adverse environmental conditions) cause substantial increase of force of relations between separate components of productivity (morphological characters). It is assumed that changes of structure of relations reflect possibility of switching of regulatory mechanisms directing processes of growth and morphogenesis which in turn are additional way of adaptation to changing conditions of environment. Transformations of structure of interrelations reflect specific features of realization of adaptive reactions of different parameters and of different genotypes. Breeding of cultivated plant species according to Ref. [13] conducts to formation of rigid, rather stable system of interrelations which changes at external influences are limited with fluctuations of level of these relations.

However influence of ions of heavy metals on mode of interrelation of elements of yield structure in cereal crops having different level of the general resistance to abiotic factors from this point of view was not studied yet. In this connection, studying of character of changes of directions and force of correlations between elements of yield structure in oats and barley under the influence of heavy metals (cadmium, lead, iron, and manganese) and aluminum was the purpose of the given study.

7.2 MATERIALS AND METHODOLOGY

Considerable intravarietal variability of a studied trait is explained with polygene control of any quantitative parameter. It is possible to assume that the plants having different level of resistance to abiotic stressors (ions of aluminum, iron, manganese, cadmium, and lead) will differ on a mode of action of genetic systems under conditions of absence of stressful influence too. Treating of plants with stressors during short time it is possible to divide initial sample of plants into groups of resistance and at the further growth in non-stressful conditions to find out how these distinctions affect character of interrelation of formation of elements of yield structure. Study has been spent in two series of experiments.

1. Influence of aluminum ions on structure of complex of correlations in oats and barley plants
For decrease in influence of intravarietal variability the plants having equal length of a germinal root were used. For this purpose about 400 seeds of oats (varieties Krechet and Ulov) and barley (varieties Dina and Elf) have been put into rolls of a filter paper for germination during 4 days. Upon termination of this time 100 seedlings of each variety with average for a variety length of

germinal root have been selected. This initial length of a root has been noted for each individual seedling. Further these seedlings have been placed for three days in a solution of 1 mM aluminum in the form of sulfate at pH 4.3. Upon termination of stressful influence the length of a root of each individual seedling has been measured again for calculation of a re-growth and, accordingly, level of resistance of a given seedling. All seedlings of each variety have been divided into 4 groups: I – without increase in root length in aluminum solution; II – increase of root length is 1.0–4.9%; III – increase of root length is 5.0–9.9%; IV – increase of root length is more than 10%. All seedlings have been sowing under conditions of microplot trial on neutral soil for the further growth and development. Upon termination of growth season elements of yield structure have been estimated at each individual plant by a technique [14]. In Tables 7.2–7.4 elements of yield structure are designated as follows: 1 – "plant height," 2 – "number of stems," 3 – "productive tillering capacity," 4 – "length of an ear/panicle," 5 – "main ear/panicle mass," 6 – "lateral ear/panicle mass," 7 – "number of grains per main ear/panicle," 8 – "grain mass per main ear/panicle," 9 – "number of grains per lateral ear/panicle," 10 – "grain mass per lateral ear/panicle," 11 – "sheaf mass".

2. Influence of heavy metals on structure of complex of correlations in oats and barley plants

As in our previous researches it has been found out that the same varieties of cereal crops differently reacted to presence of ions of various heavy metals in root growth media [15] it was not possible to select for research in conditions of microplot trials two varieties of each species contrast reacting on character of development of root systems to all used stressors. Therefore, varieties bred in the North-East Agricultural Research Institute (Kirov, Russia) shown the greatest and least degree of reaction to all investigated heavy metals as a whole (though their degree of reaction of root systems on separate stressor may not differ) have been selected for researches: barley Kupets as most resistant as a whole to action of heavy metals, and barley selection line 406–99 as least resistant against them; oats Krechet as resistant variety, and oats Butsefal – as variety sensitive to action of ions of heavy metals.

Seeds have been presoaked in solutions of salts of metals (test treatment) or in the distilled water (control treatment) within 4 hours before sowing. The used concentrations of salts were: cadmium – 100 mM; lead – 500 mM; iron – 120 mg/L; manganese – 160 mg/L. The given concentrations of acting substances have been picked up experimentally in greenhouse and laboratory trials spent earlier. Further 400 seeds of each variant have been sowing under conditions of microplot trials. Plants grew up to a stage of full maturing of seeds. With 1 week interval after germination plants were processed with corresponding solutions of salts of heavy metals (a control variant – with distilled water) in the morning by means of a manual sprayer.

At the end of growth season degree of development of elements of yield structure was estimated [14] on an example of 20 plants of each variant; values of coefficients of pair correlations between separate elements of yield structure were counted; stability of system of similar interrelations for each individual variety was estimated. A designation of variants in Tables 7.5 and 7.6 and in Figures 7.1 and 7.2: 1 – "plant height," 2 – "number of stems," 3 – "productive tillering capacity," 4 – "length of an ear/panicle," 5 – "main ear/panicle mass," 6 – "lateral ear/panicle mass," 7 – "number of grains per main ear/panicle," 8 – "grain mass per main ear/panicle," 9 – "number of grains per lateral ear/panicle," 10 – "grain mass per lateral ear/panicle," 11 – "1000 grains mass".

Data were processed statistically with use of software Statistica 10 (StatSoft) and Microsoft Office Excel 2007. Average arithmetic means of studied traits ± average error are presented in tables.

7.3   RESULTS AND DISCUSSION

1. Influence of aluminum ions on structure of complex of correlations in oats and barley plants

The average root length of initial plants has made on varieties: oats Ulov – 70.8 ± 0.8 mm, oats Krechet – 64.9 ± 0.7 mm, barley Dina – 71.0 ± 0.7 mm, barley Elf – 81.3 ± 0.8 mm; an average increase of root systems for three days of influence of aluminum has made accordingly 3.6 ± 0.3 mm (5% to initial length), 6.2 ± 0.4 mm (10% to initial length), 3.7 ± 0.3 mm (5% to initial length), and 1.7 ± 0.3 mm (2% to initial length). High intravarietal variability of size of increase is noted: zero increase and increase more than 10% have shown accordingly 18 and 15% of plants of oats Ulov, 9 and 41% of plants of oats Krechet, 16 and 16% of plants of barley Elf, and 53 and 5% of plants of barley Dina (Table 7.1).

Thus, in each investigated variety there are plants with different level of aluminum resistance. Accordingly, it is necessary to expect changes in structure of correlations between elements of yield structure at plants with different level of aluminum resistance.

Similar changes should occur owing to that the set of genes participating in the control of a trait "aluminum resistance" in plants of different groups of resistance will be a little distinguished. If any gene influences development both resistance trait and elements of yield structure it is possible to expect increase of value of coefficients of pair correlations between level of resistance and a level of development of these elements.

The matrix of pair correlations between values of development of elements of yield structure in oat and barley and values of increase in root length of plants under influence of 1mM aluminum is presented in Table 7.2.

The data of Table 7.2 indicate that as a whole for the investigated barley varieties relations of aluminum-resistance level with development of elements of yield structure in variants I-III it is not observed practically. For variety Elf significant negative correlations between value of root re-growth and "number of stems," "lateral ear mass," "number of grains per lateral ear," "grain mass per lateral ear," and "sheaf mass" are shown only in the plants having the greatest increase of root length in aluminum solution. In other groups of plants and in whole sample statistically significant relations between increases of root length in aluminum solution and values of elements of yield structure is not revealed. Possibly genetic

TABLE 7.1 Intravarietal Variability of Oats and Barley on Increase of Root Length of Plants (1 mM aluminum, 3 days of influence), % from sample

Variety	Increase in root length, %			
	0	1.0–4.9	5.0–9.9	More than 10
Oats Ulov	18	39	28	15
Oats Krechet	9	21	29	41
Barley Elf	16	42	26	16
Barley Dina	53	31	11	5

TABLE 7.2 A Matrix of Pair Correlations Between Values of Elements of Yield Structure in Oat and Barley and Values of Increase in Root Length of Plants at Growing in 1mM Aluminum Solution

Group of plant*	Elements of yield structure*										
	1	2	3	4	5	6	7	8	9	10	11
Barley Elf											
I	0.007	−0.070	−0.083	−0.084	−0.150	−0.184	−0.079	−0.123	−0.134	−0.168	−0.187
II	−0.253	−0.147	−0.145	0.114	−0.326	−0.307	−0.190	−0.363	−0.212	−0.302	−0.295
III	−0.112	0.124	−0.141	0.170	0.040	−0.090	0.190	0.034	−0.064	−0.087	0.112
IV	0.022	*−0.646*	*−0.594*	−0.236	0.079	*−0.657*	−0.083	0.188	*−0.612*	*−0.647*	*−0.712*
Barley Dina											
I	0.014	−0.103	−0.053	−0.092	−0.007	−0.040	0.073	0.000	−0.043	−0.039	−0.050
II	−0.377	−0.334	−0.296	−0.301	−0.347	−0.320	−0.329	−0.346	−0.344	−0.354	−0.285
III	0.475	0.695	0.506	0.383	−0.265	0.492	0.012	−0.071	0.428	0.462	0.692
IV	0.729	*0.990*	*0.990*	0.793	*0.838*	*0.952*	0.396	0.768	*0.972*	*0.958*	0.785
Oats Ulov											
I	*0.257*	*−0.261*	−0.218	−0.001	−0.055	−0.169	0.096	0.060	−0.128	−0.161	−0.084
II	0.092	−0.191	0.136	−0.188	−0.049	0.261	0.086	0.053	0.313	0.281	−0.012
III	−0.233	−0.395	*−0.663*	−0.049	−0.128	*−0.677*	−0.377	0.349	*−0.745*	*−0.724*	−0.357
IV	0.488	0.204	0.177	0.474	0.309	0.229	0.388	0.286	0.199	0.199	*0.645*
Oats Krechet											
I	0.016	−0.107	0.036	0.123	0.121	0.003	0.100	0.160	0.038	0.013	−0.164
II	0.146	−0.040	−0.139	0.062	−0.193	−0.082	−0.161	−0.201	−0.141	−0.075	−0.100
III	−0.131	0.173	0.087	−0.289	−0.099	0.133	−0.057	0.086	0.188	0.161	−0.033
IV	0.088	−0.115	0.067	0.278	0.094	0.119	0.099	0.152	0.134	0.106	0.101

Note: * see "Material and methodology"; Bold italics indicate correlations significant at $p < 0.05$.

systems participating in control of aluminum resistance of the given variety negatively influence development of lateral shoots. In barley Dina, on the contrary, level of interrelation of aluminum resistance of plants of IV group with development of lateral shoots is strong positive, as well as with mass of the main ear.

Oats varieties, contrary to barley, do not show practically interrelations between re-growth of roots in aluminum of the most resistant plants (with increase of root length more than 10%) and elements of yield structure. Only in one case significant relations of value of this increase with mass of sheaf (variety Ulov) is found out. At the same time middle-resistant plants of variety Ulov (with increase in root length in aluminum from 5 to 10%) differ with presence of correlations with such traits, as "productive tillering capacity," "lateral panicle mass," "number of grains per lateral panicle," and "grain mass per lateral panicle". As well as in a case of barley Elf these relations are negative that is middle-resistant plants have the lowered ability to development of lateral shoots.

Thus, most aluminum-resistant plants of barley Elf and middle-resistant plants of oats Ulov under conditions of neutral soil have low potential of development of lateral shoots. The most resistant plants of barley Dina – on the contrary, have high potential. The oats Krechet at any grouping of plants has not shown significant relations between level of aluminum resistance and degree of development of elements of yield structure in neutral conditions of a soil solution.

As a whole obtained data can testify to considerable varietal distinctions in structure of the genetic control of aluminum resistance both in oats and in barley.

Comparison of matrixes of pair correlations (Tables 7.3 and 7.4) for two sets of plants – having zero re-growth of roots in aluminum and having the greatest re-growth (more than 10%) has given the basis to assume that in most aluminum-resistant plants of barley (both variety Dina and variety Elf) high level of aluminum resistance is associated with destruction of many interrelations, characteristic for aluminum sensitive plants.

If compare these two groups of plants of barley Elf by significance of correlations for 55 pairs of traits the following facts can be pointed out: in resistant plants in 30 cases correlations have ceased to be statistically significant, in two cases new significant correlations are revealed, and only in 19 cases

TABLE 7.3 A Matrix of Coefficients of Pair Correlations for Elements of Yield Structures in Plants of Barley Contrast on Aluminum Resistance Level

Trait	1	2	3	4	5	6	7	8	9	10	11
Variety Elf											
1		0.52	0.47	-0.13	0.79*	0.57	0.76*	0.79*	0.54	0.57	0.37
2	0.95*		0.78*	0.17	0.34	0.88*	0.32	0.29	0.84*	0.90*	0.85*
3	0.82*	0.89*		0.07	0.51	0.95*	0.48	0.47	0.98*	0.94*	0.70*
4	0.79*	0.77*	0.65		0.15	-0.01	0.08	0.07	0.06	-0.01	0.00
5	0.93*	0.91*	0.85*	0.73*		0.45	0.91*	0.99*	0.50	0.44	0.22
6	0.92*	0.97*	0.92*	0.82*	0.90*		0.48	0.40	0.99*	1.00*	0.81*
7	0.91*	0.92*	0.89*	0.63	0.97*	0.88*		0.89*	0.50	0.48	0.37
8	0.93*	0.91*	0.87*	0.69	0.99*	0.89*	0.99*		0.45	0.40	0.16
9	0.85*	0.92*	0.99*	0.69	0.85*	0.96*	0.88*	0.87*		0.98*	0.74*
10	0.91*	0.96*	0.92*	0.82*	0.89*	1.00*	0.87*	0.88*	0.96*		0.82*
11	0.87*	0.89*	0.68	0.82*	0.87*	0.84*	0.78*	0.83*	0.70	0.82*	
Variety Dina											
1		0.77	0.77	0.47	0.75	0.76	0.46	0.78	0.85	0.79	0.53
2	0.42*		1.00*	0.80	0.80	0.93*	0.38	0.74	0.96*	0.94*	0.74
3	0.54*	0.91*		0.80	0.80	0.93*	0.38	0.74	0.96*	0.94*	0.74
4	0.35*	0.52*	0.39*		0.73	0.58	0.65	0.69	0.70	0.60	0.27
5	0.64*	0.54*	0.51*	0.75*		0.81	0.78	0.99*	0.89*	0.83	0.57
6	0.52*	0.87*	0.94*	0.46*	0.59*		0.28	0.75	0.98*	1.00*	0.92*

TABLE 7.3 Continued

Trait	1	2	3	4	5	6	7	8	9	10	11
7	*0.56**	*0.62**	*0.57**	*0.77**	*0.90**	*0.60**		0.84	0.46	0.32	−0.06
8	*0.66**	*0.53**	*0.54**	*0.71**	*0.98**	*0.61**	*0.89**		0.85	0.77	0.47
9	*0.52**	*0.90**	*0.97**	*0.43**	*0.54**	*0.99**	*0.59**	*0.56**		0.98*	0.79
10	*0.52**	*0.86**	*0.96**	*0.44**	*0.57**	*1.00**	*0.59**	*0.59**	*0.99**		0.90*
11	*0.47**	*0.86**	*0.76**	*0.64**	*0.72**	*0.83**	*0.69**	*0.69**	*0.79**	*0.80**	

Note: Above a diagonal – the data for group "increase of root length in the presence of aluminum more than 10%," below a diagonal – the data for group "without increase in root length in aluminum solution"; cells in which correlations of traits at two groups of plants differ significantly are indicated in bold italics. Numbers of elements of yield structure – see "Material and Methodology."

* statistically significant at p < 0.05.

TABLE 7.4 A Matrix of Coefficients of Pair Correlations for Elements of Yield Structures in Plants of Oats Contrast on Aluminum Resistance Level

Trait	1	2	3	4	5	6	7	8	9	10	11
Variety Ulov											
1		0.44	0.26	0.57	0.91*	0.23	0.83*	0.88*	0.24	0.21	*0.79**
2	*0.19*		0.73*	0.35	*0.64**	*0.68**	*0.77**	*0.65**	0.67*	*0.65**	*0.61**
3	0.32	0.81*		0.17	0.34	0.80*	0.43	0.35	0.79*	0.78*	0.32
4	*0.70**	0.06	0.12		0.54	*0.78**	0.54	0.49	*0.89**	0.48	0.34
5	0.74*	0.28	0.30	*0.90**		0.34	0.90*	0.99*	0.34	0.31	*0.75**
6	0.50	0.61	0.88*	0.30	0.35		0.53	0.36	0.99*	1.00*	0.16
7	0.69*	0.45	0.46	*0.87**	0.95*	0.53		0.89*	0.55	0.50	*0.75**

TABLE 7.4 Continued

Trait	1	2	3	4	5	6	7	8	9	10	11
8	0.73*	0.36	0.40	0.90*	0.98*	0.46	0.99*		0.34	0.33	0.74*
9	0.44	0.41	0.76*	0.33	0.31	0.96*	0.48	0.42		0.99*	0.13
10	0.47	0.45	0.78*	0.37	0.37	0.96*	0.55	0.49	0.99*		0.11
11	0.53	0.83*	0.85*	0.24	0.42	0.83*	0.55	0.49	0.65	0.65	
Variety Krechet											
1		0.16	0.05	0.75*	0.78*	0.21	0.72*	0.74*	0.30	0.30	0.48*
2	-0.29		0.79*	0.32	0.12	0.69*	0.22	0.08	0.74*	0.72*	0.75*
3	-0.58	0.58		0.32	0.09	0.85*	0.19	0.04	0.86*	0.86*	0.75*
4	-0.16	0.16	-0.38		0.70*	0.35	0.64*	0.60*	0.45*	0.43*	0.60*
5	0.66	-0.45	-0.83*	0.50		0.26	0.95*	0.96*	0.32	0.31	0.47*
6	-0.36	0.41	0.71*	-0.70*	-0.85*		0.34	0.18	0.98*	0.98*	0.86*
7	0.19	-0.59	-0.31	0.14	0.43	-0.68		0.96*	0.40*	0.40*	0.48*
8	0.67	-0.61	-0.42	-0.20	0.60	-0.49	0.77*		0.24	0.24	0.38
9	-0.54	0.64	0.85*	-0.51	-0.94*	0.90*	-0.55	-0.56		0.99*	0.89*
10	-0.48	0.55	0.80*	-0.60	-0.94*	0.96*	-0.60	-0.55	0.98*		0.89*
11	0.26	0.67	-0.10	0.39	0.26	-0.05	-0.41	-0.15	0.04	-0.02	

Note: see Table 7.3.

coefficients of correlation were similar at both sets of plants. For barley Dina it is also possible to notice that in 41 case coefficients of correlation between pair of traits, significant in group of sensitive plants, have lost the reliability (have ceased to correlate) in group of highly resistant plants, thus only 14 pair of traits have kept statistically significant relations among themselves. It is interesting to notice that 29 pairs of trait relations between which has ceased to be significant at transition from sensitive to the most resistant plants, coincide for both grades; 14 pairs of the correlating traits which have remained significant are also identical to both varieties of barley.

As to oats varieties it is possible to note the following. First, the number of the correlations ceased to be significant at transition from sensitive to the most resistant plants is much less – six for variety Ulov and five for variety Krechet, thus in each grade it is different pairs of traits. Secondly, much more correlations became significant: in variety Ulov – 12, and in variety Krechet – 24; only in 5 cases correlations were common for both varieties. Thirdly, significant relation was kept among traits in 14 cases in variety Ulov and in 7 cases in variety Krechet (these seven cases coincide for both varieties).

Also that fact is of interest that 6 pair of traits has significant correlations in both sets of plants of all four studied varieties. These traits are linked with development of lateral shoots – "productive tillering capacity," "lateral ear/panicle mass," "number of grains per lateral ear/panicle," and "grain mass per lateral ear/panicle".

In special experiments on sunflower, rice, soft spring wheat and other crops it has been established [13] that adverse conditions of environment cause substantial increase of force of relations between morphological traits and components of productivity; similar distinctions are found out at comparison of correlations in genotypes of different degree of adaptedness at cultivation in identical conditions [16].

It is possible to assume that changes of structure of relations reflect possibility of switching of regulatory mechanisms directing processes of growth and morphogenesis which, in turn, are additional way of adaptation to changing conditions of growth [11]. Transformations of structure of interrelations reflect specific features of realization of adaptive reactions of different traits and genotypes.

In our study it was showed species-specifity in change of structure of correlations of the investigated traits in barley and oats. If for both varieties of barley the complex of correlations of traits of grain mass with other elements of yield structure was much more complex in group with absence of re-growth of root system in stressful conditions then for both tested varieties of oats, on the contrary, more resistant plants had more complex structure of correlations.

Distinctions of two used varieties of barley were showed only for a complex of correlations of a trait "grain mass per main ear": variety Elf has not shown significant relations with trait "length of an ear" for sensitive group and, on the contrary, has shown significant relations with traits "plants height" and "number of grains per main ear". In the rest the investigated complex of correlations of these two varieties coincided at both groups of plants.

Distinctions between oats varieties are more distinct and concern traits "grain mass" both per lateral and per main panicles. In variety Krechet the complex of correlations for the main panicle is poorer than in variety Ulov, and for lateral panicles, on the contrary, more richly. Thus distinctions on the main panicle are more strongly shown for sensitive plants, and on lateral – for aluminum resistant groups.

As a whole obtained data indicate considerable morpho-physiological distinctions of groups of plants with different level of aluminum resistance.

2. Influence of heavy metals on structure of complex of correlations in oats and barley plants

According to the theory of the organization of quantitative trait [17] genetic control of a quantitative trait in various ecological conditions is carried out by a various quantitative and qualitative set of the genes influencing development of the given trait. Reorganization of the genetic control of separate trait and metabolic processes occurs without change of a genotype of a plant, as is one of the reasons of modification variability.

Data on degree of development of each of elements of yield structure received as a result of the study is shown in Tables 7.5 and 7.6.

Calculations of coefficients of pair correlations between separate elements of yield structure in the tested varieties in different variants of treatment have shown that all four stressors used in research made significant impact on development of separate elements of yield structure in both varieties of oats and barley and change of system of correlations between these elements (Figures 7.1 and 7.2).

TABLE 7.5   Change of Elements of Yield Structure in Barley Under Influence of Heavy Metals

Variety	Treatment	Elements of yield structure										
		1	2	3	4	5	6	7	8	9	10	11
Kupets	Control	68.5	7.2	6.7	8.9	1.3	5.7	25.7	1.2	118.2	4.7	45.62
	Fe	70.1	7.4	6.3	8.9	1.3	5.5	24.9	1.2	113.3	4.7	48.09
	Mn	65.9*	6.6	6.1	8.3*	1.2*	4.4*	22.9*	1.1*	95.3	3.9	46.78
	Cd	67.5	5.1*	4.6*	8.7	1.2*	3.4*	24.7	1.1*	71.7*	3.0*	44.53
	Pb	67.4	7.1	6.2	9.1	1.4*	5.4	25.4	1.2	105.3	4.6	75.18*
Line 406–99	Control	62.7	4.8	4.6	8.4	1.6	4.0	22.7	1.4	63.7	3.6	62.51
	Fe	67.0*	4.8	4.7	8.9	1.6	4.2	23.6	1.5	67.7	3.8	62.03
	Mn	70.0*	6.3*	6.0*	7.8*	1.5*	4.9	22.3	1.4	79.4	4.5	60.86
	Cd	69.4*	5.5	5.4	8.7	1.6	5.1	23.6	1.5*	75.1	4.6	62.81
	Pb	75.0*	7.7*	7.1*	9.7*	1.6	7.7*	24.4*	1.5	123.5*	6.9*	62.38

Note: * differences from the control are statistically significant at $p < 0.05$. Numbers of elements of yield structure – see "Material and Methodology."

TABLE 7.6 Change of Elements of Yield Structure in Oats Under Influence of Heavy Metals

Variety	Treatment	Elements of yield structure										
		1	2	3	4	5	6	7	8	9	10	11
Butsefal	Control	85.9	3.3	2.6	18.2	2.5	3.0	68.6	2.3	81.5	2.7	33.52
	Fe	85.9	3.1	2.5	17.5	2.7	3.3	108.8*	2.4	89.8	3.0	30.40*
	Mn	83.0	3.5	2.9	20.0*	4.2*	4.8*	111.4*	3.7*	138.6*	4.5*	35.10
	Cd	88.1	3.1	2.5	18.5	3.2*	3.4	92.2*	2.8*	93.2	2.9	30.95*
	Pb	91.5*	2.7*	2.2*	18.7	2.8*	2.6	86.7*	2.4	70.8	2.0*	27.64*
Krechet	Control	77.6	2.7	2.3	16.8	2.4	2.3	59.2	2.0	60.6	1.9	34.54
	Fe	83.5*	2.7	2.5	17.7	2.9*	2.7	73.1*	2.6*	69.1	2.4	36.35
	Mn	78.8	3.1	2.9*	19.8*	3.6*	4.4*	94.6*	3.3*	108.1*	3.8*	35.19
	Cd	81.1*	2.7	2.7*	18.2*	3.2*	3.2*	81.4*	2.9*	79.7*	2.8*	35.76
	Pb	82.1*	3.4*	2.6	18.9*	3.5*	4.1*	89.2*	3.1*	92.4*	3.5*	34.95

Note: * see Table 7.5.

As it is possible to see from the resulted figures, in many variants of study it is observed both complication and simplification of system of interdependence in development of separate trait of plants in comparison with control variants. In other words, there is a considerable reorganization of genetically caused relations in plants under the influence of stressful abiotic factors.

If to consider separate species and varieties of plants used in microplot trials it is possible to note considerable distinctions both between varieties, and between different species on a studied question.

As a whole barley were more resistant against ions of heavy metals than oats. For each of barley variety similar qualitative influence of iron, manganese, cadmium, and lead only on two pairs of the interrelated traits is noted. In case of variety Kupets all four metals have led to strengthening of correlation in development of pairs traits "length of main ear – main ear mass" (in the control the interrelation was statistically doubtful, $r = 0.434$; in test treatment $r = 0.644$–0.797, for example, it is statistically significant at $p < 0.05$) and "length of the main ear – grain mass per main ear" ($r = 0.320$ and $r = 0.515$–0.721 accordingly).

For selection line 406–99 more susceptible to stressful influence, on the contrary, destruction of interdependence in development of following pairs of traits is noted: "plant height – length of the main ear" (relation is statistically significant in the control $r = 0.712$ and is doubtful at stressful influence), "plant height – number of grains per main ear" (in the control $r = 0.624$).

For both investigated varieties of barley the strongest change of system of correlations under the influence of ions of iron and lead and minimum – under the influence of manganese ions (see Figure 7.1) was characteristic. Cadmium ions have four times more strongly action on a complex of correlations of variety Kupets than of selection line 406–99; in selection line 406–99 ions of cadmium have destroyed 4 of significant relations existing in the control, and in variety Kupets, on the contrary, 13 pairs of interactions became significant there where in the control it was not found out of these genetic interrelations. If in selection line 406–99 ions of iron have led to destruction of existing interrelations in 32 pairs of elements than in variety Kupets the basic influence of iron was showed an increase of reliability of interaction of 14 pairs of elements. It is possible to name action of ions of lead equally destructive for both samples – in variety Kupets they have broken 14 of existing relations, in selection line 406–99–25 interdependences.

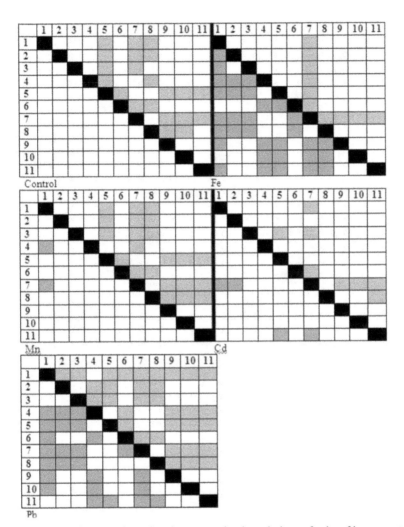

FIGURE 7.1 Influence of top-dressing processing by solutions of salts of heavy metals on structure of correlations between elements of yield structure of plants of barley Kupets (above a diagonal) and selection line 406–99 (below a diagonal). Pair of traits between which coefficients of correlations are statistically doubtful at $p < 0.05$ are painted. Numbers of elements of yield structure – see "Material and Methodology."

Thus variety resistant to stressors at cultivation in the environment, containing ions of heavy metals, has shown ability to more coordinated action in the conditions of stressful influence than sensitive variety of barley.

For both varieties of oats (see Figure 7.2) manganese ions in a greater degree have created new significant correlative relations (8 – in variety

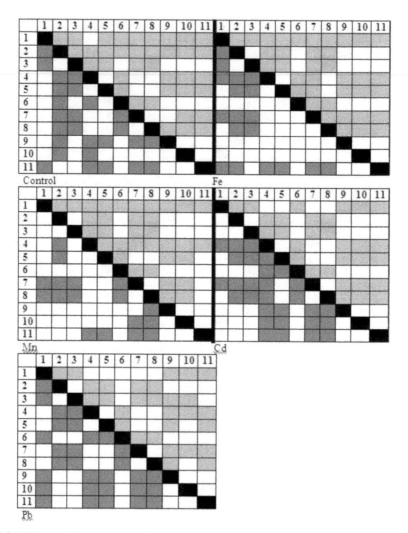

FIGURE 7.2   Influence of top-dressing processing by solutions of salts of heavy metals on structure of correlations between elements of yield structure of plants of oats Butsefal (above a diagonal) and oats Krechet (below a diagonal). Pair of traits between which coefficients of correlations are statistically doubtful at $p < 0.05$ are painted. Numbers of elements of yield structure – see "Material and Methodology."

Butsefal and 13 – in variety Krechet) than have destroyed existing at control variants (5 and 4 accordingly). Iron ions have created 11 new significant communications in plants of variety Krechet having destroyed only 3 whereas in variety Butsefal the number of newly arisen and destroyed relations was equal (on 5). For the tested varieties of oats it is possible to name action of ions of cadmium and lead opposite on mode. In variety Butsefal ions of these metals promoted formation of new interdependences more likely than destruction of old (8 against 2 for cadmium and 11 against 3 for lead). In variety Krechet on the contrary cadmium has destroyed 9 old interrelations, having formed only 6 new, and lead, accordingly, 7 and 3.

In resistant oat variety Krechet stressful influence has led to occurrence of statistically significant relations between pairs of traits which development has not been coordinated in control conditions: "general tillering capacity – lateral panicle mass," "general tillering capacity – number of grains per lateral panicle," and "general tillering capacity – grain mass per lateral panicle" (coefficients of pair correlations were statistically significant at $p < 0.05$ and have made accordingly 0.535–0.923; 0.551–0.862; 0.546–0.926).

Sensitive variety Butsefal has shown changes of correlations in six pairs of traits, thus four pairs interactions became statistically significant unlike the control: "general tillering capacity – lateral panicle mass," "general tillering capacity – number of grains per lateral panicle," "general tillering capacity – grain mass per lateral panicle" (as well as in variety Krechet), and also has amplified relations between the general and productive tillering capacities (corresponding coefficients of pair correlations have made 0.680–0.872; 0.630–0.747; 0.619–0.745, and 0.743–0.889). Two other pairs of traits, on the contrary, have lost co-ordination in development (relations between them became statistically insignificant): "general tillering capacity – 1000 grains mass" and "number of grains per main panicle – number of grains per lateral panicle".

Thus, resistant variety of oats, as a whole, also was capable to more coordinated growth in stressful conditions, than sensitive variety of this cereal crop.

7.4 CONCLUSIONS

Thus, the obtained data testifies to considerable infringement of a metabolism of plants under the influence of heavy metals. Each of the studied metals differs from others on influence on development of separate elements of yield structure of plants. For varieties of oats, contrast on resistance to abiotic factors it was possible to reveal three pairs of elements of yield structure which change coordinately under the influence of all four studied metals. The varieties of barley taken into study sharply differ from each other in this point of view – in each of them the coherence of different elements of yield structure qualitatively changes under the influence of all metals. Most likely, it indicates their higher genetic distinction than in studied varieties of oats.

As a whole in study it was showed species-specifity in change of structure of correlations of the investigated traits in barley and oats. If for both varieties of barley the complex of correlations of traits of grain mass with other elements of yield structure was much more complex in group of least aluminum resistant plants than for both tested varieties of oats, on the contrary, more aluminum resistant plants had more complex structure of correlations.

Distinctions of two used varieties of barley were showed only for a complex of correlations of a trait "grain mass per the main ear": variety Elf has not shown significant relations with trait "length of an ear" for aluminum sensitive group and, on the contrary, has shown significant relations with "plant height" and "number of grains per main ear". In the rest the investigated complex of correlations of these two varieties coincided at both groups of plants.

Distinctions between oats varieties are more distinct and concern "grain mass" both per lateral, and per main panicles. In variety Krechet the complex of correlations for the main panicle is poorer, than in variety Ulov, and for lateral panicles, on the contrary, more richly. Thus distinctions on the main panicle are more strongly shown for aluminum sensitive plants, and on lateral – in aluminum resistant group.

As a whole it is possible to point out considerable morpho-physiological distinctions of intravarietal groups of plants with different level of aluminum resistance. Our research allows to add the received [13] conclusion that selection at cultural crops conducts to formation rigid rather stable system of the interrelations which changes at external influences are limited to fluctuations of level of relations. Addition it consists that similar fluctuations can indicate intravarietal heterogeneity on separate traits.

The data received during performance of study, allows to draw a conclusion that under the influence of ions of aluminum and heavy metals there is a considerable reorganization of genetically caused system of interrelations of development of elements of yield structure of plants. In the majority of variants of treatment effects of destruction of the relations existing in control variants prevailed. As a whole for all studied metals and both investigated cereal crops resistant varieties differed with more coordinated growth in stressful conditions than sensitive varieties of these crops.

ACKNOWLEDGEMENTS

We would like to thank breeders DrSci G.A. Batalova and PhD I.N. Shchennikova (North-East Agricultural research Institute, Kirov, Russia) for seed material and critical remarks on the manuscript.

KEYWORDS

- **aluminum**
- **barley**
- **heavy metals**
- **intravarietal heterogeneity**
- **oats**
- **resistance**

REFERENCES

1. Ryan, P. R., Raman, H., Gupta, S., Horst, W. J., Delhaize, E. A second mechanism for aluminum resistance in wheat relies on the constitutive efflux of citrate from roots. Plant Physiol. 2009, Vol. 149. N. 1. 340–351.
2. Raman, H., Ryan, P. R., Raman, R., Stodart, B. J., Zhang, K., Martin, P., Wood, R., Sasaki, T., Yamamoto, Y., Mackay, M., Hebb, D. M., Delhaize, E. Analysis of *TaALMT1* traces the transmission of aluminum resistance in cultivated common wheat (*Triticum aestivum,* L.). Theor. Appl. Genet. 2008, Vol. 116. N. 3. 343–54.

3. Sasaki, T., Yamamoto, Y., Ezaki, B., Katsuhara, M., Ahn, S. J., Ryan, P. R., Delhaize, E., Matsumoto, H. A wheat gene encoding an aluminum-activated malate transporter.. Plant, J. 2004, Vol.37. 645–653.
4. Delhaize, E., Gruber, B. D., Ryan, P. R. The roles of organic anion permeases in aluminum resistance and mineral nutrition. FEBS Letters. 2007, Vol. 581. N. 12. 2255–2262.
5. Houde, M., Diallo, A. O. Identification of genes and pathways associated with aluminum stress and tolerance using transcriptome profiling of wheat near-isogenic lines. BMC Genomics. 2008, Vol. 9. 400.
6. Ermolayev, V. Genes differentially expressed in soybean lines sensitive and tolerant to aluminum stress. Dissertation Zur Erlangung des akademischen Grades doctor rerum naturalium (Dr. rer. nat.). Halle (Saale), 2001, (In German)
7. Ma, J. F., Furukawa, J. Recent progress in the research of external Al detoxification in higher plants: a mini review. J. Inorg. Biochem. 2003, Vol.97. N. 1. 46–51.
8. Ma, J. F., Hiradate, S., Matsumoto, H. High aluminum resistance in buckwheat. II. Oxalic acid detoxifies aluminum internally. Plant Physiol. 1998, Vol. 117. 753–759.
9. Li, X. F., Ma, J. F., Hiradate, S., Matsumoto, H. Mucilage strongly binds aluminum but does not prevent roots from aluminum injury in *Zea mays*. Physiol. Plant. 2000, Vol. 108. 152–160.
10. Delhaize, E., Ryan, P. R. Aluminum toxicity and resistance in plants. Plant Physiol. 1995, Vol.107. 315–321.
11. Rostova, N. S. Correlation analysis in population investigations. Ecology of populations. Moscow: Nauka Publ., 1991, 69–86 (In Russian).
12. Lisitsyn, E. M., Batalova, G. A., Tiunova, L. N. Intravarietal variability of the structure of correlations of yield components in oat and barley plants. Russian Agricultural Sciences. 2011, Vol. 37. N. 4. 280–282.
13. Rostova, N. S. Correlations: structure and variability. St. Petersburg: Unipress, 2002, 308 p (In Russian).
14. Methodic of State varietal testing of agricultural crops. Moscow. Ministry of Agriculture, 1989, 250 p (In Russian).
15. Zubkova O.A., Russkikh E.A., Shikhova, L. N., Lisitsyn, E. M. Influence of heavy metal ions on source-sink system of oat and barley plants. Leguminous and groat crops. 2012, N. 3. 42–47 (In Russian).
16. Rostova, N. S., Koval, S. F. Structure of correlations of productivity elements in low-height isogenic lines of spring soft wheat. Agricultural biology. 1986, N. 8. 61–67 (In Russian).
17. Dragavtsev, V. A., Litun, N. P., Shkel, I. M., Nechiporenko, N. N. Model of ecological-genetic control of plant quantitative traits. Report of Russian Academy of Sciences. 1984, Vol. 274. N. 3. 720–723 (In Russian).

PART II

HORTICULTURAL CROP SCIENCE

CHAPTER 8

BIOLOGICAL FEATURES OF THE NEW PEAR CULTIVARS (*PYRUS COMMUNIS* L.)

MYKOLA OL. BUBLYK, OKSANA IV. MYKYCHUK,
LIUDMYLA A. FRYZYUK, LIUDMYLA M. LEVCHYK, and
GALYNA A. CHORNA

CONTENTS

ABSTRACT

The important source of the pear assortment renovation in most of Ukraine's regions is introduction of new cultivars. When establishing pear orchards a special attention should be paid to the selection of cultivars because it is just they that play a decisive role in the creation of a high-productive orchard and the entire complex of measures concerning tending, protection from pests and diseases is directed to support and securing of optimum conditions contributing to the display of all the cultivar potential possibilities. Therefore, without detailed study of the cultivars economic and biological characteristics in research institutions it is impossible to introduce them in production under the concrete soil and climatic conditions.

8.1 INTRODUCTION

Prydnistrovya one of the best regions in Ukraine for the cultivation of high-quality early winter and late-winter pear cultivars. This crop takes the second place after apple. Pear fruits have high nutritive value. Their taste dietetic and medicinal properties are conditioned with high content of sugars (6–16%), organic acids (0.1–0.3%), presence of the vitamins A, B, P, PP, C, microelements, nitrogenous and biologically active substances.

The intensification of the horticultural branch, high requirements to the environment protection makes it necessary to review the assortment and select high-quality cultivars. According to the market requirements the assortment of pear, like other fruit crops, is constantly improved and renovated. The former cultivars are substituted for new ones, which thanks to the efforts of breeders obtain better qualities as compared to parental forms surpassing them considerably by many parameters [1, 2].

The modern intense orchards need cultivars with low and average low trees and dense or average dense, usually compact crowns, early ripening, with high and regular yield, high fruit quality, resistant to all the unfavorable environmental conditions and diseases and, besides, contribute to mechanization of the operations in orchards.

The role of a cultivar and requirements to it increases in the modern intensive horticulture even more. Under the similar conditions only by the selection of cultivars it is possible to achieve a 1.3–1.5 higher yield from the area unit and increase significantly the economic efficiency of the fruits production [3, 4].

One of the important biological peculiarities influencing the limits of the pear spread is winter-hardiness of cultivars. Its degree depends on the cultivar origin, age, weather conditions of summer and autumn, general readiness for hibernation, presence of the snow coverage during the winter period, yield, etc. [5, 6].

Early ripening is one of the most important economic and biological properties of the fruit crop. It is just early maturing cultivars that return capital costs for the establishment and care of orchards more rapidly than those which begin fruit-bearing in later terms [7, 8].

For the creation of early ripening high productive pear orchards with low trees it is necessary to select dwarf and semidwarf cultivars. Trees having those properties may be planted densely (1000 per 1 ha and more), which reduces the labor intensity [9, 10].

Thus the increase of the yield and profitableness of pear orchards is connected closely with the assortment improvement. The horticulture intensification and new orchard constructions cause increased require-ments to pear cultivars. For the establishment of intense orchards cultivars are necessary with a not high compact crown of trees, early beginning of fruit bearing, and high yield of high quality fruits as well as high resistant to the main fungous and bacterial diseases.

8.2   MATERIALS AND METHODOLOGY

Proceeding from the tasks the researches are directed to studying the biological peculiarities of new inland and introduced pear cultivars in an orchard and determination of their economic favorability for growing in the soil and climatic conditions of the Prydnistrovya and Naddnestrovschina.

15 cultivars were studied: 11 autumn cultivars ('Vrodlyva,' 'Madam Balle,' 'Bilka,' 'Oksamyt,' 'Krasa Kubani,' 'Kyrgyzka Zymova,' 'Bristol Cross,' 'Legenda Karpat,' 'Saiva,' 'Bukovynka' – control and 4 winter cul-tivars ('Monzana,' 'Yakymivska,' 'Gurzufska' and 'Talgarska Krasunya,' 'Yablunivska' – control).

The orchard was planted in 2002. Nutrition area was 4.5 × 3 m, wild pear rootstock served. The crown was flat, formation volumetrical. The soil management is bare fallow; in a row herbicides are applied.

The following methods are used in the process of the researches:

- program and methods of the strain investigation of the fruit, small fruit and nuciferous crops (methods of ARRIBFC) [11];
- methods of making examinations of the cultivars of the fruit, small fruit and nuciferous crops and grape [12];
- methods of carrying out field researches with fruit crops [13]; Methods of estimating the fruit and small fruit products quality [14].

The statistical processing of the research results is carried out by the method of the disperse analysis [15].

The following estimates and observations are conducted.

1. The trees height is measured after the finish of growth by means of a measuring rule [16].
2. The phenological phazes are studied with the use of the methods of the Uman National University of Horticulture (former Uman Agricultural Institute – UAI)) [16] and the above mentioned programs and methods of ARRIBFC [11].
3. The yield is calculated per each registration tree by the gravimetric method [16].
4. The net photosynthesis productivity is determined by means of ringing fruit-bearing branches [16].
5. The specific productivity is calculated as the ratio of yield per tree to the crown projection surface (m^2), crown volume (m^3) and leaf area (m^2) by the methods of UAI [16].
6. The average fruit mass is determined by the methods of UAI [16].
7. The appearance and taste qualities of fruits are estimated on the basis of the degustation estimation of fresh fruits by the methods of ARRIBFC [11].
8. The dry substances content is determined by the arbitrary method, total sugar content by the colorimetric method, total acidity by the titration with the NaOH solution (0.1 n), vitamin C content by the photocolorimetric method, phenol compounds content by the method of Folin–Denis, pectic substances content by the carbozolic method [14].
9. The drought-resistance and winter-hardiness are determined by the method of V.K. Smykov [17].

Peculiarities of the technology of the pear fruits cultivation.

In the experiment of studying the economic and biological peculiarities of the pear growth and fruit-bearing in an orchard the technology is applied according to the recommendations of IH NAAS. The soil management is bare fallow; additional fertilization of the orchards is carried out proceeding from the presence of accessible nutritive elements in the soil. The crown is flat and volumetric. The trees protection from pests and diseases is conducted.

While using bare fallow not deep hoeing of the upper soil layer 4–5 times is applied alternating the use of disc harrows and cultivators. Weeds near with a tree were exterminated to manually.

The orchard pruning is carried out annually after frosts. The trees growth is limited up to 2–2.5 m for the cultivars with short-holed trees, to 3.0 m for average and to 3.5–4.0 m for high ones.

Fruits are collected in the period of table ripeness, as e rule, with a fruit stem.

The plant protection from pests and diseases is conducted according to the recommendations of the Prydnistrovska RSH (Prydnistrovska Research Station of Horticulture).

The climate in the region of the researches is temperate continental. Accumulated effective temperatures above 10°C vary by years within 3169–3643°C. The frost-free period lasts 249–284 days. According to the average multiyear data autumn frosts begin in the third decade of November – first decade of December and spring ones finish in the second-third decades of March.

The temperature peaks were in the year 2012: the absolute maximum +38°C and absolute minimum –32.8°C (February 12) (Figure 8.1).

Concerning the precipitation amount this region is that with unsteady moistening. During the vegetation period the average precipitation sum is 507.1 mm.

The most droughty year was 2012 when the precipitation amount was 453.2 mm, the humidest month was April (the precipitation sum was 90.3 mm), and the highest humidity deficiency was observed in July – 38 mm (norm is 94 mm) (Figure 8.2).

It should be noted that the highest average summer temperature (22.1°C) was in the droughtiest periods – June–August.

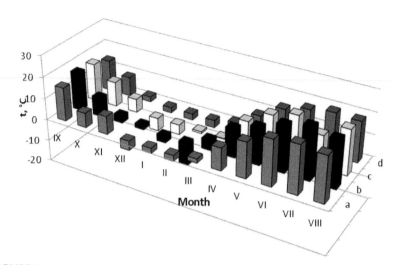

FIGURE 8.1 Average monthly air temperature (°C) during through years of research [Notes: a – average monthly temperature, 2010–2011; b – average monthly temperature, 2011–2012; c – average monthly temperature, 2012–2013; d – average multiyear temperature, °C].

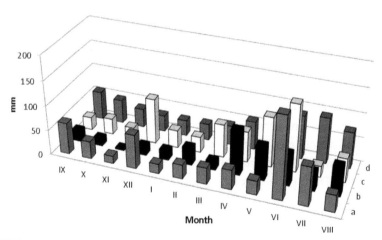

FIGURE 8.2 Precipitation amount (mm) during the research years [Notes: a – monthly sum, mm, 2010–2011; b – monthly sum, mm, 2011–2012; c – monthly sum, mm, 2012–2013; d – Average multiyear precipitation amount, mm].

8.3   RESULTS AND DISCUSSION

The collection pear orchards of the Prydnistrovsky Research Station of Horticulture (Prydnistrovska RSH) contain 159 strain samples.

In order to select cultivars for the introduction into production and amateur gardening as well as into the breeding programs as parental forms the researched samples are estimated according to their characteristics biological and valuable for economy.

Winter-hardiness is one of the main indices, which determines the possibility of growing cultivars under concrete conditions, their productivity and value for production. The productional pear development is limited by the presence of cultivars that are high winter-hardy with marketable fruit qualities. Therefore, one of the major tasks is study and selection of new winter-hardy cultivars, which meet the climatic conditions of the given zone and estimation of cultivars as initial forms in the breeding programs concerning the concrete sign. For instance, one of the factors that determine the future yield is the reproductive buds resistance to low air temperatures in winter. Different resistance degree may be explicated by different water-holding capacity and differentiation of flower buds [8].

The cause of the flower embryos destruction in the winter period is freezing of deeply overcooled water in cells. The critical temperature of the destruction of flower buds which are in the state of dormancy is –23 to –27°C. Certain European cultivars are damaged even under –18 to –20°C. Especially dangerous for pear are lasting winter thaws after which the temperature lowering to –9 to –14°C can be critical [18].

The critical destruction temperature of flower buds of both different cultivars and the same cultivar is not steady. Under the temperature fluctuations in the second half of winter or when early frosts come in the autumn-winter period the temperature causing the flower buds damage lowers considerably [19]. In the collectional pear orchards the destruction was revealed in the winter period of 2011–2012. For example, in 2012 the reproductive buds damage was caused by different air temperature fluctuations in the second half of winter. The fruit-buds of the cultivars

'Bukovynka' (control), 'Gurzufska,' 'Madam Balle' (3 points each), 'Bilka' and 'Oksamyt' (5 points) were damaged by low temperatures (−32.8°C, February 12) to the greatest degree. 'Yablunivska' (control), 'Krasa Kubani,' 'Kyrgyzka Zymova' displayed the high fruit-buds resistance to slight freezing (0 points).

Characterizing slight freezing of reproductive buds it should be noted that not all the cultivars have their buds damaged by winter temperatures in full during the research years. The analysis of the obtained data shows that this phenomenon is connected on the whole with general state of trees. If leaves and branches are affected with fungous diseases the frost-resistance of both fruit buds and of the whole tree reduces acutely.

Since different extent of the reproductive buds resistance can be explicated by different water-holding capacity the important task is to study the drought-resistance of the investigated cultivars.

Drought-resistance is one of the significant economic and biological characteristics of a fruit crop. It is ensured by the high water-holding capacity of the cells of leaves and shoots, presence in their vacuoles and cytoplasm low-molecular compounds with high hydrophily as the drought in the summer period causes reducing of the shoots and roots increase, early summer wilt and fall of leaves, disturbance of the CO_2 assimilation as well as weakens the leaf apparatus development. Those phenomena influence negatively the formation of the productivity and fruits quality. Besides, the unfavorable drought effect can be the reason of the considerable winter-hardiness decrease and lack of moisture in the soil and high temperatures cause functional and parasitic diseases. The moisture deficiency influences transpiration, growth, development and metabolic processes (depending on the lasting of its effect) as well. Low content of all the forms of sugars in fruits is also explicated by lack of moisture [20, 21].

A comparative estimation of the researched pear cultivars drought-resistance was carried out in 2011–2013. Shoots with leaves were selected in the periods of the strongest water regime strain. 2011–2012 summer was rather hot. Especially droughty appeared June and July. During a year precipitations are distributed irregularly. According to the average multi-year data their sum in winter is only 19.1% of the annual amount, in spring – 16.5%, in autumn 18.6% and in summer 45.8%. During the vegetation period the average amount precipitations is 507.1 mm.

In order to determine the cultivars drought-resistance the laboratory-field method is applied which includes the study of the leaves water regime: determination of the water content in tissues, water deficiency as well as the capacity of leaves to hold water and renovate turgor. The experiment for the definition of drought-resistance is carried out in dynamics (in 2, 4 and 24 hours).

Among the autumn cultivars it is 'Gurzufska' that has the lowest index of water deficiency – 31.1%, amongst winter ones 'Kyrgyzka Zymova' (21.8%) and 'Yablunivska' (control) – 28.0%. Most of the investigated cultivars have this index temperature (within 31.1–42.2%). Under this index the greater part of researched cultivars (93.3%) is high and average resistant to drought, the rest (6.7%) low resistant (Figures 8.3 and 8.4).

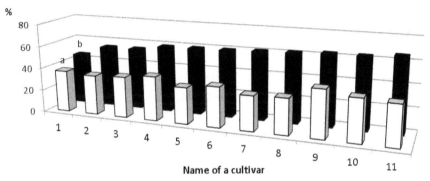

FIGURE 8.3 Average indices of the pear autumn cultivars drought-resistance [Notes: a – water deficiency, %; b – moisture content in the leaves, %; cultivars: 1 – Oksamyt, 2 – Bristol Cross, 3 – Talgarska Krasunya, 4 – Vrodlyva, 5 – Krasa Kubani, 6 – Yakymivska, 7 – Gurzufska, 8 – Monzana, 9 – Bukovynka (c.), 10 – Madam Balle, 11 – Legenda Karpat].

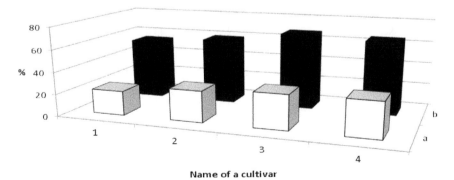

FIGURE 8.4 Average indices of the pear winter cultivars drought-resistance [Notes: a – water deficiency, %; b – moisture content in the leaves, %; cultivars: 1 – Kyrgyzka Zymova, 2 – Yablunivska (c.), 3 – Saiva, 4 – Bilka].

The obtained results show that the content of a certain water amount in leaves is connected in the main with the individual characteristics of cultivar. The investigated cultivars have it 45.9–68.2%. For instance, 'Gurzufska,' 'Monzana,' 'Bukovynka' (control), 'Madam Balle,' 'Bilka,' 'Legenda Karpat' and 'Saiva' distinguish themselves for high water content in tissues – 61.3–68.2%. The cultivar 'Oksamyt' has the lowest content of moisture in the leaves – 45.9%. The water content in the plants of the other researched cultivars is 54.1–60.2% (see Figures 8.3 and 8.4).

The determination of the changes concerning the leaves water-holding capacity shows that during the first 2 hours of the exposition the difference among cultivars as about the water loss is not significant (within 14–40%). But in 4 hours already most of the cultivars lost more than 50% of water. For example, among the autumn cultivars 'Talgarska Krasunya' and 'Gurzufska' distinguish themselves for the highest water-holding capacity. Their average loss of water for 2 hours is 11.1–17.5%. Among the winter cultivars it is 'Kyrgyzka Zymova' that has the lowest water loss for two hours during the three years – 10.5% (Table 8.1).

TABLE 8.1 Water-Holding Capacity in Leaves of the Pear Cultivars, 2011–2013, Average

Cultivar	Loss of water, %, in			Turgor renovation, %
	2 hours	4 hours	24 hours	
Autumn cultivars				
Bukovynka (control)	32.5	37.4	59.5	68.9
Vrodlyva	30.6	33.3	53.9	69.0
Gurzufska	17.5	22.7	43.8	80.1
Krasa Kubani	31.5	34.6	64.5	81.6
Madam Balle	22.7	34.1	63.8	84.7
Monzana	26.1	29.3	47.5	82.8
Oksamyt	22.9	27.3	47.5	69.3
Talgarska Krasunya	11.1	20.1	32.2	89.6
Legenda Karpat	30.1	35.4	52.6	91.2
Bristol Cross	24.8	27.1	43.3	57.8
Yakymivska	27.6	37.1	58.1	76.3
LSD_{05}	3.07	2.72	4.24	6.59

TABLE 8.1   Continued

Cultivar	Loss of water, %, in			Turgor renovation, %
	2 hours	4 hours	24 hours	
Winter cultivars				
Yablunivska (control)	17.9	32.0	60.8	84.8
Bilka	29.9	37.3	67.8	65.1
Kyrgyzka Zymova	10.5	26.2	36.3	92.7
Saiva	18.7	30.6	49.9	86.4
LSD_{05}	2.68	2.70	3.28	3.81

On the basis of the results of the field evaluation of the pear drought-resistance and laboratory analysis of the water regime certain parameters the drought-resistant cultivars have been selected for growing in the regions with non-sufficient humidity with a complex of characteristics valuable for economy – 'Gurzufska,' 'Talgarska Krasunya,' 'Kyrgyzka Zymova' and 'Yablunivska.'

The estimation of the researched cultivars resistance to fungous diseases is conducted by the field method. On the background of the plant protection system received at the Prydnistrovska RSH that foresees 6–8 sprayings against a complex of fungous diseases, namely: monilial blight, the symptoms of its affection displayed in 2012. This disease injured pear cultivars considerably during the research years. It is caused by the fungus Monilia fructigena, Monilia cydoniae, Monilia cinerea. Its most spread form is brown rot. The favorable weather conditions in the spring-summer period of 2012 brought in the progress of this disease in the middle of summer. It should be remarked that the cultivars which were damaged by frost in winter appeared less resistant to monilial blight.

Highly resistant to this disease proved 'Vrodlyva,' 'Krasa Kubani,' 'Kyrgyzka Zymova' and 'Yablunivska' (control), low resistant 'Oksamyt' (35.5% of the disease spread), somewhat higher resistant 'Gurzufska,' 'Bilka,' 'Monzana,' 'Madam Balle' and 'Yakymivska' (19.0–34.0% of the disease spread).

Other diseases were observed on certain cultivars but a considerable affection was not detected.

Pear psylla (Prulla puril) is a small gray or yellow-brown insect 3 mm long with four transparent wings. Its larvae discharge a great amount of transparent sweet liquid – 'honey dew.' It protects a larva against unfavorable environmental conditions [22]. Among the introduced cultivars the most susceptible to this pest have appeared 'Yakymivska,' 'Bristol Cross,' 'Oksamyt' (affection is 5 points or 31.5–34.5% of spread), rather high resistant 'Vrodlyva,' 'Krasa Kubani,' 'Madam Balle,' 'Legenda Karpat,' 'Saiva,' and 'Yablunivska' (control) (1 point). Concerning the cultivar Kyrgyzka Zymova the symptoms of its affection by pear psylla have not been detected at all.

The analysis of the cultivars resistance to gall mite shows that high resistant to this pest are 'Vrodlyva,' 'Kyrgyzka Zymova' (0 points). The affection of most of the investigated cultivars does not exceed 1–3 points. Low resistant are 'Oksamyt' (15.0% of the pest spread) and 'Bristol Cross' (14.0%).

Net photosynthesis productivity. Photosynthesis is major process of the plant vital activity that results in the creation of the whole fruit tree biological mass, in particular, yield through the synthesis of the organic substance from the carbonic acid and water. Therefore, one of the tasks of our researches was determination of the net photosynthesis productivity (NPP).

The difference between the cultivars as for this index is significant. For instance, autumn cultivars with average trees ('Talgarska Krasunya' and 'Madam Balle' – 3.8 m each, 'Vrodlyva' – 3.5 m and 'Bukovynka' (control) – 3.9 m) distinguish themselves for the greatest organic substance accumulation and their NPP is 9.9, 8.7, 8.6 and 8.4 g/m^2 per day respectively (Figure 8.5).

Besides, the cultivars that accumulate much organic substance have also rather high average yield indices (8.8–16.6 kg/tree).

The investigations have detected the dependence between the indices of height, NPP and yield among winter cultivars as well. For example, the highest NPP and yield (9.6 g/m^2 per day and 31.1 kg/tree – the average yield per three years) are characteristic for the cultivar 'Kyrgyzka Zymova' with high trees (4.07 m). The rest of cultivars has this index within 5.6–7.9 g/m^2 (Figure 8.6).

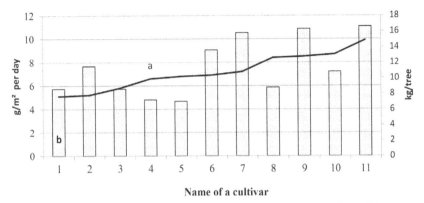

FIGURE 8.5 Net photosynthesis productivity of pear autumn cultivars [Notes: a – average index of the net photosynthesis productivity, g/m² per day; b – average index of yield, kg/tree; cultivars: 1 – Monzana, 2 – Yakymivska, 3 – Krasa Kubani, 4 – Oksamyt, 5 – Gurzufska, 6 – Bristol Cross, 7 – Legenda Karpat, 8 – Bukovynka (c.), 9 – Vrodlyva, 10 – Madam Balle, 11 – Talgarska Krasunya].

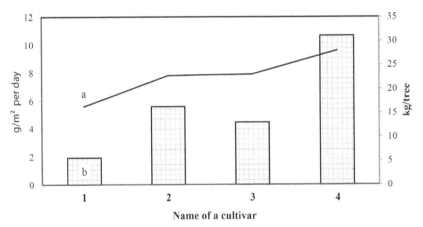

FIGURE 8.6 Net photosynthesis productivity of pear winter cultivars [Notes: a – average index of the net photosynthesis productivity, g/m² per day; b – average index of yield, kg/tree; cultivars: 1 – Bilka, 2 – Saiva, 3 – Yablunivska (c.), 4 – Kyrgyzka Zymova].

One of the most important cultivar characteristics is its yield. Harvest determined by genetics cultivar and increased at a rational planting of trees, appropriateness of applying agricultural technology, from pruning and from ways protect them from pests and diseases [23].

Yield is an integrated index which characterizes the amount of the laid reproductive buds, the number of flowers in a flower cluster, amount of infructescene fruits and their average mass [24].

Growth in the present day conception is rapidity of accumulating the mass of the plant structural substance that synthesizes from the fund of free assimilants created in the process of photosynthesis as well as the processes of storing the organic substances of the plant reproductive and storing organs which determine the level of the most valuable part of the yield.

A great role in the non-specific plant resistance to stress factors belongs to the root-leaf interrelations. They determine the plants water regime and effectivity of using mineral nutritive elements from the soil. Non-specific plant organism resistance also depends on the buffer volume of the vegetative organs and root system for the utilization the excess of the assimilants appearing in the stress conditions when growth is inhibited. Metabolites is plastic and energetic material for the growth processes and for the organogenesis respectively – process of the formation of the biological and net yield [24].

The researched cultivars yield is significantly different (Figures 8.7 and 8.8). Among autumn cultivars 'Talgarska Krasunya' has high average

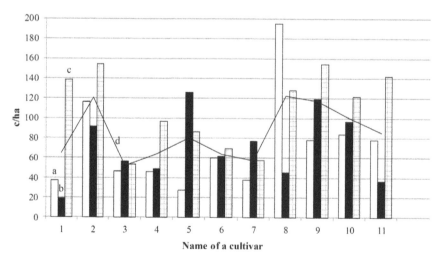

FIGURE 8.7   Yield of the pear autumn cultivars [Notes: a – yield, c/ha, 2011; b – yield, c/ha, 2012; c – yield, c/ha, 2013; d – average yield for 2011–2013, c/ha; cultivars: 1 – Bukovynka (c.), 2 – Vrodlyva, 3 – Gurzufska, 4 – Krasa Kubani, 5 – Madam Balle, 6 – Monzana, 7 – Oksamyt, 8 – Talgarska Krasunya, 9 – Legenda Karpat, 10 – Bristol Cross, 11 – Yakymivska].

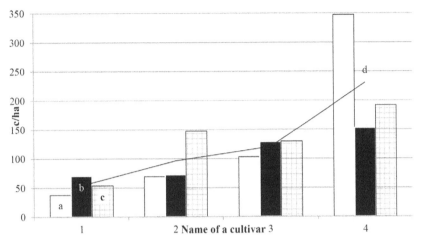

FIGURE 8.8   Yield of the pear winter cultivars [Notes: a – yield, c/ha, 2011; b – yield, c/ha, 2012; c – yield, c/ha, 2013; d – average yield for 2011–2013, c/ha; Cultivars: 1 – Bilka, 2 – Yablunivska (c.), 3 – Saiva, 4 – Kyrgyzka Zymova].

yield – 16.6 kg/tree (122.7 c/ha) as well as 'Vrodlyva' – 16.3 kg/tree (120.5 c/ha), 'Legenda Karpat' – 15.8 kg/tree (117.1, c/ha), 'Bristol Cross' – 13.6 kg/tree (100.5 c/ha) (see Figure 8.7). These cultivars exceed the control 'Bukovynka' by 54.8–88.9%.

High-yielding among winter cultivars both in the current year and for the three investigation years has been 'Kyrgyzka Zymova' – 31.1 kg/tree (230.6 c/ha) that has exceeded the control one by 1.2 times. 'Saiva' and 'Yablunivska' (control) have moderate yield – the average one for the three years is 16.3 kg/tree (120.5 c/ha) and 13.0 kg/tree (96.5 c/ha) (see Figure 8.8). The yield variation by years is connected with the capacity of laying reproductive buds for the future yield and of holding fruits on a tree under unfavorable factors (disease, drought).

Yield per 1 ha characterizes the orchard productivity more fully. For instance, the cultivars 'Bristol Cross,' 'Legenda Karpat,' 'Vrodlyva,' 'Talgarska Krasunya,' 'Saiva' and 'Kyrgyzka Zymova' have the highest average yield for three years – 100.5–230.6 c/ha.

Such regularity is also observed as concerns gross yield. For example, 'Legenda Karpat,' 'Saiva,' 'Talgarska Krasunya' and 'Kyrgyzka Zymova' have the heaviest gross yield (351.2–619.9 c/ha).

The analysis of the research years data shows that autumn cultivars exceed significantly the control one as about yield.

Specific productivity. The investigated cultivars had not any high indices of the fruit load per volume and projection surface of a tree since the low temperatures of the winter (−32.8°C) and high ones of the summer periods (+38°C) in the years 2011–2012 caused considerable decrease of yield. Among the autumn cultivars 'Yakymivska' has the above mentioned indices moderate (4.9 kg/m² and 4.7 kg/m³, respectively). 'Vrodlyva' (4.4 kg/m² and 4.0 kg/m³) and 'Krasa Kubani' (4.5 kg/m² and 4.8 kg/m³) have somewhat lower indices (Figure 8.9). It is 'Saiva' that has the highest fruit load per unit of volume and projection surface of a tree (4.2 kg/m² and 3.6 kg/m³, respectively) among the winter cultivars (Figure 8.10).

Early ripening. Cultivars of each fruit crop are divided into three groups: early, average and late. Pear cultivars on vigorous rootstocks are referred to early-ripening if they begin fruit-bearing on the fifth-seventh year after planting. Under the conditions of the Prydnistrovya those terms move to earlier fruit-bearing. Our researches show that most of the modern

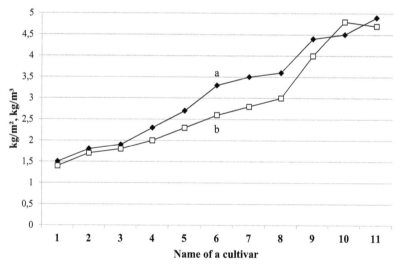

FIGURE 8.9 Specific productivity of the pear autumn cultivars [Notes: a – average load with fruits per unit of a tree projection surface, kg/m² (2011–2013); b – average load with fruits per unit of a tree projection volume, kg/m³ (2011–2013); cultivars: 1 – Talgarska Krasunya, 2 – Gurzufska, 3 – Oksamyt, 4 – Monzana, 5 – Bukovynka (c.), 6 – Legenda Karpat, 7 – Madam Balle, 8 – Bristol Cross, 9 – Vrodlyva, 10 – Krasa Kubani, 11 – Yakymivska].

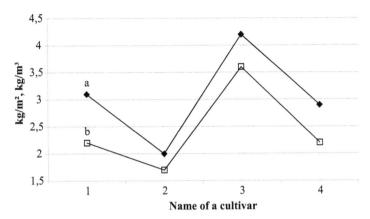

FIGURE 8.10 Specific productivity of the pear winter cultivars [Notes: a – average load with fruits per unit of a tree projection surface, kg/m² (2011–2013); b – average load with fruits per unit of a tree projection volume, kg/m³ (2011–2013); cultivars: 1 – Yablunivska (c.), 2 – Bilka, 3 – Saiva, 4 – Kyrgyzka Zymova].

pear cultivars are early-ripening. 'Bukovynka', 'Yablunivska', 'Vrodlyva', 'Bilka', 'Krasa Kubani', 'Kyrgyzka Zymova' begin to bear fruits on the fourth year after planting.

Besides, 'Yablunivska' and 'Kyrgyzka Zymova' appeared capable of laying fruit buds on one-year planting tree already therefore we consider that their feeding areas may be decreased (to 3.0–3.5×1.0–1.5 m). Such orchards on the rootstocks quince MA, VA–29, Sido are traditional for the countries of Europe today and though they are seldom presented practically by the cultivar Conference only their yield is 50–80 t/ha.

For the modern farm orchards the most valuable cultivars are those which distinguish themselves not only for early-ripeness but also for high marketable and taste fruit quality that is a rather important economic characteristic. Marketable fruits qualities are determined by their size, dimensional homogeneity, form and coloring. The fruit size that effects considerably the cultivar marketability is determined by mass, diameter and height.

The researched cultivars differ in the fruit mass which varies from 172 to 340 g (Table 8.2). Within one cultivar difference as for the average fruit mass is also observed depending on the tree load with fruits. Most of the cultivars have fruits of the size above average with the average mass 172–185 g. 'Bristol Cross', 'Bukovynka', 'Oksamyt', 'Madam Balle',

TABLE 8.2 Characteristics of Pear Fruits By Their Size and Qualitative Estimation

Cultivar	Degustation estimation, point, 2011–2013 (average)		Average fruit mass, g			
	appearance	taste	2011	2012	2013	average
Autumn cultivars						
Bukovynka (control)	7.6	8.3	233	255	246	245
Vrodlyva	8.6	7.0	296	334	390	340
Gurzufska	6.7	6.9	167	182	176	175
Krasa Kubani	8.2	7.2	226	165	125	172
Madam Balle	8.7	8.4	287	259	310	285
Monzana	8.7	7.8	179	169	208	185
Oksamyt	7.3	7.2	236	253	328	272
Bristol Cross	7.3	7.2	210	222	281	238
Talgarska Krasunya	8.6	7.6	193	169	184	182
Legenda Karpat	7.4	8.9	361	185	323	290
Yakymivska	8.3	7.7	198	138	198	178
LSD_{05}			33.9	20.4	17.8	24.0
Winter cultivars						
Yablunivska (control)	8.7	8.6	217	268	319	268
Bilka	8.1	7.7	221	198	228	216
Kyrgyzka Zymova	8.3	7.6	205	156	163	175
Saiva	8.7	8.3	287	182	339	269
LSD_{05}			44.5	15.3	11.7	23.8

'Legenda Karpat', 'Vrodlyva' have very big fruits (their average size is 238–340 g). Among winter cultivars 'Yablunivska' and 'Saiva' have fruits of big size (286 and 269 g respectively). The fruits of most of the winter cultivars do not exceed the control one as regards size but, on the contrary, are smaller by 27–50%.

On the whole the pear cultivars concerning the fruit size were divided into 3 groups:

- very big fruits (272–340 g) – 'Oksamyt,' 'Madam Balle,' 'Yablunivska', 'Saiva', 'Legenda Karpat', 'Vrodlyva';
- big (216–245 g) – 'Bilka', 'Bristol Cross', 'Bukovynka';
- fruits with the size above average (172–185 g) – 'Krasa Kubani', 'Kyrgyzka Zymova', 'Gurzufska', 'Yakymivska', 'Talgarska Krasunya', 'Monzana' (see Table 8.2).

Depending on the fruit diameter cultivars were divided according to the average data as follows:

- with a big fruit diameter (8.3–8.8 cm) – 'Legenda Karpat', 'Saiva', 'Vrodlyva';
- with average (7.1–7.9 cm) – 'Madam Balle', 'Yablunivska', 'Bristol Cross', 'Gurzufska', 'Bukovynka', 'Oksamyt';
- with small (5.9–6.8 cm) – 'Talgarska Krasunya', 'Krasa Kubani', 'Monzana', 'Yakymivska', 'Kyrgyzka Zymova'.

The majority of the researched cultivars have 54.5% of fruits with a big and average diameter, the rest (45.5%) with a small one.

So the qualitative evaluation of investigated cultivars shows that the considerable amount of the autumn cultivars is characterized by the fruit mass very big (272–340 g) and above average (172–185 g). The number of cultivars with big fruits is 20%. The autumn cultivars 'Vrodlyva,' 'Oksamyt', 'Legenda Karpat', 'Madam Balle' according to the average fruit mass exceed the control one by 11–38.8%. Among winter cultivars 'Kyrgyzka Zymova' and 'Bilka' have this index as lower than the control 'Yablunivska' by 21–34%.

The marketable value of fruits is undoubtedly influenced by their coloring and taste. Especial importance in the formation of the fruits marketable qualities belongs to the cellular structure. Thick and firm walls of cells that attach staunchness to fruits can increase their marketable value despite the worsening of taste properties, the fruits of 'Kyrgyzka Zymova,' 'Talgarska Krasunya', 'Bristol Cross' being an example [25].

The best appearance (8.7 points) is characteristic for the fruits of 'Madam Balle', 'Monzana', 'Yablunivska', 'Saiva', excellent taste qualities for 'Legenda Karpat', 'Yablunivska', 'Saiva', 'Bukovynka', 'Madam Balle'. On the basis of the obtained data the investigated cultivars were divided in keeping with the fruits appearance into three groups: first – 'Madam Balle', 'Monzana', 'Yablunivska', 'Saiva' (8.7 points each), 'Talgarska Krasunya' and 'Vrodlyva' (8.6 points each), second – 'Bilka,' 'Krasa Kubani', 'Kyrgyzka Zymova', 'Yakymivska' (8.1–8.3 points), third – 'Gurzufska', 'Oksamyt', 'Bristol Cross', 'Legenda Karpat', 'Bukovynka' (6.7–7.6 points) (see Table 8.2).

The important index of the pear fruits quality is their taste. The data of the organoleptic estimation show that it is the fruits of 'Bilka,' 'Yakymivska', 'Monzana', 'Saiva', 'Bukovynka', 'Madam Balle', 'Yablunivska', 'Legenda Karpat' which are the best ones concerning this index (7.7–8.9 points). Of these fruits sweet or sourish-sweet taste is characteristic and of those of the rest of the researched cultivars sourish-sweet as indicated by the degustation estimation (6.9–7.6 points) (see Table 8.2).

The attractiveness of the fruits appearance is determined considerably by the main coloring, especially by the intensity of the surface one. There is a great consumer demand for the bright-red pear fruits. The study of the fruits appearance and taste qualities help us to select cultivar samples with high indices and to include them into the breeding work for obtaining hybrids that would inherit those characteristics.

Pear fruits taste qualities, their nutritive, medicinal and prophylactic value are determined by the chemical composition. Dry soluble substances (DSS) include mainly sugars, organic acids, water-soluble vitamins, coloring and pectic substances.

The chemical composition is known to depend on the cultivar peculiarities, climatic conditions and other cultivation factors. Proceeding from this the indices studied are different among the cultivars by years but the average data characterize the cultivars biological peculiarities and their possibilities [26].

Among the researched autumn cultivars Bristol Cross has the best indices of the DSS content (15.4%) and among the winter ones it is 'Saiva' (18.2%). By our data 'Vrodlyva' and 'Saiva' have the highest sugar content (9.6–10.2%). The pear sugars include in the main glucose and fructose. The sugar content in fruits is also effected by the interrelation between them and acids. Acidity varies by cultivars and depends on the conditions and degree of ripeness. Among investigated cultivars it is 'Saiva' and 'Bukovynka' fruits that contain the biggest amount of organic acids (0.44–0.46%).

The pear fruits are rich in phenol compounds that play an important role in the oxidative and regenerative processes which the rapidity of the human constitution cells is connected with. For instance, the highest level of the above mentioned compounds has been detected in the fruits of 'Bukovynka,' 'Gurzufska,' 'Kyrgyzka Zymova' (185–281 mg/100 g of raw mass).

The pectic substances are closely connected with the metabolic processes in the human constitution. 'Yablunivska,' 'Krasa Kubani' and 'Yakymivska' have the highest pectines content among the researched cultivars (0.8–1.0%).

The vitamin C content depends on a cultivar and ripeness degree. Growing conditions effect its accumulation more than that of other chemical composition components. The biggest amount of the ascorbic acid is contained in the fruits of 'Saiva' and 'Gurzufska' (4.9–6.4 mg/100 g of raw mass) (Table 8.3).

The pears biochemical analysis shows that the fruits of the investigated cultivars distinguish themselves for high taste qualities and technological properties. They are characterized by the harmonic acids and sugars interrelation, suitable for using as fresh and are good raw material for processing.

Thus the results of the chemical and technological analysis and evaluation of the taste qualities of pear fruits show that under the conditions of the Prydnistrovya it is possible to grow cultivars which distinguish themselves for large-fruitedness, attractive appearance and, above all, high taste qualities and technological properties of fruits. The pears of 'Madam Balle,' 'Legenda Karpat,' 'Yablunivska,' 'Saiva' have been recognized as

TABLE 8.3 Chemical Composition and Average Mass of Fruits Pear By Cultivars

Cultivar	Dry soluble substances, %	Total sugar, %	Total acidity, %	Vitamin C, mg/100 g	Pectines, % per raw mass mas soluble	Pectines, % per raw mass mas protopectine	Pectines, % per raw mass mas total amount	Phenol compounds, mg/100 g
Autumn cultivars								
Bukovynka (control)	14.9	9.0	0.46	2.0	0.28	0.53	0.82	185
Vrodlyva	14.3	9.6	0.26	1.1	0.15	0.38	0.53	143
Gurzufska	15.2	6.4	0.40	6.4	0.24	0.64	0.89	195
Krasa Kubani	13.8	6.0	0.26	2.4	0.30	0.64	0.94	104
Oksamyt	13.1	8.4	0.30	1.1	0.21	0.47	0.68	169
Bristol Cross	15.4	7.6	0.26	2.0	0.21	0.45	0.66	165
Talgarska Krasunya	14.6	5.7	0.10	2.1	0.09	0.47	0.57	142
Legenda Karpat	15.3	7.8	0.18	3.5	–	–	–	–
Yakymivska	15.2	7.5	0.10	3.2	0.30	0.72	1.02	126
Winter cultivars								
Yablunivska (control)	16.8	9.9	0.26	1.3	0.34	0.47	0.81	128
Bilka	15.2	9.4	0.20	1.3	0.21	0.43	0.65	248
Kyrgyzka Zymova	12.4	6.8	0.20	1.3	0.04	0.57	0.61	281
Saiva	18.2	10.2	0.44	4.9	–	–	–	–

the tastiest ones. Valuable are also the cultivars 'Bukovynka,' 'Monzana,' 'Bilka.' Their fruits are characterized by average taste qualities but the taste is harmonic, sour-sweet.

8.4   CONCLUSIONS

The following conclusions can be made as a result of investigating 15 pear cultivars (10–11 – year trees) in the conditions of the Prydnistrovya and Naddnistryanshchyna in 2011–2013.

1. The high resistance to the fruit buds freezing was displayed by the cultivars 'Yablunivska' (control), 'Krasa Kubani,' 'Kyrgyzka Zymova' (the damage is 0 points). The strongest fruit buds damage by low temperatures was observed among 'Bukovynka' (control), 'Gurzufska,' 'Madam Balle' (3 points each), 'Bilka,' 'Oksamyt' (5 points each).

2. Concerning drought-resistance among the autumn cultivars 'Gurzufska' (31.1%) and among winter cultivars 'Kyrgyzka Zymova' (21.85%) and 'Yablunivska' (28.0%) were characterized by the lowest water deficiency. The majority of the researched cultivars (93.3%) was high and average drought-resistant, the others (6.7%) low resistant. The determination of the dynamics of the leaf water-holding capacity showed that during the first 2 hours of the exposition the difference in the water loss was not significant (within 14–40%). But in 4 hours already most of cultivars lost more than 50% of water. Among the autumn cultivars 'Talgarska Krasunya' and 'Gurzufska' distinguished themselves for the highest water-holding capacity – their loss of water during two hours was the least (11.1–7.5%). Concerning winter cultivars it is 'Kyrgyzka Zymova' that lost during two hours only 10.5% of water average for 3 years.

3. Among diseases and pests the considerable damage during the research years was caused by monilial blight, especially to the cultivars 'Oksamyt' (35.5% of the disease spread). 'Gurzufska,' 'Bilka,' 'Monzana,' 'Madam Balle' and 'Yakymivska' turned

out somewhat more resistant to this disease – 19.0–34.0% of spread. The most dangerous pest appeared psylla, especially for 'Yakymivska,' 'Bristol Cross,' 'Oksamyt' – 5 points, or 31.5 – 34.5% of spread.

4. The autumn cultivar 'Talgarska Krasunya' distinguished itself for high average yield – 16.6 kg/tree (122.7 c/ha) as well as 'Vrodlyva' – 16.3 kg/tree (120.5 c/ha), 'Legenda Karpat' – 15.8 kg/tree (117.1 c/ha). They exceeded control cultivar 'Bukovynka' by 54.8–88.9%.

5. Among winter cultivars 'Kyrgyzka Zymova' proved high-yielding during the three investigations years −31.1 kg/tree (230.6 c/ha) that exceeded the control one by 1.2 times. 'Saiva' and 'Yablunivska' were characterized by moderate indices – their average yield for three years was 16.3 and 13.0 kg/tree respectively (120.5 and 96.5 c/ha).

6. The cultivar 'Yakymivska' had the moderate index of the trees load with fruits per unit of the tree volume and projection surface (4.9 kg/m^2 and 4.7 kg/m^3 respectively). 'Vrodlyva' and 'Krasa Kubani' had somewhat lower indices (4.4 kg/m^2, 4.0 kg/m^3 and 4.5 kg/m^2 and 4.8 kg/m^3 respectively). Among winter cultivars it is Saiva that had the highest fruits load per unit of a tree volume and projection surface (4.2 kg/m^2 and 3.6 kg/m^3).

7. The autumn cultivars 'Oksamyt,' 'Madam Balle,' 'Yablunivska,' 'Saiva,' 'Legenda Karpat,' (272–340 g), 'Bilka,' 'Bristol Cross,' 'Bukovynka' (216–245 g) distinguished themselves for the fruit mass. Besides, 'Vrodlyva,' 'Oksamyt,' 'Legenda Karpat,' 'Madam Balle' exceeded control cultivar by 11–38.8%.

8. The cultivars 'Madam Balle,' 'Monzana,' 'Yablunivska' and 'Saiva' are characterized by the best appearance and 'Legenda Karpat,' 'Yablunivska,' 'Saiva,' 'Bukovynka,' 'Madam Balle' by excellent taste qualities of fruits.

9. The strategic assessment of the economic and biological characteristics carried out in 2011–2013 resulted in selecting autumn cultivars 'Legenda Karpat,' 'Monzana,' 'Madam Balle,' 'Talgarska Krasunya,' 'Vrodlyva,' and the winter ones ('Saiva,' 'Yablunivska' and 'Kyrgyzka Zymova').

KEYWORDS

- **degustation estimation**
- **drought-resistance**
- **early ripening**
- **specific productivity**
- **winter-hardiness**
- **yield**

REFERENCES

1. Dronyk, N. I., Saiko, V. I., Satina, L. F., Stronar, O. A. Bukovynian pear cultivars and promise of their breeding for the creation of new cultivars high resistant to fire blight. Sadivnytstvo (Horticulture), 2006, N 58. 11–19 (in Ukrainian).
2. Saiko, V. I. Bukovynian pear cultivars. Sadivnytstvo (Horticulture), 2000, N 51. 63–66 (in Ukrainian).
3. Matviyenko, M. V., Babina, R. D., Kondratenko, P. V. Pear in Ukraine. Kyiv: Agrarna dumka (Agricultural thought), 2006, 320 p. (in Ukrainian).
4. Rulyev, V. A. Competity of fruits and small fruits. M. F. Sydorenko Institute of Irrigated Fruit Growing of UAAS, 2007, 361 p. (in Ukrainian).
5. Duganova, E. A. Pear winter-hardiness. Sadovodstvo (Horticulture), 1978, N 2. 25–26 (in Russian).
6. Solovyova, M. A. Atlas of the damage of fruit and small fruit crops by frosts. Kiev: Urozhai (Harvest), 1988, (in Russian).
7. Sukholytky, M. D., Gomol's'ky, M. I. State and promises of the horticulture development in Naddnistryanshchyna and Prycarpathya/Intensive technologies in the horticulture of Naddnistryanshchyna and Peredcarpathya of Ukraine. Chernivtsi, 1995, 5–7 (in Ukrainian).
8. Khodakivs'ka, J. B. Early maturity and yield of the pear cultivars and forms in the conditions of the Ukraine's Lisosteppe. Sadivnytstvo (Horticulture), 2009, N 62. 32–39 (in Ukrainian).
9. Shipota, S. Y. Donors and sources of the restrained growth, crown compactness and other characteristics useful in economy and perspective in the pear breeding/Intensive technologies in the horticulture of Naddnistryanshchyna and Peredcarpathya of Ukraine. Chernovtsy, 1995, 50–57 (in Russian).
10. Polevoi, V. V. Plant physiology. Moscow: Vysshaya shkola (High school), 1989, 464 p (In Russian).
11. Sedov, E. N. Ogoltsova, T. P. Program and methods of the strain investigation of the fruit, small fruit and nuciferous crops. Oryol: All-Russian Research Institute of the Breeding of Fruit Crops, 1999, 608 p (In Russian).

12. Methods of expertizing cultivars of fruit, small fruit, nuciferous crops and grape. State Service of the Protection of the Rights for the Plant Cultivars, 2005 (in Ukrainian).
13. Kondratenko, P. V., Bublyk, M. O. Methods of field researches with fruit crops. Institute of Horticulture of the Ukrainian Academy of Agrarian Sciences. Kyiv: Agrarna nauka (Agricultural science), 1996, 95 p. (in Ukrainian).
14. Kondratenko, P. V., Shevchyk, L. M., Levchuk, L. M. Methods of the fruit and small fruit products quality estimation. Kyiv: Institute of Horticulture of the Ukrainian Academy of Agrarian Sciences, 2008, 80 p. (in Ukrainian).
15. Fisher, R. A. Statistical methods for research workers. New Delhi: Cosmo Publications, 2006, 354 p.
16. Karpenchuk, G. K., Mel'nyk, A. V. Estimates, observations, analyzes, processing of the data in the experiments on fruit and small fruit plants: methodical recommendations. Uman: Uman' Agricultural Institute, 1987, 115 p (In Russian).
17. Smykov, V. K, Smykov, A. V., Avanova, A. S. Evaluation the biological potential of fruit plants and its implementation. Biological potential of garden plants and ways of its realization. Moscow: VSTISP (All-Russian Institute of plant breeding of gardening and nursery), 2000, 129–141 (in Russian).
18. Duganova, E. A., Grinenko, N. N., Ananyeva, G. K. Diagnosis of the pear frost-resistance in Crimea. Papers on the applied of botanics, genetics and breeding, 1977, Vol. 59. Issue 2. 161–162 (in Russian).
19. Opalko, A. I., Kucher, N. M., Opalko, O. A. Method for evaluation of regeneration potential of pear cultivars and species (Pyrus, L.). Ecological Consequences of Increasing Crop Productivity: Plant Breeding and Biotic Diversity [Eds. Anatoly, I. Opalko et al.]. Toronto New Jersey: Apple Academic Press, 2014, 141–154.
20. Skryaga, V. A., Bublyk, M. O., Moiseichenko, N. V., Kitayev, O. I. Complex estimation of the cherry cultivars drought- and heat resistance in the Ukraine's Northern Lisosteppe. Sadivnytstvo (Horticulture), 2005, Issue 57. 480–486 (in Ukrainian).
21. Kushniryenko, M. D. Physiology of the fruit plants water change and drought resistance. Chisinau: Stiinta (Science), 1975, 216 p (In Russian).
22. Matviyenko, M. V., Strel'nikov, V. O. The results of the investigation of pear autumn and winter cultivars under the conditions of the Ukraine's Northern Lisosteppe. Sadivnytstvo (Horticulture), 2000, Issue 51. 59–63 (in Ukrainian).
23. Kichina, V. V. Productivity and its biological limits in the modern conception 'Science about a cultivar in horticulture.' Problem of the fruit and small fruit crops productivity: Report of the Scientific and Productional Conf. (Moscow, Zagorye, September 9–12, 1996). Moscow: All-Russian Breeding Technological Institute of Horticulture and Nursery Practice, 1996, 79–86 (in Russian).
24. Gulyayev, B. I., Rozhko, I. I., Rogachenko, A. D. and others. Photosynthesis, productional process and plants productivity. Academy of Sciences of the Ukrainian SSR. Institute of the Plants Physiology and Genetics. Kyiv: Naukova Dumka (Scientific though), 1989, 159 p (In Russian).
25. Babina, R. D., Untilova, A. E., Gorb, N. N. Organoleptic estimation and biochemical composition of regional and promise pear cultivars in Crimea. Sadivnytstvo (Horticulture), 1998, Issue 46. 39 (In Russian).
26. Bublyk, M. Influence of weather factors on the stone crops productivity in the Ukraine. Fruit, Nutand Vegetable Production Engineering. Proceeding of the 6th International Symposium held in Potsdam, 2001, Potsdam–Bornim, 2002, 117–121.

CHAPTER 9

AMINO ACID COMPOSITION OF STRAWBERRIES (*FRAGARIA* × *ANANASSA* DUCH.)

IRINA L. ZAMORSKA

CONTENTS

ABSTRACT

The composition and content of amino acids in strawberry fruits were studied using High Performance Liquid Chromatography. Total content of amino acids in strawberry fruits – cultivars 'Ducat,' 'Pegas,' 'Rusanovka,' 'Honey' and 'Polka' – was 3297.9–5218.9 mg/100 g of dry weight. The highest content of amino acids was found in "Ducat" fruits – 5218.9 mg/100 g of dry weight; it can be recommended as an initial material for strawberry breeding.

Twenty amino acids were identified in a chemical composition of strawberry fruits, including eight out of nine irreplaceable ones: leucine, lysine, threonine, phenylalanine, histidine, valine, isoleucine, and methionine. The main acids in fruits are aspartic acid, glutamic acid, arginine and alanine. The amount of irreplaceable amino acids is 22.4–28.8% of their total content in fruits. The largest share – 21.0–23.5% is that of leucine.

Replaceable acids are presented by aspartic acid, glutamic acid, alanine, serine, glycine, proline, γ-amino-butyric acid, 4-hydroxyproline, tyrosine, 2-ethanolamine, cysteine. The amount of aspartic acid is 25.3–36.4% of the total content. The main amino acids in strawberry fruits are aspartic acid, glutamic acid, arginine and alanine.

9.1   INTRODUCTION

Strawberries belong to the most popular and valuable small berry crops due to their high gustatory properties, fast and early ripening, simple growing conditions and high yielding capacity. Strawberry fruits contain large amounts of sugars, organic acids, vitamins, phenol compounds, and mineral substances. The content and composition of amino acids are of great interest.

Amino acids are known to be the main materials for the photosynthesis of proteins, enzymes, organic acids, vitamins and other compounds [1]. Their content depends on the fruit ripeness stage, climatic conditions, farm practices, irrigation, and fertilizers. Amino acid composition can help determine the origin of the output and its adulteration [2, 3].

Amino acids add a lot to fruit flavor: arginine and asparagine contribute to sweetness, aspartate – to acidity [4]. They play an important role in developing fruit aroma [5, 6]. When strawberries ripen alanine participates in forming ethyl ethers (methyl and ethylhexanol) [7], acids, spirits [8]. A. Moing et al. [9] state that the most aromatic strawberry cultivars have the highest alanine concentration during ripening.

According to J. Zhang et al. [1] during ripening the content of free amino acids decreases gradually until strawberry fruits become red and it increases considerably during ripening. In particular, aromatic amino acids phenylalamine and tyrosine are predecessors for the biosynthesis of antocyanines and flavonoids. In the course of strawberry ripening phenylalamine amount is very high at early and late stages, which enables the development of antocyanine coloring [10–12].

H. Zhang et al. [13] identified eleven free amino acids in strawberry fruits. A.G. Perez et al. [14] consider aspartic acid, glutamic acid and ala-nine to be the main strawberry amino acids. Stój, Targonski [3] received the same data studying chemical composition of strawberry juices. Aspartic acid content in them ranges from 1.68 mmol l^{-1} to 6.64 mmol l^{-1}, and that of glutamic acid is 0.33–2.32 mmol l^{-1}. Proline, valine, methio-nine, isoleucine, leucine, tyrosine, lysine and arginine were determined in small concentrations. J.S. Elmore, D.S. Morton [15] consider strawberry juices to be the products with high content of asparagine and glutamine. The content of these amino acids is 56.8 and 13.9%.

H. Zhang et al. [13], studying biochemical changes in strawberry fruits during their ripening, found out the ways of amino acid formation: four central amino acids glutamine, glutamate, aspartate and asparagine (Gln, Glu, Asp, and Asn) are initially formed from a-oxoglutarate and oxaloac-etate in the tricarboxylic acid cycle, then they change into all other amino acids during various bio-chemical processes [16].

Our work was aimed at identifying quantitative and qualitative amino acid content of various strawberry cultivars.

9.2   MATERIALS AND METHODOLOGY

The work was done with strawberry fruits 'Ducat,' 'Honey,' 'Polka,' 'Pegas' and 'Rusanivka' in the laboratory of the department of the technology of storage and processing of fruits and vegetables at Uman national university of horticulture and at the experimental center of food quality control at the National institute of grape and wine "Magarach."

Quantitative analysis of strawberry amino acid content was made with High Performance Liquid Chromatography. To make an analysis, a chro-matographic column 4.6×50 mm filled with octadecylcylil sorbent, graini-ness 1.8 mkm, "ZORBAX" SB–C18, was used.

1.5–2.0 g of homogenized strawberry were weighed in a vial on analyt-ical scales. Then 3 mL 6 n. of water solution of hydrochloric acid, contain-ing 0.4% β-merkaptoethanol, was poured into the vial. The vial was sealed and exposed at 110°C for 24 hours. After centrifugation and filtration of hydrolyzate/digest, 100 mcl was put into a vial for analysis. After that its content was dried in vacuum at 20–40°C until hydrochloric acid was completely removed.

200 mcl of 0.8M borate buffer pH 9.0 and 200 mcl of 20 мМ solution 9-chloride fluorenilmethoxycarbonil in acetonitrile were gradually added into the vial with dried digest for analysis; after 10-minute exposure 20 mcl of 150 mM of amantadine hydrochloride in 50% water acetonitrile were added into the vial.

The following detection parameters were established:

- measurement scale 1.0
- scanning time 0.5 sec
- a wave length of detection 265 nm

Amino acid identification was carried out according to standard time exposure [17, 18].

Statistic analysis was made using StatSoft STATISTICA 6.1.478 Russian, Enterprise Single User (2007).

9.3   RESULTS AND DISCUSSION

Total amino acid content in strawberries 'Ducat,' 'Pegas,' 'Rusanovka,' 'Honey' and 'Polka' was 3297.9–5218.9 mg/10g of dry weight (Table 9.1). Higher amino acid content was observed in 'Ducat' fruits – 5218.9 mg/100 g of dry weight. Amino acid content in 'Pegas' fruits was 4153.0 mg/100 g of dry weight. 'Polka' fruits showed the lowest amino acid content – 3297.9 mg/100 g of dry weight.

Twenty amino acids were identified in a chemical composition of strawberry fruits, including eight out of nine irreplaceable ones: leucine, lysine, threonine, phenylalanine, histidine, valine, isoleucine, and methionine. The amount of irreplaceable amino acids ranged within 738.2–1278.9 mg/100 g of dry weight, which was 22.4–28.8% of their total amino acid content in fruits (Figure 9.1).

The largest share – 21.0–23.5% of the total amount of irreplaceable amino acids was that of leucine. Lysine content in fruits was equal to 131.0–230.2 mg/100 g of dry weight depending on a cultivar. Threonine content ranged from 114.7 in 'Polka' fruits to 196 in 'Ducat' fruits, that of

TABLE 9.1 Composition and Content of Amino Acids in Strawberry Fruits (mg/100 g of dry weight)

Amino acids	Cultivar				
	Ducat	Pegas	Rusanovka	Honey	Polka
Irreplaceable					
leucine	293.7	248.8	236.6	221.0	169.5
lysine	230.2	182.6	209.8	180.8	131.0
threonine	196.0	161.5	188.8	166.6	114.7
phenylalanine	171.3	143.3	132.4	129.9	95.8
histidine	137.5	113.3	116.9	120.4	80.2
valine	104.8	85.5	112.0	84.3	61.1
isoleucine	104.1	85.3	95.5	75.6	53.8
methionine	41.3	36.8	34.1	25.5	32.1
Amount	*1278.9*	*1057.1*	*1126.1*	*1004.1*	*738.2*
LSD_{05}			4.0		
Replaceable					
aspartic acid	1259.5	940.2	796.6	696.8	931.3
glutamic acid	996.3	736.4	694.1	671.9	490.6
arginine	416.5	350.3	276.9	357.8	256.8
alanine	398.7	251.1	306.4	281.5	288.3
serine	272.6	228.1	217.4	226.4	164.0
glycine	232.3	210.8	182.5	205.4	155.4
proline	210.1	184.8	167.7	170.6	136.8
γ-aminobutyric acid	49.2	67.1	56.8	67.4	55.8
4-hydroxyproline	35.9	44.6	23.7	30.1	25.5
tyrosine	32.4	51.9	22.9	19.0	28.5
2-ethanolamine	26.0	21.7	26.1	21.5	19.8
cysteine	10.5	8.9	9.6	7.7	6.9
Amount	*3940.0*	*3095.9*	*2780.7*	*2756.1*	*2559.7*
Total amount	*5218.9*	*4153.0*	*3906.8*	*3760.2*	*3297.9*
LSD_{05}			4.0		

phenylalanine – 95.8–171.3, and histidine content was 80.2–137.5 mg/100g of dry weight. Valine and isoleucine content was almost at the same level, namely 7.3–8.2% of the total amount of irreplaceable amino acids. Methionine content appeared to be the lowest in fruits – 32.1–41.3 mg/100 g

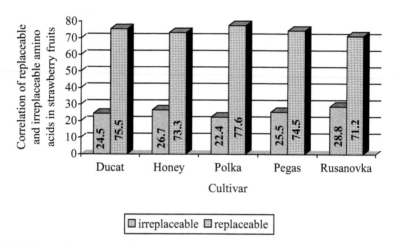

FIGURE 9.1   Correlation of replaceable and irreplaceable amino acids in strawberry fruits, %.

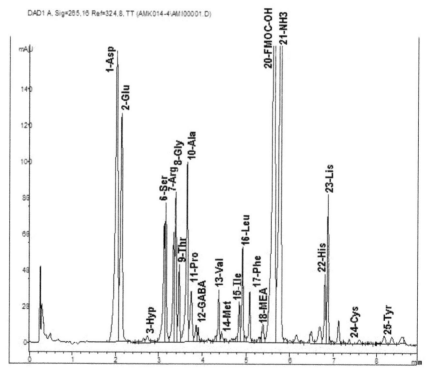

FIGURE 9.2   Chromatography of amino acid composition of 'Ducat' strawberry fruits.

of dry weight, which was 3.2–4.3% of the total content of irreplaceable amino acids in strawberry fruits.

Twelve replaceable amino acids (Figures 9.2–9.6) – 2559.7–3940.0 mg/100g of dry weight – were identified in strawberry fruits, including aspartic acid, glutamic acid, arginine, alanine, serine, glycine, proline, γ-aminobutyric acid, 4-hydroxyproline, tyrosine, 2-ethanolamine, and cysteine.

The statistics received match the results of the research done by Zhang et al. [1] and Perez et al. [7]. The share of aspartic acid is 25.3–36.4% of the total amino acid content.

Glutamic acid amount was 490.6–996.3 mg/100 g of dry weight, which was 19.1–25.3% of the replaceable amino acid content. The share of arginine was 10.0–10.6% and that of alanine – 10.1–11.3%.

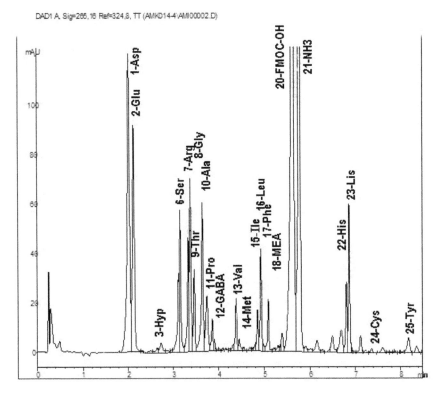

FIGURE 9.3 Chromatography of amino acid composition of 'Pegas' strawberry fruits.

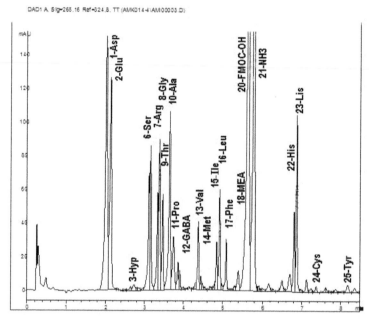

FIGURE 9.4   Chromatography of amino acid composition of 'Rusanovka' strawberry fruits.

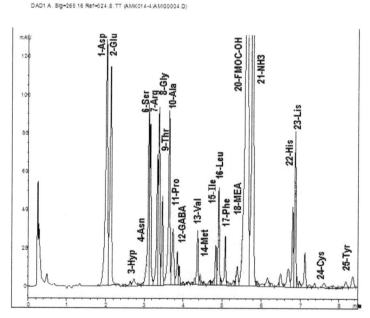

FIGURE 9.5   Chromatography of amino acid composition of 'Honey' strawberry fruits.

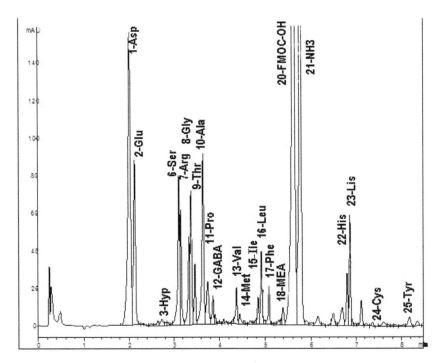

FIGURE 9.6   Chromatography of amino acid composition of 'Polka' strawberry fruits.

The content of serine, glycine, and proline was 136.8–272.6 m g/100 g of dry weight. The share of other replaceable amino acids did not exceed 2.4% of their total amount.

9.4   CONCLUSIONS

Strawberry fruits are the source of amino acids including eight of nine irreplaceable ones: leucine, lysine, threonine, phenylalanine, histidine, valine, isoleucine, and methionine. Total amino acid content in strawberry fruits was 3297.9–5218.9 mg/100 g of dry weight.

The main acids in fruits are aspartic acid, glutamic acid, arginine and alanine. Higher amino acid content – 5218.9 mg/100 g of dry weight – was found in 'Ducat' fruits; this cultivar can be recommended as an initial material for strawberry breeding.

KEYWORDS

- amino acids
- antocyanines
- chromatography
- fertilizers

REFERENCES

1. Zhang, J., Wang, X., Yu, O., et al. Metabolic profiling of strawberry (*Fragaria* × *ananassa* Duch.) during fruit development and maturation. Journal of Experimental Botany. 2010, Vol. 62, №3, 1103–1118.
2. Di Maro, A., Dosi, R., Ferrara, L. et al. Free amino acid profile of *Malus domestica* Borkh cv. Annurca from the Campania Region and other Italian vegetables. Australian Journal of Crop Science. 2011, Vol. 5, №2, 154–161.
3. Stój, A., Targonski, Z. Use of amino acid analysis for estimation of berry juice authenticity. Acta Scientiarum Polonorum Technologia Alimentaria. 2006, №5 (1). 61–72.
4. Jia, H. J., Okamoto, G., Hirano, K. Effect of amino acid composition on the taste of 'Hakuho' peaches (*Prunus persica* Batsch) grown under different fertilizer levels. Journal of the Japanese Society for Horticultural Science. 2000, Vol. 69, №2, 135–140.
5. Kader, A., Stevens, M., Albright, M., Morris, L. Amino acid composition and flavor of fresh market tomatoes as influenced by fruit ripeness when harvested. Journal of the American Society for Horticultural Science. 1978, Vol. 103, №4, 541–544.
6. Sugimoto, N., Jones, A. D., Beaudry, R. Changes in free amino acid content in 'Jonagold' apple fruit as related to branched–chain ester production, ripening, and senescence. Journal of the American Society for Horticultural Science. 2011, Vol. 136, №6, 429–440.
7. Perez, A. G., Rios, J. J., Sanz, C., Olias, J. M. Aroma components and free amino acids in strawberry variety Chandler during ripening. Journal of Agricultural and Food Chemistry. 1992, Vol. 40, №11, 2232–2235.
8. Handbook of Fruit and Vegetable Flavors. Strawberry Flavor. [Ed. Y. H. Hui]. John Wiley and Sons, Inc., Hoboken: New Jersey, 2010, 1118 p.
9. Moing, A., Renaud, C., Gaudillère, M. et al. Biochemical changes during fruit development of four strawberry cultivars. Journal of the American Society for Horticultural Science. 2001, Vol. 126, №4, 394–403.
10. Aharoni, A., Ric de Vos, C. H., Verhoeven, H. A. et al. Non-targeted metabolomes analysis by use of Fourier transform ion cyclotron mass spectrometry. Omics: a journal of integrative biology. 2002, Vol. 6, №3, 217–234.

11. Halbwirth, H., Puhl, I., Haas, U. et al. Two-phase flavonoid formation in developing strawberry (*Fragaria* × *Ananassa*) fruit. Journal of Agricultural and Food Chemistry. 2006, Vol. 54, №4, 1479–1485.

12. Hanhineva, K., Kärenlampi, S., Aharoni, A. Recent advances in strawberry metabolomics. Genes, Genomes and Genomics. Global Science Books, Ltd., 2011, 65–76.

13. Zhang, H., Wang, Z. Y., Yang, X. et al. Determination of free amino acids and 18 elements in freeze-dried strawberry and blueberry fruit using an Amino Acid Analyzer and ICP-MS with micro-wave digestion. Food chemistry. 2014, Vol. 147. 189–194.

14. Pérez, A. G., Olías, R., Luaces, P., Sanz, C. Biosynthesis of strawberry aroma compounds through amino acid metabolism. Journal of Agricultural and Food Chemistry. 2002, Vol. 50, №14, 4037–4042.

15. Elmore, J. S., Morton, D. S. Compilation of free amino acid data for various food raw materials, showing the relative contributions of asparagine, glutamine, aspartic acid and glutamic acid to the free amino acid composition. 2002, URL: http://jifsan.umd.edu/docs/acrylamide2002/wg1_aspargine_in_foods.pdf) (Accessed 2 October 2014).

16. Galili, S., Amir, R., Galili, G. Genetic engineering of amino acid metabolism in plants. Advances in Plant Biochemistry and Molecular Biology. 2008, Vol. 1, 49–80.

17. Jámbor, A., Molnár-Perl, I. Quantitation of amino acids in plasma by high performance liquid chromatography: Simultaneous deproteinization and derivatization with 9-fluorenylmethyloxycarbonyl chloride. Journal of Chromatography, A. 2009, Vol. 1216, №34, 6218–6223.

18. Jámbor, A., Molnár-Perl, I. Amino acid analysis by high-performance liquid chromatography after derivatization with 9-fluorenylmethyloxycarbonyl chloride. Literature overview and further study. Journal of Chromatography, A1. 2009, Vol. 1216, №15, 3064–3077.

CHAPTER 10

THE VIRAL DISEASES OF THE *CORILUS* SPP. BIOTECHNOLOGY OF PRODUCTION OF IMPROVEMENT PLANT MATERIAL

GALINA A. TARASENKO, IVAN SEM. KOSENKO, OLGA A. BOYKO, and ANATOLY IV. OPALKO

CONTENTS

ABSTRACT

Investigations of the representatives of the genus *Corylus* L., its viral diseases and the methods of their detection were done. The diagnostic techniques of pathogens at the species, forms and cultivars of the genus *Corylus* were examined. The observation results as for the efficient of the different *methods* of the initial explants sterilization before an introduction into in vitro culture in order to get the improvement plant material were given.

The application results of traditional and modern plant growth promoters, so as the usage of biochemical compounds from the fungi Basidiomycetes which had been used in the various investigations and industrial conditions, were discussed in the article.

10.1 INTRODUCTION

Viral diseases, which cause significant plant pathologies obtained a great expansion in the different ecological regions of Ukraine these days. As a result these destructive processes became the reason of considerable economic wastes in the crop capacity of this valued culture and attract attention of botanists, phytopathologists and virologists all over the world.

We know more than 600 pathogens, which can parasitize on the plants while the harmfulness of the some of them should reach 20–60% of the crop losses. Hereby, the substantial economic losses inflict not only certain viruses but their complex infection as well. Under the circumstances the viral pathogens have an effect on the most of physiological processes of infected plant. Viruses unlike the other infectious diseases have the line of features: viruses have been reproduced during all the life of the infected plant, as well as in its vegetative (for some pathogens in generative also) race. It brings to the storage of viruses in the plant production, seed-cultural material, so as in the environment (agrocoenosis) increasing the contagious field. In addition, the infected seedlings leaving the nursery expend the area of causative agents of a disease and the range of plants infected by them [1]. That's why we should pay a great attention to the phyto-viral state uterine-seminal and graft gardens as well as to the queen cell of clonal wilding. The main method of virus propagation is their transmission from plant to plant together with the pollen and seed, at the same time some of them have the vehicles (insects, mites, nematodes) and the aphides play an important role among them [2, 3].

The symptoms, which cause the virus of definite taxonomic group vastly, depend from the genotype and physiological state of plant, virus strain and environmental factors (temperature, light intensity, and others) [4, 5].

The processes, which cause virus propagation, wrecking and occurrence are very important either as the occurrence of their new "variants" and the impact of such anthropogenic factors on them such as:

- the destruction of natural ecosystems.
- intense modification/maintenance of the natural environment.
- irresponsible relation to the biotechnological processes of different trend and others.

Sometimes the plants infected with viruses cannot to have any visible symptoms. Besides, such factors as imbalance of the elements of mineral nutrition, its deficiency, high light intensity, insects and mites invasions, bacterial or fungus damage or genetic disturbances should cause the symptoms which are look like the viral infection. That is why the diagnosis "viral infection" should be confirmed with the modern diagnostics methods and identification of its pathogens [4].

We should admit that though the plants of the genus *Corylus* are rather stable to the diseases and pests, its activity and distribution increased these days. Considering the losses, which cause the viral diseases we think that the main trend of horticulture development in Ukraine is its removal to the basis without viruses [5].

Therefore, the aim of our investigation was to examine the phyto-virology state of fruit and ornamental plantings of the genus *Corylus* in the ecological conditions of the National Dendrological Park "Sofiyivka" of NAS of Ukraine (NDP "Sofiyivka") with the help of the different modern diagnostics methods.

10.2 MATERIALS AND METHODOLOGY

Experiments were conducted in the years 2010–2014 at the base of the laboratory of microclonal propagation of the NDP "Sofiyivka," the laboratory of virology on the base of the Institute of Horticulture NAAS of Ukraine, and the laboratory of virology on the base of the viral department in the Taras Shevchenko National University of Kyiv.

The samples of plants of the genus *Corylus* at the territory of NDP "Sofiyivka" were the object of the investigation. A lot of explorers suppose that the main reason of plantings weakening is the damage or them by different viral diseases [6].

It is admitted, that the heaviest results were noticed during the leaf damage at the first period of spring. Moreover, the plantings of natural origin consist from the attendant species of trees and bushes, which are serving as a reservoir for virus's conservation [7, 8].

That's why we select the samples by visual symptoms and without any viral signs—randomized as well [3]. The samples selection and visual observation of the ornamental and fruit plants of the genus *Corylus* we did according to the standard methods [5, 9, 10].

Field inspections and collection of samples were conducted in two-way diagonal and over the four sides of the observable area. The uterine-grafts, collection and varietal plantings were examined twice a year: in spring, at the development of the three or four leaves, and at the end of summer, during the second wave of fruit plants growing and fruit ripening [10].

During the tree inspections we inspected the crown of the every tree from the north side especially, as symptoms are seen better from this side and preserved longer at the time of "dog days". The samples were selected evenly from the fourth sides of the tree and from the branches placed inside the crown closer to the trunk at the height no more than 2 meters. The upper parts of the shoot with the young leaves were selected. The plants without any symptoms of viral infection were tested on the latent viruses. Selected samples were packed into the polyethylene packets with the labels where the cultivar of tree, place and date of selection were indicated. Later all the samples were tested by enzyme-linked immunosorbent assay (ELISA). In order to conserve the leaves and their suspensions we kept the temperature in the refrigerator +2°C to +49°C.

The next stage of our work was biological testing of the selected samples on viruses with the help of herbaceous hosts. It was done by the method of bioassays on the detector plants. The young leaves were the material for testing. Biological testing of samples for the presence of *Apple mosaic ilarvirus* (ApMV) was done with the help of the plants *Cucumis sativus* L. and *Phaseolus vulgaris* L., *Nicotiana occidentalis* H.-M. Wheeler, *Nicotiana tabacum* L. cv. Samsun; the presence of *Prunus necrotic ringspot virus* (PNRSV) was tested with the help of the plants *Chenopodium quinoa* Willd. [10].

As we found the symptoms of *tobacco mosaic virus* (TMV) [11] we decided to test it with the help of plant which are receptive to this virus they were: *Nicotiana tabacum, Datura stramonium* L., *Nicotiana glutinosa* L. Tested plants had been growing in the hothouse at the usual parameters of humidity, temperature, photoperiod and light intensity and in the boxes with the universal soil after autoclave. Plants were planted with the distance 10 cm.

Inoculation samples were prepared by grinding in a mortar which was rinsed with 0.01 M phosphate buffer (pH 7.2) containing 0.1% nicotine 0.25% sodium dietildytiocarbomat 0.2%, $NaSO_3$ and diluted 1:3 (wt/vol) with buffer [5]. The plants were inoculated mechanically by rubbing with a glass spatula dipped in inoculum and using abrasive (carborundum) or giving an injection into the midrib by the juice from the plant with the viral symptoms. The test plants were inoculated with buffer, which we used for plant homogenization. In order to increase the plant sensitivity to virus after the injection they were kept in the dark place nearly 24–48 hours. Symptoms were later observed in the greenhouse.

In order to detect and fix the intracellular inclusions we generally used the method of luminescent microscope. Having prepared the samples we took the low streak epidermis of plants of the genus *Corylus* with the symptoms of disease. They were colored at the method of Gimza and other methods for luminous microscope in dilution (1:1000), acrylic-orange and other inks were used. During the experiment we removed off the lower epithelium from the plant with the symptoms of viral infection and immersed it into 1.5% solution of trichloracetic acid on 3 minutes to fix the sample. Then we rinsed it and examined at the glass plate under the microscope. The inclusions were fixed with the help of video camera, which is specially arranged, and after that the picture was transferred on to the computer monitor. Presence of ApMV and PNRSV was tested by ELISA test. The viruses were assayed using the Agdia protocol for double antibody sandwich enzyme-linked immunosorbent assay (DAS-ELISA) [12–15]. For DAS-ELISA the sample was ground with mortar and pestle. Identification of viruses was done with the help of test-system LOEWE (made in Germany) according to the producer introduction [9].

The results were registered at the automatic ELISA-analyzer Start Fax: 2100 (Awarennes Technology, USA). The index E_{405} was taken as positive and it was three times bigger for control index. The sample in suspension was at a dilution—1:100.

The suspension was centrifuged in 5000 r/m during the 10 minutes. The refined supernatant was applied on to the immunological plate.

The method of electronic microscopy was used for virus detection in the refined samples. We selected the samples with the symptoms of mosaic, chlorosis, rolling and necrosis for the researches. The diluted suspension of plant sap (the dilution method) was researched and pathogen preparation production from refinement suspension, which we got by the method of virus purification with differential centrifuged (1–2 cycles, 10,000–12,000 rt/min. and 28,000–32,000 rt/min. pH of solution 7.2 in the phosphate buffer 1/15 M). Besides, the samples were detected in the electronic microscopes JEM-1400 (Japan) та EM-125 (Sumy, Ukraine) during some of tests we used the preparation tenuous suspension using the method of "immersion" [16].

In order to do it we took the cuts of young leaves and located the shoots into the drop of the salt of heavy metal (contrast), which was situated at the liner with tape for electronic microscopes. We drained the samples in the sterile Petri dish [1]. We had been working a lot of with the methods of sterilization of plant material before the research as we all know that the representatives of the genus *Corylus* as well as the other wood plants are rather difficult for in vitro propagation and very often they are fall out even during the first passage through the lesion by the infectious diseases, fungi and bacteria. During the experiment of micropropagation for sterilization of starting materials (shoots 10–15 cm long from chosen foundation stock) we used methods of sterilization with different function substance; the seven of them are following:

Treatment 1:
- starting materials were washed with water for 10 min.;
- submersion into antibacterial solution of "Manorm" for 5 min.;
- submersion into water with 5% solution of NaOCl+Tween 20 for 10 min.;
- rinsing in distilled sterile water thrice (for 3 min. each time).

Treatment 1a:
- starting materials were washed with water for 10 min.;
- submersion into antibacterial solution of "Manorm" for 5 min.;
- submersion into water with 0.1% solution of $HgCl_2$+Tween 20 for 10 min.;
- rinsing in distilled sterile water thrice (for 3 min. each time).

Treatment 2:
- starting materials were washed with water for 1 hour;
- submersion into antibacterial solution of "Manorm" for 5 min.;
- submersion into water 70% solution of $C_2H_5(OH)$ for 5 seconds;

- submersion into water 0.05% solution of $C_9H_9HgNaO_2S$+Tween 20 for 10 min.;
- rinsing in distilled sterile water thrice (for 3 min. each time);
- submersion into water with 5% solution of NaOCl+Tween 20 for 10 min.;
- rinsing in distilled sterile water thrice (for 3 min. each time).

Treatment 3:
- starting materials were treated with antibacterial solution of "Manorm" for 1 hour;
- submersion into water 70% solution of $C_2H_5(OH)$ for 5 seconds;
- rinsing in distilled sterile water thrice (for 3 min. each time);
- submersion into water with 15% solution of NaOCl+Tween 20 for 5 min.;
- rinsing in distilled sterile water for 5 min.;
- submersion into water with 0.2% solution of NaOCl for 5 min.;
- submersion into water with 100 mg in 100 mL (volume) solution of $C_6H_8O_6$ for 10 min.;
- rinsing in distilled sterile water thrice (for 3 min. each time).

Treatment 4:
- starting materials were washed with water for 10 min.;
- submersion into antibacterial solution of "Manorm" for 5 min.;
- treatment water with 1.0% solution of $AgNO_3$ for 2 min.;
- rinsing in distilled sterile water thrice (for 3 min. each time).

Treatment 5:
- The starting materials were washed with water for 10 min.;
- treatment water with 1 mg/L solution of fundazol for 5 min.;
- submersion into antibacterial solution of "Manorm" for 5 min.;
- rinsing in distilled sterile water thrice (for 3 min. each time).

Treatment 6:
- starting materials were washed with water for 10 min.;
- submersion into antibacterial solution of "Manorm" for 5 min.;
- submersion into water 0.05% solution of $C_9H_9HgNaO_2S$ for 10 min.;
- rinsing in distilled sterile water thrice (for 3 min. each time).

The next stage of our research was the test with plant growth promoters, which have immune-incentive effect. The usage of these preparations during the cultivation of the planting stocks of the woody species is still not very popular and widespread. That is why we decided to give estimation of their practical perceptiveness using the literary and practical data.

As long ago, as 70–80 years of the last century, some fundamental works devoted to the investigations of the plant growth promoters and their influence on to the plant growth, development and the explication of its work facility were published [2].

Eventually the additional data as for the positive activity of the plant growth promoters on to the woody species were developed. It was confirmed that under the effect of these promotoring agents the processes of albuminous substances and sugar synthesis were intensified and protoplasm tenacity decreased, its penetrability improved, the tissues renovated, the chlorophyll contents so as the photosynthesis activity increased, the development of the root system and adventives roots especially intensified [17, 18].

At the same time, many questions are still unknown. It concerns hardly implanted woody species especially.

We used three different plant growth promoters ACTO, BIOLAN which are widely known at the market of these preparations and BiOECOFUNGE-1 (bio-specimen which was engineered on the base of refined biochemical compounds from the Basidiomycetes fungi and vegetative combinations from plants of genus Polygonaceae, Betulaceae, Cannabaceae, Caprifoliaceae, Scorphulariaceae, Asteraceae) [19].

The preparation BiOECOFUNGE-1 was also added to the composition of the nutrient medium for in vitro propagation in the dilution 0.1–0.5% (the general contents in the medium 0.01%). Traditional nutrient mediums so as the mediums with our receptive modification were used during the research. This preparation was also used for plant treatment in the field conditions (0.1–0.5%). The plants were sprayed and treated in the different variants during the adaptation to *ex vitro* conditions.

10.3   RESULTS AND DISCUSSION

During the period of the survey and inspection of experimental, production and ornamental plantings of the genus *Corylus* at the territory of NDP "Sofiyivka," as well as owing to the used methods of investigation we established that the most widespread viruses among the plants of the genus

Corylus are: *Apple mosaic virus* (ApMV), *Prunus necrotic ringspot virus* (PNRSV). Besides, there were some symptoms of *Tobacco mosaic virus* (TMV) and had a place mixed infection.

According to the literary data *C. avellana* L. and *C. maxima* Mill. are the most sensitive to the viral infection [9]. We observed this tendency also.

The mosaic symptoms provoked by ApMV were evinced in the form of general yellowing, yellow rings and lines and yellow flecking (Figure 10.1).

The agent could have latent form at some species during the long period (even to 2–3 years). As soon as the leaves appear, the mosaic symptoms are turned up. The most distinct symptoms were registered in May–June, and during the hot days the symptoms were masked [20]. The symptoms, which are appeared during the viral infection, we may confuse with the symptoms, which turn up under the toxic materials (herbicides, promotoring agents), or the lack of some microelements (zinc, boron, iron). Through the likeness of symptoms which ensure from different reasons we cannot get objective rating of the tree condition during the visual observation. The main reason of this problem is latency—the plants should be infected and in the risk group for using them as uterine trees, without any symptoms of infection [9].

Prunus necrotic ringspot virus (PNRSV) is characterized by rings, strips and broad veinbanding. Sometimes the narrow chlorotic trimming is created. The symptoms are occurred at the leaves of all the stages, to the extent of plant development the size of necrosis increase and it leads up to the slot formation [21].

FIGURE 10.1 Mosaic symptoms on hazelnut leaves caused by ApMV.

The typical features of the diseases are the presence of bright green rings and strips. The rings diameter is about 1–2 mm to 1 cm upon to the cultivar. The tissue died in the middle of the ring and then fall out. The leaves become full of holes. Sometimes only the central and lateral ribs are kept (Figure 10.2). But these symptoms you should differ from the pest damage and injury by parasitical fungi [20].

The testing results by the method of biotests on the herbaceous indicator-plants indicated that during the test on ApMV and latent ring spotted virus symptoms appear promptly and safely on *Chenopodium quinoa* Willd. while during the test on PNRSV the symptoms were clearly seen on *Cucumis sativus* L. [4]. The symptomless course of infection was seen on the next plants: *Beta vulgaris* L., *Datura stramonium* L., *Helianthus annuus* L., *lactuca sativa* L., *Nicotiana talacum* cv. Samsun

As a result of our investigations we revealed that the best plants for biotest were: *Chenopodium quinoa, Nicotiana tabacum* cv. Samsun and *Cucumis sativus* L. [22].

Upon the infection we observed the specific response on the indicator-plants, namely the small chlorosis local lesions, systematic strips, yellow spots on the seed-lobes, deformation of the real leaf and the depression of growth.

With the help of the method of luminescent microscope we had an opportunity to observe intracellular inclusions. Owing to this, we did the first conclusion about the potential presence of viral infection in the some samples of hazelnut, attendant plants and indicator-plants (Figures 10.3 and 10.4).

A total of 73% of the ELISA-tested hazelnut trees (110–150) in the plantings of the NDP "Sofiyivka" were found to be infected with viral

FIGURE 10.2   The symptoms of PNRSV on hazelnuts.

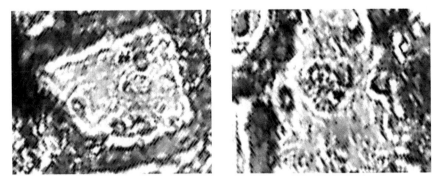

FIGURE 10.3   The intracellular inclusions in the samples of leaves of *C. colurna* infected with *TMV.*

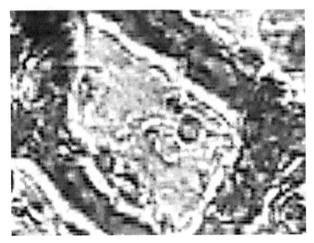

FIGURE 10.4   The intracellular inclusions in the samples of leaves *C. avellana* infected with *TMV.*

infection. The virus antigen of PNRSV was found in 68,4% tested samples, almost 23,3% tested samples of hazelnuts were infected with ApMV, and all the trees that had been observed with mosaic symptoms tested positive by ApMV, while 20% gave positive result for the compound viral infection. The highest extinction values were recorded in the young leaves of the infected hazelnut plants collected in spring.

Thus, the results of testing the hazelnuts samples by ELISA test showed the wide spreading of the viral infection among the testing samples. There was not any regularity in the distribution of the viral infection among the species, forms or cultivars of the representatives of the genus *Corylus*.

The investigations with electron microscope revealed the presence in the samples which we got from the sap of the tested plants of hazelnuts the flexible stick-shaped parts ≈ 550–640 nm (*Carlavirus*) by length and isometric parts with the diameter 29–32 nm (*Ilarvirus*). Herewith, during the inoculation of the indicator-plants with the sap of these plants the symptoms typical for the representatives of the genus *Ilarvirus, Tobamovirus, Carlavirus*.

At the Figure 10.5 you should clearly seen the short and stick-shaped viruses, which harm the plant in the compound infection and were extracted from the plants of hazelnut.

Thus, the testing results of the indicator-plants, ELISA test and electron microscopy revealed the wide spreading of the viral infections among the tested samples of the genus *Corylus*. At the same time, there was not any regularity in the distribution of the viral infection among the species, forms or cultivars of the representatives of the genus *Corylus*.

As a result of the conducted investigations as for the explants sterilization before the in vitro culture we think that the most suitable was that one from the treatment №5 which included the treatment with antifungal preparation — fundazol and treatment №6 where the main operate substance was natrium merthiolate. Result of these two modes of sterilization we got the biggest number of the sterile and viable explants (Figure 10.6).

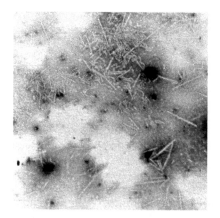 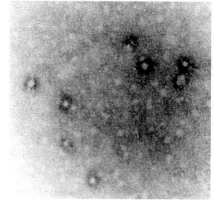

FIGURE 10.5 Electromyography of lengthy (on the left) and spherical (on the right) parts extracted from the plants of the genus *Corylus*.

FIGURE 10.6   The viable explants of the plants of the genus *Corylus.*

During the last inspection of the plants of genus *Corylus* which took the part in our experiment and were treated with the plant growth promoters we detected that disregard the morning frosts up to $-10°C$ and the strong snowfall in the beginning of November the tested plants treated with the bio-preparation Bioecofunge-1 and Biolan were going on vegetation. However, it was characteristically just for one-year seedlings of *C. colurna,* and at the same time there were not any changes on the other plants (Figures 10.7 and 10.8).

As the usage of plant growth promoters during the growth of the planting material of the wood species is not very popular and widespread.

FIGURE 10.7   The condition of the one–year seedlings of *C. colurna* and five-year old hazelnut plants treated with promotoring agents in three weeks after treatment.

FIGURE 10.8   The condition of the one–year seedlings of *C. colurna,* which were not treated with the bio-preparation at the end of November.

We tried to give the advanced estimation to it having used the literary data and practical observations. There were some positive results with the usage of the bio-preparation "Bioecofunge-1" (which has the biochemical compounds of the fungi Basidiomycetes at its basis) as the addition agent to the nutrient medium. The propagation factor of the explants on the nutrient medium with the bio-preparation "Bioecofunge-1" in its composition was biggest in comparison to the other mediums and the number of viable explants also varied to extension (Figure 10.9).

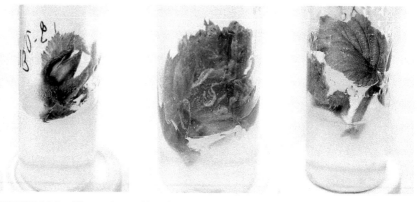

FIGURE 10.9   The explants of hazelnuts planted in vitro on the nutrient medium with the bio-preparation "Bioecofunge-1" in its composition.

10.4   CONCLUSIONS

Owing to the usage of the complex of methods of virus fixing we can identify the pathogen, which is the reason of the certain outsides plant's anomaly and investigate the viral diseases of this plant. Taking into consideration that the each pathogen has the sharp and specific character in the infected plants and indicator-plants has specific reaction it is possible to identify the pathogens having got the results of these researches. Besides, we should observe the intracellular inclusions using luminescent microscope. The analysis of literary data as for the usage of plant growth promoters with synthetic and natural origin testified availability of this technological action for acquisition of the qualitative planting material of the woody species. The integrated plant growth promoters had the positive effect on to the ground microbes and plants, as well as accompanied the enhancement of the ground-forming process.

The application of traditional and modern plant growth promoters, so as, the use of biochemical compounds of the fungi Basidiomycetes at its basis, is very perspective for the next development of plant biotechnology and transference of nurseries on healthy plant material without viral diseases, as well as the creation of propagation system of its propagation and certification.

ACKNOWLEDGEMENTS

We thank member of NAS of Ukraine DSc of Biology Anatoly Boyko for sharing information and observations with us. This information was immensely helpful in preparing this summary. We also grateful to PhD Oleksandr Balabak for kindly providing the samples and to PhD Myhaylo Nebykov for his valuable help and contributions.

KEYWORDS

- cultivar
- explant

- hazelnut
- herbicides
- immunosorbent
- plant growth promoters

REFERENCES

1. Boyko A. L. The ecology of plant viruses. Kyiv: Higher School, 1990, 164 p (In Russian).
2. Hamburg, K. Z., Kulaeva, O. N., Muromtsev, G. S. et al. The plant growth regulators. [Ed. G. S. Muromtsev]. Moscow: Kolos (The publishing house of Ears in Rus), 1979, 248 p (In Russian).
3. Lakin, G. F. Biometry. Teaching book [3-th edit. Revised and completed.]. Moscow: Higher School, 1980, 293 p (In Russian).
4. Kashin, V. I., Borisova, A. A., Prikhodko, Y. N. et al. Technological process for acquisition of planting materials of fruit and berry crops without viruses: methodological guidelines. Moscow, All Russia Selection Technological Institute of Horticulture and Nursery of RAAS, 2001, 109 p (In Russian).
5. Tarasenko, G. A. Serological detection of the genus *Corylus, L.* representatives using enzyme-linked immunosorbent assay (ELISA). Autochthonous and alien plants. The collection of proceedings of the National Dendrological park "Sofiyivka" of NAS of Ukraine. 2013, Vol. 9. 137–141. (in Ukrainian)
6. Tarasenko, G. A., Kosenko, I. S., Nebykov, M. V., Boyko A. L. Contamination of plants of the genus *Corylus, L.* during the period of their ontogeny by the viruses of the various taxonomic groups. The role of the botanical gardens and dendroparks in the conservation and enrichment of the biological multiformity of the urban lands: Materials of the international scientific conference (Kyiv, 28–31 May 2013) [Ed.-in-Chief, V. G. Radchenko]. Kyiv: SCEBM (Scientific center of ecomonitoring and biodiversity of megapolisis of NAS of Ukraine), PJSC "Vipol" (Privat Joint-Stock Company), 2013, 304 p. (In Ukrainian)
7. Šutic, D., Ford, R. E., Tosic, M. T. Handbook of Plant Virus Diseases. Boca Raton: Chemical Rubber Company (CRC) Press, 1999, 584 p.
8. Sökmen, M. A. The occurance of apple mosaic virus (ApMV) at the hazelnuts (*Corylus avellana, L.*) in Samsun. Department of plant protection. Samsun. 2005, №3, 22–24. (In Turkish)
9. Metiyuz, R. The viruses of plants. [Ed. J. G. Atabekov]. Moscow: Mir (The publishing house of World – in Rus), 1973, 469 p (In Russian).
10. Verderevskaya, T. D., Marinesku, V. G. Viral and mycoplasmal diseases of fruit cultures and grape. Kishinev: Shtiintsa, 1985, 311 p (In Russian).

11. Shmyglia, V. A. The types of the infectious process of tobacco mosaic virus and the diagnostics of the plant material contamination. Scientific reports of the high school of Biological science. 1987, №6, 22–28 (In Russian).

12. Clark, M. F. Characteristics of the microplate method of the enzyme – linked immunosorbent assay for the detection of plant virus. J. Gen. Virol. 1977, Vol. 34, №3, 475–483.

13. Fleg, C. L. The detection of Apple chlorotic leaf spot virus by a modified procedure enzyme-linked immunosorbent assay (ELISA). Ann. Appl. Biol. 1979, 61–65.

14. Jelkmann, W. An immunocapture-polymerase chain reaction and plate trapped ELISA for detection of apple stem pitting virus. J. Phytopathol. 1997, 145. 499–504.

15. Voller, A. The detection of viruses by enzyme-linked immunosorbent assay (ELISA). J. Gen. Virol. 1976, 33. 165–167.

16. Boyko A. L. Virus electronography: Album. Kyiv: DIA. 2012, 56 p. (In Ukrainian)

17. Shternshys, M. V., Dzhalilov, F. S., Andreeva, I. V., Tomilova, O. G. Biological preparations in the complex of plant protection. Teaching book. Novosibirsk: Novosibirsk State Agricultural University, 2000, 128 p (In Russian).

18. Shevchuk, V. K., Doroshenko, O. L. Biostimulants – against diseases. Plant protection. 2000, №4, 7 (In Russian).

19. Boyko, O. A., Veselskiy, S. P., Grygoryuk, I. P., et al. The biochemical evaluation of drug that are developed on the basis of Basidiomicetes. Ukrainian Biochemical Journal XI Int. Bich. Kong. 2014, Vol. 86, №5, 174–175.

20. Ryzhkov, V. L., Protsenko A. E. The atlas of viral diseases of plants, Moscow: Nauka (The publishing house of Science – in Rus.), 1968, 136 p (In Russian).

21. Moskovets, S. M., Bobyr A. D., Glushak, L. Y., Onyshchenko A. M. Viral diseases of agricultural cultures. [Ed. A.D. Bobyr]. Kyiv: Urozhay (The Publishing House of Harvest – in Ukr.), 1975, 152 p. (In Ukrainian)

22. Fulton, R. W. Ilarvirus group. CMI/AAB Descriptions of Plant Viruses. Association of Applied Biologists, Wellesbourne, 1983, №. 275, 39–48.

CHAPTER 11

PHYLOGENETIC CONNECTIONS BETWEEN REPRESENTATIVES OF THE GENUS *AMELANCHIER* MEDIK

ANATOLY IV. OPALKO, OLENA D. ANDRIENKO, and OLGA A. OPALKO

CONTENTS

ABSTRACT

Within the frames of retrospective discourse, the information concerning breeding value of a representative of the genus *Amelanchier* Medik. — for national pomiculture, decorative gardening, and pharmacy – is integrated. This article characterizes their: biological peculiarities, ecological adaptiveness, palatability traits and cooking qualities of their fruits, their availability for drying and processing, namely preparing juices, syrups, jams, candied fruit jellies, comfiture, and also fruit wine. The effectiveness of using Juneberry for phytomeliorative is mentioned, some ethnobotanical aspects are discussed. Data about chromosome numbers and the geographical origin of the genus *Amelanchier* representatives cultivated in Ukraine and their closest congeners from the family *Rosaceae* Juss. are cited. Controversial questions of the genus *Amelanchier* system were discussed from the classical and molecular genetic approaches. The results of phylogenetical and molecular genetic researches made by scientists of different countries offer a possibility to specify the systematic position of the genus *Amelanchier* representatives of the family *Rosaceae* Juss. grown in Ukraine, and to place them temporarily in a big subfamily *Amygdaloideae* Arn., which combines the former subfamilies *Amygdaloideae*, *Spiraeoideae*, and *Maloideae*, tribe – *Maleae* Small, subtribe – *Malinae* Rev.

11.1   INTRODUCTION

Among the decisive premises of successful conservation of biotic diversity (biodiversity), and the enrichment of local diversity of any plant, including representatives of the genus *Amelanchier* Medik. under certain conditions, one should determine their systematic position, ascertainment of geographical origin and peculiarities of phylogenetical connections on the interfamily and interspecies levels. Such information will be favorable to scientifically ground a planning of their introduction, prevention from invasion, and also create sources of outgoing material to conduct their breeding.

According to the classification of Armen Takhtadzhan [1], representatives of the genus *Amelanchier* (Juneberry) belong to the family *Rosaceae*, subfamily *Pyroideae* (former *Maloideae*), tribe *Maleae*.

In Ukraine, representatives of the genus *Amelanchier* are considered unconventional for growing, but interest in Juneberry and many other promising, but currently undervalued plants (mostly known to a narrow range of wildlife lovers) increases with increasing population welfare. First of all, it is referred to the species *Hippophae* L. (sea buckthorn), *Lonicera* L. (honeysuckle), *Sorbus* L. (mountain ash) and *Viburnum* L. (viburnum), which, now together with Juneberry, are gaining more popularity due to the decorativeness, and high taste, remedial, and dietic qualities of their fruits [2, 3].

Juneberry is a very flexible and unpretentious plant. In the culture, and as well as in the natural state, it grows in the form of a large shrub, sometimes a tree. It can be used as an ornamental, nectareous, phyto-reclamative, and medicinal plant. It is valued as a fast-growing, fast-fetal, and perennial fruit crop. It has a number of other benefits. According to the degree of resistance to unfavorable conditions, Juneberry is a unique plant. It has a great tolerance to winter conditions. The plant itself is capable of withstanding temperatures of 40–50° C below zero, and the flowers that blossomed – to minus 5–7°C [2, 4, 5].

Juneberry successfully grows on the soils of different mechanical makeups and acidity. It thrives on the rather moisty light soil, sometimes even on the marshy ones. At the same time, it withstands drought well and can grow on rocky and sandy dry areas. However, it can't withstand poorly drained clay soils with low humus content [2, 5].

Juneberry is a photophilous plant, but it can grow in the shade. Burmistrov [5] mentions an interesting biological feature of young Juneberry plants (under 5 years), that it has an ability to withstand relatively intense shading. Juneberry plants are distinguished by being sufficiently fast growing and by the age of ten they reach their full development. Duration of the yielding period for the bush is 60–70 years (some shoots can grow up to 15 years) [5, 6].

Juneberry is a hermaphrodite plant, outcrossing (self-fertilizes rarely). The Juneberry can berry on last years' shoots even if it had a single area that was self-fertilized [3–5]. It starts to berry rather early, it produces crops in 3–4 years. It is characterized by annual and abundant berrying, which reaches 3–5 kg from wild-growing bushes of 5 years, and up to 10–12 kg from 10 or more year-old ones [4, 5].

A positive feature of Juneberry is that diseases rarely affect it. Sometimes, there can be powdery mildew, fruit rot, leaf blight, and leaf rust. However, many pests willingly settle on it. Among them are: the green apple aphid (*Aphis pomi* Deg.), the apple blossom weevil (*Anthonomus pomorum* L.), the garden chafer (*Phyllopertha horticola* L.), caterpillars of different species of butterflies (*Operophtera brumata* L., *Euproctis chrysorrhoe* L., *Orgyia antiqua* L., *Dasychira pudibunda* L., *Acronicta tridens* (Den. and Schiff.), *Melanchra persicariae* L.), leaf-rolling moths of the genus *Pandemis,* and others. Forming the complex of phytophages of this genus takes place mainly due to broad polyphages and olygophages connected primarily with apples, hawthorns and some other representatives of *Maleae* [6–8]. Besides, sparrows, thrushes, and robins like juneberries, so it is sometimes necessary to scare these birds to preserve the harvest [6, 7].

Juneberries are sweet, exquisitely tender, and very useful. Their sugar content is 8–12% with a prevalence of fructose and glucose; organic acids 0.4–0.7 (preferably apple); tannins and dyes 0.5–0.8; 1.5–2.5% of pectin; 0.4–0.7 mg/100 g of carotene; 35–45 of vitamin C, from 7 to 12 mg/100 g of vitamin B_2, 0.2–1.0 mg/100 g of provitamin A; to 100 mg/100 g to anthocyanins; among these trace constituents, there is also copper, lead, and cobalt. Beta-sitosterol can also be found in them. It is a substance that helps to reduce cholesterol, and coumarins, which are characterized by an anti-sclerotic effect [2, 4, 9]. Its fruit – in the fresh, dried, frozen, and processed form – is consumed. Juice, syrup, wine, liqueur, jam, confiture, jelly, and marmalade can be prepared. While processing juneberries, other berries (e.g., black currant) can be mixed together, but only 300 grams of sugar per 1 kg of fruit is used (due to the high sugar content of juneberries). A peculiarity of freshly picked juneberries is the fact that they are very difficult to squeeze juice from, but if they are left for about a week, then 70% of juice, from the total mass, accumulates in them [6].

The value of the Juneberry fruit as a fine, raw material for producing fruit wines was first emphasized by the academician V.V. Pashkevych who initiated introducing juneberry into the Ukrainian culture. It was V.V. Pashkevych who launched its plantation while establishing an arboretum in the territory of modern NDP "Sofiyivka" NAS of Ukraine, now known as Pashkevich Arboretum [5, 10].

The fruit of Juneberry dries quickly in the bright sun, as well as, in the home dryers and, by its appearance, are similar to raisins – dried berries of seedless grape cultivars. Fruits that have just ripened contain more vitamin C and are better for freezing and conservation; completely ripe fruits contain more sugar and can be used to make juice and wine. While cooking wine, the juice is fermented without adding sugar. The wine has a pleasing savor, nice dark-ruby color, and its strength is 8–10% [6].

Juneberry, as a fruit crop, is grown on the industrial scale in the USA and Canada. Accordingly, much attention is paid to the breeding work. There are cultivars grown for fruit: 'Beaverlodge', 'Bluff', 'Buffalo', 'Elizabeth', 'Idaho Giant', 'JB30', 'Killarney', 'Lee № 3', 'Moonlake', 'Sturgeon', 'Thiessen', 'Thiessen RS', 'Timm' and so on, and also cultivars, which combine its high productivity and quality with decorative value: 'Altaglow', 'Gypsy', 'Honeywood', 'Martin', 'Nelson', 'Northline', 'Parkhill', 'Pembina', 'Regent', 'Smoky', 'Success' and so on. There are purely ornamental cultivars: 'Altaglow', 'Autumn Brilliance', 'Autumn Sunset', 'Ballerina', 'Carleton', 'Cumulus', 'Fergi', 'Forest Prince', 'Helvetia', 'Hollandia', 'Jennybelle', 'Lustre', 'Prince Charles', 'Prince William', 'Princess Diana', 'Rainbow Pillar', 'Reflection', 'Robin Hill', 'Silver Fountain', 'Tradition', 'White Pillar', etc. Among them there are representatives of the different species: *A. alnifolia*, *A. bartramiana*, *A. canadensis*, *A. laevis*, *A. spicata*, *A. stolonifera*, and a number of interspecies and hybrids between other species [6, 11, 12].

Juneberry, as an ornamental plant, is suitable for the arboretums, Dendrological parks, and settlement gardening. It is possible to form alleys, delicate hedges (well-tolerated to a cut), Juneberry is effective in group plantings and solitaires. It gives off a pleasing effect when placed in the background of other plants or along buildings. At the same time, due to the abundant frondescence, blossoming and fruiting Juneberry plants are ornamental throughout the year. In spring, at the time of blossoming, its inflorescences are light and delicate against the background of the young leaves, and its white and cream-colored flowers have a light pleasant scent. In early summer, because it is still ripening, the fruit is first green. Then, on the one side of the little fruit, there is a pink erubescence and the ripe fruit is usually blue and purple, but the color can vary from cream to almost black. The juneberries' leaves display a special decorativeness throughout the growing season: when blooming, they are white-tomentose, later – green, green-gray, green-red, in autumn – yellow, orange, red, purple. In winter Juneberry shoots can be graphically distinguished above the snow cover [5, 6, 12]

During blossoming, Juneberry is eagerly visited by bees; providing them with an early spring honey gathering (lots of pollen and little nectar) and in gardens it attracts them to other fruit crops, thus increasing their productivity [13–15].

Juneberry is used for fixing gullies and eroded slopes. While phytomelioration of recreational and devastated forest areas, it can even be used as an attractive factor for forming forest environment [14, 16]. It is recommended that Juneberry be planted in multifunctional shelter belts, namely in forest shelter and snow shelter belts along railways or highways. It can also be planted in different rows and tiers of wind belts as an orchard-protecting belt that would protect field crops from winds – both dry and hot, as well as, help capture snow and use its meltwater better. Under its protection, currants, gooseberries, raspberries, strawberries, and others can be grown, while simultaneously being capable of capturing snow. If fruit crops need protection from the winter cold, the location in areas that are blown by the wind does not matter for Juneberry (due to its high degree of frost hardiness). Besides, heavy beds of Juneberry as orchard-protecting belts are a great place for nesting insectivorous birds [4–6, 7, 17]. Also, according to A.D. Burmistrov, the offer to plant Juneberry on the edge of the garden is not devoid of practical sense. During the Juneberry fruiting, which lasts about a month, its non-simultaneous ripening coincides with the time of fruiting strawberries, black currants, and sometimes, cherries. Because of this, attacks from birds (starlings, blackbirds) on these baseline fruit cultures are strongly reduced [5].

Juneberry wood is solid and resilient; with gray, reddish or reddish-brown color with slightly visible beams and annual rings. It has a silky surface; it can be easily bent and polished. It doesn't have a timber industry value because of the small diameter. In the past Juneberry wood was used for making ramrods and canes. It is perfectly suited for wickerwork, industrial and domestic and art objects; delicate holders for climbing plants can be made of it [14, 18, 19].

Juneberry fruits are used with the purpose of treatment for athero-sclerosis (due to the content of beta-sitosterol, which is an antagonist of cholesterol); for different diseases of the gastrointestinal tract (as an astringent); for the prevention of hypo-and avitaminosis C and B (as multivitamin remedy). Tincture of Juneberry flowers is used as antihypertensive and cardiotonic remedy. Decoction of the rind and leaves has astringent and coating properties and is used for the prevention of gastrointestinal diseases and for septic wounds epulosis [9, 17].

A synonymous name for Juneberry was given to name to the City of Saskatoon – the largest in the Canadian province of Saskatchewan. It is derived from "mis-sask-quah-toomina," which is how the aboriginal inhabitants called the most wide-spread, local berries [20].

The importance of plants is proved by the fact that the Indian tribes distinguished between 8 individual species based on morphological differences in the plants. Juneberry flowers and fruits were used in ceremonial rites, and the beginning of harvest was celebrated by solemn feasts. Some tribes believed that even the first humans were created from Juneberry bushes [20].

Juneberry was widely used in the everyday life of the aboriginal inhabitants, and subsequently of the first settlers. Fruits were one of the staple foods, and often the only kind of fruit in sufficient quantity. They were consumed fresh, cooked, and dried. They were part of the ethnic dishes – pemykan. Young cut shoots, dried fruits, and leaves were used for making drinks and treatment remedies for children, adults, and animals. Arrows and household tools were made from Juneberry solid wood [20].

The value of the genus *Amelanchier* representatives and some problems concerning their classification, especially their place in the family, led to an active search for phylogenetic connections between cultivated species and close families.

11.2 MATERIALS AND METHODOLOGY

Considering the importance of the problem of the starting material for breeding and taking into account the data obtained from the analysis of experimental and theoretical studies performed in different countries over a long historical period by scientists from different scientific schools [1, 4, 11, 12, 20–28, 33–42], the attempt to generalize available information is made. In this study the quota sampling method was used, which allowed to eliminate dubious publications using the criteria in peer-reviewed publication citing and giving priority to research that is carried out by international programs. Works on the domestication of the genus *Amelanchier* and their nearest families published in different years, were analyzed, summarized [3, 9, 12, 20–34] and supplemented with the results of our study in the preparing process of the article.

11.3 RESULTS AND DISCUSSION

11.3.1 THE ORIGIN OF THE GENUS NAME AMELANCHIER AND ETNOBOTANICHNI ASPECTS OF THE SPECIES EPITHETS

Genus *Amelanchier* Medik. (Juneberry) was described in 1789 by Friedrich Casimir Medicus [35], a German botanist and physician, director of the botanical garden in Mannheim. One of the first records about the plant dates back to the year 1581 [36]. However, before singling out *Amelanchier* as a separate genus (probably due to the similarity of its morphological features) Joseph Pitton de Tournefort referred its species to the genus *Mespilus* [37] and Carl Linnaeus – to the genus *Chionanthus* [38].

The origin of the international name genus *Amelanchier* have several versions that are associated with taste or size of the fruit. According to one that is presented in a botanical dictionary by M.I. Annenkov edited in 1878 [39], the name *Amelanchier* is derived from the Greek words melea – apple and anchein – astringe, due to the astringent flavor of the fruit. According to another version, A.I. Poyarkova, while describing the genus in the flora of the USSR, associates the name with Provencal *amelanche*, which indicates the honey taste of the fruit [21].

The version proved by Caden and Terentyeva [40] also explains the origin of the plant name from the Provencal amelanche, but as a fruit name of only one type of Juneberry, such as *Amelanchier vulgaris* Moench. Referring to a number of sources, they suggest a Celtic origin of the word *Amelanchier*. Besides, the genus *Amelanchier* species are characterized by a large number of epithets that indicate their popularity and are usually associated with morphological features, habitat characteristics, fruit taste, etc.

Thus, among the common American names of Juneberry species, G.N. Jones names: serviceberry, sarviceberry, sarvis, maycherry, june-berry, shadblow, shadbush, shadberry, shadblossom, shadflower, shad-wood, sugar pear, wild pear, lancewood, boxwood, Canadian medlar [41]. In Canada Juneberry is known as saskatoon, originating from the Indian missask-quah-too-min [5].

Attention is drawn to ethnobotanical and symbolic aspects of the genus *Amelanchier* application by indigenous peoples of North America, emphasizing the value of the plant. G.N. Jones [41] gives an interesting interpretation of certain species epithets of American Juneberry species by associating them with the botanical characteristics and value of the plant for the indigenous population. Thus, the name Juneberry is stipulated by ripening fruit in early summer (from the month name June); in the eastern United States the names shadblow, shadberry, shadblossom, shadflower and shadwood are stipulated by the period of blossoming plants in early spring, which is an indicator of the breeding beginning of the shad run (river herring), which begins spawning migration from oceanic salt into fresh water rivers; the names lancewood and boxwood are stipulated by the use of dense wood as a part of tools (handle).

According to the botanical dictionary by M.I. Annenkov [39], the genus *Juneberry* is called in Polish – Swidośliwka, in Czech – Muchownik, in Serbian – Grašac, Irga, in German – Fluhbirne, Beermispel, Felsenbirnbaum, in French – Amelanchier, in English – The Medlar. The dictionary of Ukrainian scientific and vernacular names of vascular plants compiled by Yuri Kobiv [42] suggests names – sadova irha and irha.

11.3.2   THE SYSTEMATIC POSITION OF THE GENUS AMELANCHIER

The genus *Amelanchier* in classical phylogenetic, as well as in the molecular phylogenetic (cladistic) classification system of plants, is defined as a component of the family Rosaceae Juss. of the range *Rosales* Bercht. et J. Press. [1, 43–45].

The family Rosaceae is quite a large family of angiosperms, comprising about 90–110 genera and 2000–4828 species [43, 45–49], which averages about 100 genera and 3000 species [31].

Numerous "microspecies" are distinguished in many genera of Rosaceae, morphological differences between which are slight (e.g., details of pubescence), but they are considered stable. Microspecies appear in groups where free interbreeding in populations is limited because of apomixis spread or other reasons. Therefore, if counting microspecies, the number of Rosaceae species can significantly increase [49].

Traditionally, on the basis of differences, mainly in fruit morphology and in basic chromosome numbers, the family Rosaceae were separated into 4 subfamilies: Spiraeoideae (Meadowsweet) – fruit – hose, rarely capsule, basic chromosome numbers 8 and 9; Rosoideae (Rose) – coccus, aggregate fruit, aggregate-accessory fruit, the hypanthium often takes part in the fruit formation, basic chromosome numbers 7, 9, rarely 8; Maloideae (Apple) – fruit – apple, basic chromosome number 17; Prunoideae (Plum) – fruit – drupe, basic chromosome number 8 [46, 47]. Other authors, depending on the occurrence of stipules, calyx structure, hypanthium, gynoecium, fruit, and other signs in the family Rosaceae distinguish from 3 to 12 subfamilies [43].

The genus *Amelanchier,* since the times of Adolf Engler (1903) [45], was defined within the subfamily Pomoideae (later Maloideae):

Division – Embryophyta siphonogama
Subdivision – Angiospermae
Classis – Dicotyledoneae
Subclassis – Archichlamydeae
Ordo – Rosales
Subordo – Rosineae
Familia – Rosaceae
Subfamilia – Pomoideae
Genus – *Amelanchier.*

Formed at the beginning of the last century [32], synopsis of the genera of the subfamily Maloideae as a part of the family Rosaceae with certain deviations [34] in his near-classical state is supported by many authors [22, 23, 30, 50] (Figure 11.1).

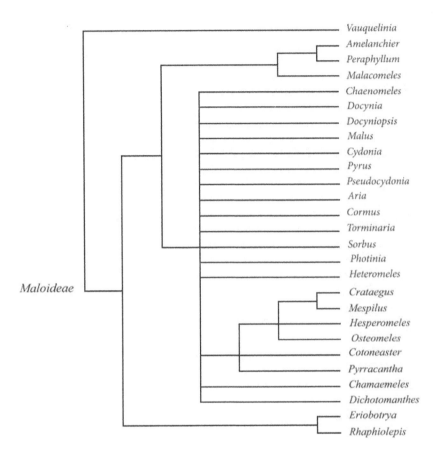

FIGURE 11.1   Simplified cladogram of the subfamily Maloideae (according to Aldasoro J.J. et al., 2005 [22] as amended [51]).

However, more evidences are provided concerning the revision of the family Rosaceae appropriateness on regrouping subfamilies, supertribes, tribes, subtribes, some genera and species with the simultaneous elimination of the subfamily Maloideae [24, 26, 31, 51].

The revision of the family Rosaceae was supported by Armen Takhtajan, who suggested a new version of flowering plants system, revised according to the latest results of molecular phylogenetics in the book "Flowering Plants" reissued in 2009 [1].

Armen Takhtajan highlights subfamily Pyroideae (formerly Maloideae) in the family Rosaceae, combining in it 27 genera in 4 tribes, defining the genus *Amelanchier* among the families of the tribe Maleae (Table 11.1):

TABLE 11.1   The Synopsis of the Pyroideae Genera (formerly Maloideae) by Armen Takhtajan (2009) [1]

Tribus	Genus
Kageneckieae	*Kageneckia*
Lindleyieae	*Vauquelinia*; *Lindleya*
Maleae	*Photinia* (у тому числі *Stranvaesia*); *Heteromeles*; *Eriobotrya*; *Rhaphiolepis*; *Sorbus*; *Chamaemespilus*; *Aronia*; *Amelanchier*; *Pyrus*; *Malus*; *Eriolobus*; *Peraphyllum*; *Docynia*; *Cydonia*; *Pseudocydonia*; *Chaenomeles*
Crataegeae	*Cotoneaster*; *Malacomeles*; *Chamaemeles*; *Pyracantha*; *Crataegus*; *Mespilus*; *Hesperomeles*; *Osteomeles*

Accordingly, the systematic position of the genus *Amelanchier,* according to Armen Takhtajan's system [1], appears as follows:

Divisio – Magnoliophyta
Classis – Magnoliopsida (Dicotyledons)
Subclass – Rosidae
Superordo – Rosanae
Ordo – Rosales
Familia – Rosaceae
Subfamilia – Pyroideae (Maloideae*)*
Tribus – Maleae
Genus – *Amelanchier.*

According to the analysis of subfamilies from the family Rosaceae, performed by a group of scholars of different universities in the USA, Canada and Sweden after six nuclear (18S, gbssi1, gbssi2, ITS, pgip, ppo) and four chloroplastic (matK, ndhF, rbcL, and trnL-trnF) segments of DNA sequences [24, 26, 31], only the subfamily Rosoideae (Juss.) Arn. turned out monophyletic, with the basic chromosome number $x=7$ or 8, except for the tribe Dryadeae ($x=9$). Instead the subfamilies Prunoideae and Maloideae in the traditional sense were paraphyletic, and Spiraeoideae – polyphyletic group. On this basis, the rank of the first two subfamilies is proposed to reduce to the tribe and together with the other related tribes to combine into one monophyletic (in a very broad sense) subfamily Spiraeoideae C. Agardh, with $x=8$, 9, 15 or 17. Therefore, the supertribe Pyrodae Camp., Ev., Morg. et Dick. with the tribe *Pyreae* Baill.

were included to the subfamily Spiraeoideae (x=17, with the exception of the genus *Vauquelinia* Correa ex Humb. Et Bonpl. with x=15), the subtribe of which Pyrinae absorbed most of the genera of the subfamily *Maloideae*, including the genus *Amelanchier*.

This extension of the subfamily Spiraeoideae enabled us to determine the systematic position of the genus *Amelanchier* within the family Rosaceae as follows [24, 51, 52]:

Familia – Rosaceae Juss.
Subfamilia – Spiraeoideae C. Agardh
Supertribus – Pyrodae Camp., Ev., Morg. et Dick.
Tribus – Pyreae Baill.
Subtribus – Pyrinae Dumort.
Genus – *Amelanchier* Medik.

However, there appeared to be a need to change the name of the subfamily Spiraeoideae due to the inclusion of the former subfamily Amygdaloideae to the newly formed subfamily Spiraeoideae. The fact is that under the International Code of Nomenclature for algae, fungi, and plants updated in 2011 [53] the taxon names must correspond to the earliest published name, so the priority name for the subfamily, which combines Spiraeoideae, Maloideae and Amygdaloideae is the name Amygdaloideae; for the tribe Pyreae – name Maleae Small; for the subtribe Pyrinae – name Malinae Rev. (Article 19.5, ex. 5).

While comparing the systematic position of the genus *Amelanchier* Medik., according to the different classification systems of plants different in time of creation and research level, the change in view on genus' phylogenetic connections can be partially traced (Table 11.2).

Herewith, the relative stability of the genus *Amelanchier* positiom within the major taxa of higher ranks should be noted. The range of fluctuations in the number of accepted species within the genus *Amelanchier* is quite wide: from 6 to 33 [30, 55], and with infraspecific taxa to 37 [48]. The number of Latin species names used by different authors is nearly ten times as much. Most of these names, which are now considered to be unresolved, have: unplaced and unassessed names, synonyms [48], provisionally accepted names, infraspecific taxa [56], interspecific hybrids, or misapplied names.

TABLE 11.2   The Systematic Position of the Genus *Amelanchier* Medik. According to Different Plant Classification Systems

Taxon	Classification systems of plants		
	Engler, 1903 [45]	**Takhtajan, 2009 [1]**	**APG III (2009) [53, 54].**
Division	Embryophyta siphonogama	Magnoliophyta	–
Subdivision	Angiospermae	–	–
Classis	Dicotyledoneae	Magnoliopsida (Dicotyledons)	–
Subclassis	Archichlamydeae	Rosidae	–
Superordo	–	Rosanae	–
Ordo	Rosales	Rosales	Rosales
Subordo	Rosineae	–	–
Familia	Rosaceae	Rosaceae	Rosaceae
Subfamilia	Pomoideae	Pyroideae (Maloideae)	Amygdaloideae
Tribus	–	Maleae	Maleae
Subtribus	–	–	Malinae
Genus	Amelanchier	Amelanchier	Amelanchier

A complex taxonomy of the genus is explained by morphological variation features of vegetative and generative organs, a large number of divergent and intermediate forms, polyploidy, hybridization, and a tendency to apomixis, causing the so-called occurring of agamospecies [57] and determining some taxonomic difficulties.

Generalized data on the genus *Amelanchier* taxonomy combine 279 names (including infraspecific). Of these 243 scientific plant names of species, 28 (11.5%) are accepted species names, 93 (38.3%) unassessed names, and 122 (50.2%) are synonyms [48].

We can assume that the ancestors of the genus *Amelanchier* modern species emerged by the end of the Cretaceous period of the Mesozoic era, or the beginning of the Paleogene period of the Cenozoic era, when there was a relatively rapid modernization of the flowering plants genus structure. So in the Eocene floras, most species can be classified within contemporary families. So, it is quite natural that certain fossilized footprints of *Amelanchier* are found in western North America, namely in the deposits of the Eocene period (48–50 million years ago) [58]. During the Neogene, in the arid regions of North America, a kind of "Madro-*Tertiary*" flora was formed, a detailed study of which [59] showed genus *Amelanchier* among other typical representatives of the fossil flora.

11.3.3 EVOLUTION DIRECTIONS OF THE GENUS AMELANCHIER

Adaptive radiation and hybridization are distinguished among the determinative evolution directions of the genus [24]. Thus, adaptive radiation most likely is caused by the formation of fleshy fruit and is related to vital functions of animals.

The conception about the growing importance of endozoochory during the process of fouling symphycarpous fruit aggregating by floral tube (apples formation), is also supported by Armen Takhtajan [60].

In general, the hypothesis about the origin of the subfamily Pyroideae (formerly Maloideae) shows the connection of this group with common ancestors of the most ancient subfamily Spiroideae. At the same time, they are close to Rosoideae according to the type of fruit-apples structure, as well as *Prunoideae*, as woody plants have a similar leaf shape, type of inflorescence, and structure of sepals and petals [47].

It should also be mentioned that the representatives of Pyroideae (Maloideae) have a basic chromosome number $x=17$ [61, 62].

Most of the other representatives of the Rosaceae family are characterized by a much lower number of chromosomes $x=7$, 8, or 9. That's why the logical assumption about the polyploid origin of chromosome number Pyroideae (Maloideae) was made. According to C.D. Darlington and A.A. Moffett (1930), Pyroideae (Maloideae) appeared from Rosoideae and is a triple trisomic tetraploid ($x=7+7+3=17$). In other words, it would double the number of chromosomes (all seven) with the addition of one more chromosome from three different pairs in one of the ancient specimens of Rosaceae with haploid set $x=7$ (very common chromosome number in the family Rosaceae) [51, 63].

However, the probability of triple trisomy is significantly lower than of amphidiploidy, so more followers supported the hypothesis of K. Sax (1931), who believed that Pyroideae (Maloideae) are alloploids arising out of doubling the number of chromosomes in hybrids between distant ancestors of two distant generic types. According to his view, this could be representatives of the subfamilies *Prunoideae*, which has a basic chromosome number $x=8$ and Spiroideae – with x=9, the uniting of which has put modern Pyroideae (Maloideae) x=17 chromosomes in a common genome [51, 61].

In those times, quoting Stebbins (1950), Armen Takhtajan [60] expressed an opinion that taking into account the data of cardiology and morphology of the flower, the most probable explanation for the origin of Pyroideae (Maloideae) is based on the fact that Pyroideae (Maloideae) is diploidizated polyploid arising out as a result of an ancient hybridization between Spiroideae and Rosoideae, which explains the basic haploid number of this subfamily $x=17$.

The fact of mainly bivalent chromosomes conjugation of Pyroideae (Maloideae) [61] gives ground to define the representatives of this subfamily as functional diploids. Although one can find tetraploid (68 chromosome) species (including the genus *Amelanchier*) near the diploid $2n=34$ in the subfamily [46, 51, 62], the proportion of functional diploids in Pyroideae (Maloideae) prevails, and it is much larger than in other subfamilies of Rosaceae [25].

The results of the research released at the beginning of the 21st century, which were carried out at comparing the DNA sequences of the subfamily Pyroideae (Maloideae) and a large number of other representatives of the family Rosaceae, shake the prestige of these hypotheses [24, 26, 31].

The analysis of the obtained materials on the genomes similarity of the subfamily Pyroideae (Maloideae) specimens and the genus *Gillenia* from the subfamily Spiraeoideae, gave good reasons to believe that probably the genus *Gillenia* is the closest relative to the apple. All nuclear and chloroplast cladograms show that the genus *Gillenia* is invariably manifested as a sister group to *Pyroideae* (*Maloideae*). Taking into consideration that *Gillenia* has a less number of chromosomes ($x=9$) than all Pyroideae (*Maloideae*) ($x=17$), the authors assumed that the genom of Pyroideae (Maloideae) was formed monophyletic as a result of autopolyploidy of the genus *Gillenia* representatives from $x=9$ to $x=18$ and subsequent aneuploidy (nullisomy) to the current number of chromosomes $x=17$. Thus, the basic chromosome number $x=17$ is common to all Pyroideae (Maloideae) and some Spiraeoideae (*Kageneckia* and *Lindleya*), although *Vauquelinia* representatives have $x=15$, which may become a reason of a system revision [27].

Flow cytometry data [64] also showed the similarity of the genomes Pyroideae (Maloideae) and *Gillenia*. The comparative analysis of the characteristics of female and male gametophytes *Gillenia* and seven genera representatives of the subfamily (*Chaenomeles, Cotoneaster, Crataegus, Mespilus, Photinia, Rhaphiolepis* and *Sorbus*) confirmed the similarity of Pyroideae (Maloideae) and *Gillenia* floral development [29].

Admitting evidences of monophyletic nullisomic origin of the subfamily Pyroideae (Maloideae) representatives one has to explain the facts of mainly bivalent conjugation of their chromosomes by a prolonged evolution, in the process of which during interspecies hybridization and polyploidization within common ancestral group with *Gillenia* took place. Such course of events is more likely than the gradual formation of a functional diploid from an autoaneuploid that perhaps arose out of an autotetraploid because of its nullisomy.

These assumptions were confirmed as a result of summarizing data of collinearity (the order of location) analysis of genes along each of the 17 chromosomes of the apple. Working according to a united international program, 86 scientists from Italy, France, New Zealand, Belgium and the United States have examined the chromosomes of the genome sequence of the apple 'Golden Delicious' [33].

They showed similar collinearities between large segments of chromosomes 3 and 11, 5 and 10, 9 and 17, 13 and 16, and between short segments of chromosomes 1 and 7, 2 and 7, 2 and 15, 4 and 12, 12 and 14, 6 and 14, 8 and 15, about which they reported in a joint publication. They found that relatively not long ago, less than 50 (approximately 30–45) million years ago, there was a spontaneous duplication (autoreduplication) of the 9 chromosome ancestor genome of the apple with subsequent loss of the eighteenth chromosome and forming 17 chromosome karyotype of modern apples (Figure 11.2).

Herewith, the first chromosome of the apple ancestor donated genetic material to 5 and 10 chromosomes of apple trees, respectively 3 and 11 chromosomes intergraded from the second chromosome, 9 and 17 – from the third one, 13 and 16 – from the fourth one, and 4, 6, 12 and 14 chromosomes of apples are combined from the fragments of the fifth and

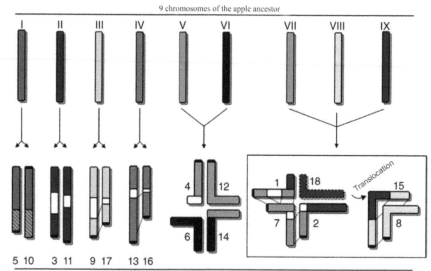

FIGURE 11.2   The scheme of forming 17-chromosome karyotype of *Malus domestica* Borkh. (the scheme could be extended to other apples, including the genus *Amelanchier*): the virtual chromosome 18 is shown as a source of genetic material for 1, 2 and 15 apple chromosomes; white areas indicate that sequences localized at them have no ancestral prototypes (according to R. Velasco et al., 2010 [33] as amended [51]).

sixth ones. The first and the second chromosomes are developed from the fragments of the seventh and ninth ancestors of the apples and from the seventh chromosome – seventh one respectively. The origin of 8 and 15 chromosomes of apples is a little more complicated. If the eighth chromosome of apples contains sequences of the eighth apple ancestor chromosome, then 15 consists of fragments of the eighth and ninth ones. But there is a reason to believe that the translocation of genetic material of the ninth chromosome took place before the above-mentioned nullisomy (loss of the 18th pair of chromosomes) [33].

It is assumed an existence (currently unidentified) of the gene regulator of conjugation of homologous chromosomes apple with functions similar to the display of the gene *Ph1*, which governs the on-goings of wheat chromosomes during meiosis, preventing from multivalent conjugation of partially homologous chromosomes poliploids. It provided mainly bivalent conjugation (pairs) and the formation of functional diploids from the ancestral autotetraploid.

11.3.4 THE AREAL OF THE GENUS AMELANCHIER

The genus *Amelanchier* representatives grow mainly in the forested areas of the moderate zone in the Northern Hemisphere mostly in light woodland slopes, light forests to an altitude of 1900 m above the sea level, and grow well on a variety of soils [21].

The areal of the genus *Amelanchier* is quite wide, it occupies the extratropical Northern Hemisphere and covers almost all of North America and Europe, partially extratropical North Africa and extratropical Asia. Some species can be found in the subtropics and occasionally in the tropical latitudes (Figure 11.3), but mainly in the mountains, where conditions are similar to moderate or subtropical climate [21, 22, 30].

The analysis of the genus *Amelanchier* existence is defined by Armen Takhtajan (1978) biogeographic regions confirms the predominant settlement in the moderate latitudes of the Northern Hemisphere. Types of *Amelanchier* occur in all regions of the Boreal subkingdom, namely in Circumboreal, East Asian, Atlantic and North-American regions and the region of the Rocky Mountains; in Mediterranean and Irano-Turanian regions of the Ancient Mediterranean Subkingdom and in Madrean area of the Madrean (Sonoran) Subkingdom in Holarctic Kingdom (Figure 11.4) [22, 66].

However, the vast majority of species of the genus grows within the original areal in the North America territory, from 18 to 26 [41, 50, 65].

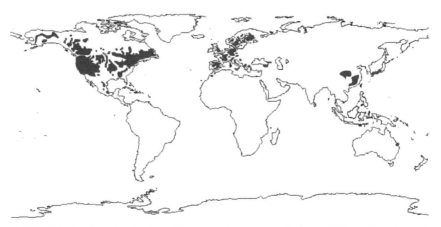

FIGURE 11.3 The distribution of the genus *Amelanchier* in the world (based on site EOL (Encyclopedia of Life) [65]).

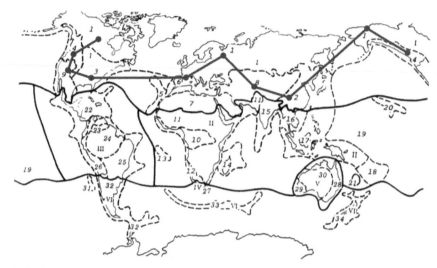

FIGURE 11.4 The distribution of the genus *Amelanchier* in floristic regions defined by Armen Takhtajan (1978) [66]. The firm line defines the conventional boundaries of floristic kingdoms, dashed – areas; point (conditional core of the floristic region) at the figures shows the distribution of the genus within the regions: 1 – Circumboreal 2 – East Asian, 3 – Atlantic, North American, 4 – Rocky Mountain region, 6 – Mediterranean, 8 – Irano-Turanian, 9 – Madrean.

One species is typical for Europe, Africa, and Turkey (*A. ovalis*) [21, 30, 67]. Another one occurs in Greece, on the island of Crete (*A. cretica*), the other – European natural hybrid of *A. lamarckii*, and 2 species grow in Turkey (*A. integrifolia; A. parviflora*) [30]. Two species grow in China, Korea, and Japan (*A. sinica; A. asiatica*) [30, 68].

11.3.5 THE INVASIVENESS PROBLEMS OF THE GENUS AMELANCHIER INDIVIDUAL REPRESENTATIVES

The genus *Amelanchier* representatives actively spread and are able to naturalize in natural phytocenoses of the second range. Thus, the distribution of some of them in the territory of some European countries and the European part of Russia can apply the character of phytoinvasion (Table 11.3).

The data in Table 11.3 show that *A. spicata* is characterized as the most aggressive in the European countries among presented species.

TABLE 11.3 The Invasion Degree of the Genus *Amelanchier* Individual Representatives in Northern and Central Europe counTries and the European Part of Russia, According to the Materials of the Website NOBANIS [69]

Species	Country										
	Belgium	Denmark	Estonia	The European part of Russia	Latvia	Lithuania	Norway	Finland	Sweden
A. alnifolia	–	n/i	–	–	–	–	i	p/i	i
the growth place: mixed coniferous and deciduous forests, open areas, damaged areas, urban areas									
A. canadensis	–	–	–	n/i	–	–	–	–	–
the growth place: mixed coniferous and deciduous forests, urban areas									
A. laevis	–	–	–	–	–	–	–	p/i	n/i
the growth place: mixed coniferous and deciduous forests, open areas, damaged areas, wetlands									
A. lamarcki	p/i	p/i	–	–	–	–	i	p/i	i
the growth place: mixed coniferous and deciduous forests, open areas, damaged areas, wetlands									

TABLE 11.3 Continued

Species	Country								
	Belgium	Denmark	Estonia	The European part of Russia	Latvia	Lithuania	Norway	Finland	Sweden
A. ovalis	–	–	–	n/i	–	–	n/i	–	–
the growth place: mixed coniferous and deciduous forests, urban areas									
A. spicata	–	i	i	n/i	i	i	i	i	i
the growth place: mixed coniferous and deciduous forests, open areas, damaged areas, agricultural areas, urban areas									

Note: n/i – non-invasive; p/i – potentially invasive; i – invasive; dash – no information available.

The invasiveness of the rest of the presented species of the genus *Amelanchier* has a non-systemic character. You can accept the opinion of A.S. Mosyakin [70] that the invasive processes are controlled by multi-directional biotic and abiotic factors, the interaction of which depends on the invasive possibility of a certain type. In particular, this invasive possibility is not some fixed trait, proper to species, it is found only in specific environmental conditions.

11.3.6 THE REPRESENTATIVES OF THE GENUS AMELANCHIER IN THE FLORA OF UKRAINE

In the flora of Ukraine, a number of the genus *Amelanchier* species is limited to two or three [3, 52, 70, 71]. These are: *A. ovalis*, *A. canadensis* and *A. spicata*. Herewith, *A. ovalis* is defined as an indigenous species, and *A. canadensis* and *A. spicata* are defined as introduced and naturalized in the secondary habitat. *A. rotundifolia*, *A. integrifolia*, *A. oligocarpa*, *A. laevis*, *A. alnifolia*, *A. florida*, *A. utahensis*, *A. asiatica*, which are unsystematically cultivated, mainly as ornamental, in private collections, botanical gardens and arboreta are referred to as a promising species for introduction, apart from *A. canadensis* and *A. spicata*.

Until recently, in the collection of the National Dendrological Park (NDP) "Sofiyivka," the genus *Amelanchier* was represented only by two species *A. ovalis* and *A. canadensis* [72].

Among the supplies of the last decade, there are representatives of the species: *A. alnifolia*, *A. asiatica*, *A. canadensis*, *A. florida*, *A. laevis*, *A. ovalis*, *A. spicata*, *A. stolonifera*, *A. utahensis* and *A. pumila*. Among them, there were plants that were delivered to "Sofiyivka" in 50–60 years of the last century, but identified only in 2004–2014, as well as new supplies from various botanical institutions. In some cases, the re-introduction of species contributed to the species specification of existing plants. Now 14 species names can be counted in the collection of the NDP "Sofiyivka." Table 11.4 introduced the species names of the genus *Amelanchier* specimens growing in the NDP "Sofiyivka," the plants of which were identified and also 28 species names, accepted in the database of Royal Botanical Gardens Kew [56]. During the time that has passed since our previous report [52], the collection of NDP "Sofiyivka" was enlarged to 14 species.

TABLE 11.4   The Genus *Amelanchier* Collection List of the NDP "Sofiyivka" of NAS of Ukraine Compared to the Catalogue of Life: 2014 Annual Checklist

Catalogue of Life., 2014 [56]	Collection list of the NDP "Sofiyivka," 2014
A. alnifolia (Nutt.) Nutt. ex M. Roem.	*A. alnifolia* (Nutt.) Nutt. ex M. Roem.
A. arborea (F. Michx.) Fernald	absent
A. asiatica (Siebold & Zucc.) Endl. ex Walp.	*A. asiatica* (Siebold & Zucc.) Endl. ex Walp.
A. australis Standl.	absent
A. bakeri Greene	absent
A. bartramiana (Tausch) M. Roem.	absent
A. canadensis (L.) Medik.	*A. canadensis* (L.) Medik.
A. covillei Standl.	absent
A. cretica (Willd.) DC.	absent
A. cusickii Fernald	absent
Provisionally accepted name	*A. florida* Wiegand
A. grandiflora Rehder	*A. grandiflora* (Wiegand) Wiegand
A. interior E.L. Nielsen	absent
A. intermedia Spach	absent
A. laevis Wiegand	*A. laevis* Wiegand
Absent	*A. lamarckii* F.G. Schroed.
A. neglecta Eggl. ex G.N. Jones	absent
A. obovalis (Michx.) Ashe	absent
A. ovalis Medik.	*A. ovalis* Medik.
A. pallida Greene	absent
A. parviflora Boiss.	absent
A. pumila (Nutt. ex Torr. & A. Gray) M. Roem.	*A. pumila* (Nutt. ex Torr. & A. Gray) M. Roem.
A. quinti-martii Louis-Marie	absent
Provisionally accepted name	*A. rotundifolia* (Lam.) K. Koch
A. sanguinea (Pursh) DC.	*A. sanguinea* (Pursh) DC.
A. sinica (C.K. Schneid.) Chun	absent
A. spicata (Lam.) K. Koch	*A. spicata* (Lam.) K. Koch
A. stolonifera Wiegand	*A. stolonifera* Wiegand
A. turkestanica Litv.	absent
A. utahensis Koehne	*A. utahensis* Koehne

*Catalogue by Royal Botanical Gardens Kew.

Besides, cited databases [48, 56] in recent years became very close to the list of species names, which gave a reason to limit them to comparing the species names of the collection NDP "Sofiyivka" from the Annual check-list of Catalogue by Royal Botanical Gardens Kew.

The species *A. alnifolia* was imported from the Krivoy Rog Botanical Garden of NAS of Ukraine. This species name was considered synony-mous with *A. sanguinea* var. *alnifolia* (Nutt.) P. Landry in the past in the working list of known plant species The Plant List Royal Botanic Gardens, Kew and Missouri Botanical Garden, but now it is recognized as a sepa-rate species name in the above mentioned catalog The Plant List [48], as well as in the Catalogue of Life, 2014 [56].

The species *A. asiatica* were delivered to the NDP "Sofiyivka" collec-tion from O.V. Fomin Botanical Garden, Taras Shevchenko Kyiv National University research institutions in 2009. This species name is accepted in all catalogs known to us.

A. canadensis plants were first imported from Minsk Botanical Garden (now the Central Botanical Garden of NAS of Belarus) in 1959, but cer-tainty about their species belonging was questioned after transferring plants from the domestic park arboretum to the active research and com-mercial arboretum, that prompted to the re-introduction of this species in 2010 from the Krivoy Rog botanical Garden. The species name is now accepted in both above mentioned databases [48, 56].

The similar story of the repeated (in 2010) introduction from the Krivoy Rog Botanical Garden of *A. florida* plants, representatives of which were first imported in 1959 from Leningrad Botanical Garden (now V.L. Komarov Botanical Garden of the Botanical Institute of RAS). In the Plant List [48], the name of *A. florida* Wiegand is considered unresolved, and *A. florida* Lindl. is synonymous with *A. alnifolia var. semi-integrifolia* (Hook.) CLHitchc., and in the Catalogue of Life [56] *A. florida* Wiegand name is given as a provisionally accepted name without the name *A. flor-ida* Lindl.

The species *A. grandiflora*, in the NDP "Sofiyivka" collection, is represented by two cultivars: 'Autumn Brilliance' and 'Forest Prince.' Its species name – *Amelanchier grandiflora* (Wiegand) Wiegand is now accepted in The Plant List [48] a synonym of *Amelanchier sanguinea* var. *grandiflora* (Wiegand) Rehder., but in the same catalog the accepted name

Amelanchier×grandiflora Rehder is given. Instead, in the Catalogue of Life [56] the accepted name *A. grandiflora* Rehder is given, whereas *A. grandiflora* Wieg. is considered a synonym for *A. sanguinea* (Pursh) DC.

A. laevis plants were first delivered in 1958 from the Botanical Garden of Uzbekistan (now the Botanical Garden of the Uzbekistan AS) and re-imported in 2010 from the Krivoy Rog Botanical Garden, which contributed to specifying the plant species. This species name is now accepted as a separate species in both above mentioned databases [48, 56].

In the NDP "Sofiyivka" collection, the species *A. lamarckii* FG Schroed. is presented as a cultivar 'Prince William.' This species name is an accepted name in the database of The Plant List [48]. But in the Catalogue of Life: 2014 Annual Checklist [56] this species name (*A. lamarckii*) is removed, although in the Catalogue of Life: 2010 Annual Checklist it was given as an accepted name.

A. ovalis Wieg. plants are classified as representatives of the long and widely spread species (in all parts of "Sofiyivka") by the NDP "Sofiyivka" catalog in 2000 [72]. This name is accepted by both the above mentioned catalogs [48, 56].

The species name *A. pumila* (Nutt. ex Torr. & A.Gray) M.Roem. is now accepted in both above mentioned databases [48, 56].

The species name *A. rotundifolia* (Lam.) K. Koch, plants of which were imported from Kaunas Botanical Garden (now the Vytautas the Great University Botanical Garden) in 1958, is included in The Plant List [48] as an unresolved name, and in the Catalogue of Life [56] is considered as a synonym for *A. ovalis subsp. ovalis* Medik.

Re-introduction (in 2010) from the Krivoy Rog Botanical Garden of *A. sanguinea* (Pursh) DC. plants, representatives of which were first imported from Leningrad Botanical Garden (now V.L. Komarov Botanical Garden of the Botanical Institute of RAS) in 1958, will contribute to specifying species belonging of existing plants. The species name is accepted by both the above-mentioned catalogs, which gives grounds for certainty in its status [48, 56].

A. spicata (Lam.) K. Koch, as well as *A. ovalis* are classified by NDP "Sofiyivka" Catalogue of 2000 as a long spread species in all parts of the park [72]. However, the features of vegetative and generative organs variation, that is referred to this species plants, prompts for further more grounded analysis. Therefore, *A. spicata* representatives delivered from the Krivoy Rog Botanical Garden in 2010 were planted in the NDP "Sofiyivka" collection to compare and specify the status of the existing plants under this plant name. In the Plant List [48] *A. spicata* is accepted as a species name. But in the Catalogue of Life [56] *A. spicata* is considered as a synonym for *A. canadensis* (L.) Medik.

The plants *A. stolonifera* Wiegand and *A. utahensis* Koehne, imported from the Krivoy Rog Botanical Garden in 2010, belong to new supplies of the genus *Amelanchier* specimens. Thus, in the Plant List … [48] and in the Catalogue of Life … [56] are both accepted species names.

The similar divergences in approaches to systematize species names and specify the composition the genus are observed when comparing the names of other *Amelanchier* species, so far absent in NDP "Sofiyivka" of NAS of Ukraine collection, but available in other catalogs [48, 56]. The desired consensus can be achieved by combining the results of the species identification by classical (morphological) and molecular genetic criteria. Today, collecting new genotypes of the genus *Amelanchier* continues, and exploring new supplies has already started.

In addition to the above-mentioned species of *Amelanchier* in the NDP "Sofiyivka" collection, a number of cultivars are researched, seedlings of which are grown from in vitro propagated plants: 'Autumn Brilliance,' 'Forest Prince,' 'Krasnojarskaja,' 'Pembina,' 'Prince William,' 'Slate,' 'Snowcloud,' 'Smoky,' including old cultivars: 'Pembina,' 'Smoky.'

The analysis of differences in species names and common names of the genus *Amelanchier* specimens in some well-known websites [48, 56], demonstrates the need for their further arranging. However, as a great advantage of these [48, 56] and other similar electronic databases of plant species names, one should accept their general availability, ease of use and, what is very important, is a constant dynamism, the ability to collect and analyze new information and arrange it.

11.4 CONCLUSION

Thus, in Ukraine, the representatives of the genus *Amelanchier* are still unconventional plants for the culture, but interest in them is constantly growing, due to their fruit ornamental value, nectareous, medicinal phytomeliorative abilities.

The results of the phylogenetic and molecular genetic studies performed by scientists from different countries give an opportunity to specify the systematic position of the representatives of the genus *Amelanchier* of the family Rosaceae Juss. grown in Ukraine, and temporarily place them in a large subfamily Amygdaloideae Arn., which unites the former subfamilies Amygdaloideae, Spiraeoideae and Maloideae, tribe – Maleae Small, subtribe – Malinae Rev.

The divergences in species and interspecies classification of the genus *Amelanchier* representatives found in various publications indicate incompleteness of the genus system and necessity for further studies by classical and molecular genetic methods.

ACKNOWLEDGEMENT

This material is partly based on the work supported by the National Dendrological park "Sofiyivka" of NAS of Ukraine (№ 0112U002032) together with Uman National University of Horticulture (№ 0101U004495) in compliance with their thematic plans of the research work. We thank corresponding members of NAS of Ukraine Ivan Kosenko and Ph.D. Galyna Chorna for consultations and discussion.

KEYWORDS

- areal
- biodiversity
- chromosome number
- DNA sequence
- family
- floristic region
- Juneberry
- phytomeliorative
- taxonomy
- tribe

REFERENCES

1. Takhtajan, A. L. Flowering plants [corr. 2nd ed.]. NY: Springer Science+Business Media, 2009, 871 p.
2. Markovskyi, V. S., Bakhmat M.I. Juneberry. Berries in Ukraine. Kamenetz-Podolskiy: Medobory-2006, 2008, 166–168. (in Ukrainian).
3. Opalko, A. I., Zaplichko, F. O. Pome fruit crops breeding. Fruit and vegetable crops breeding. Kyiv: Higher School, 2000, 345–385. (in Ukrainian).
4. Andrienko, N. V., Roman, I. S. Juneberry. Rare berry and fruit crops. Kyiv: Urozhai, 1991, 79–84, 153–154. (in Ukrainian).
5. Burmistrov, A. D. Juneberry. Berry crops. Leningrad: Agropromizdat, 1985, 240–245 (In Russian).
6. Antsiferov, A. Juneberry – honey wine. Enjoyable garden. 2011, №6 (65). 44–47 (In Russian).
7. Kuklina, A. G. Honeysuckle, Juneberry: A Manual for amateur gardeners. Moscow: Niola Press, 2007, 240 p (In Russian).
8. Sautkin, F. V. Complex structure of phytophagans – Juneberry (*Amelanchier* spp.) pests under the conditions of Belarus. Bulletin of the Belarusian State University. 2012, Edition 2, №2, 38–42 (In Russian).
9. Lim, T. K. *Amelanchier alnifolia*. Edible medicinal and non-medicinal plants. Dordrecht; Heidelberg; London; New York: Springer, 2012, Vol. 4: Fruits. 358–363.
10. Kosenko, I. S., Khraban, G. Y., Mitin, V. V. and Garbuz, V. F. Flora of "Sofiyivka." Dendrological park "Sofiyivka": 200 years. [Ed. M. A. Kokhno]. Kyiv. Naukova Dumka, 1996, 165–180. (in Ukrainian).
11. Kuklina, A. G. Naturalization of *Amelanchier* species from North America in a secondary habitat Pleiades Publishing. Russian Journal of Biological Invasions. 2011, Vol. 2, №2–3. 103–107.
12. Zatylny, A. M., St-Pierre, R. G. Revised international registry of cultivars and germplasm of the genus *Amelanchier*. Small Fruits Review. 2003, Vol. 2, №1, 51–80.
13. Gluhov, A. A. Nectareous plants. Moscow: State publishing House of agricultural literature, 1955, 512 p (In Russian).
14. Grysyuk, M. M., Yellin Yu.Ya. Wild food, technical and nectareous plants of Ukraine. Kyiv: Urozhai, 1993, 208 p. (in Ukrainian).
15. Shabarova, S. I., Tarhonskiy, P. N. Juneberry genus – *Amelanchier*. Fruit, berry and nut plants of the USSR forests. Kyiv: Higher School, 1984, 47–48. (in Ukrainian).
16. Shukel, I. V., Dyda, A. P., Nizhalovskiy, Y. V. Using *Amelanchier ovalis* Medik. in the recreational phytomelioration. Scientific Bulletin of the Ukrainian State Forestry University. 2003, Iss. 13.5. 379–383.
17. Ryazanova, O. A. New garden culture. Gardening and Viticulture. 1999, №3, 23–24 (In Russian).
18. Lukin, V. D. Best bars for handicrafts – from Juneberry. Orchard and garden. 2006, №8(89). 30 (In Russian).
19. Nekrasov, S. A. Juneberry. Berry garden. Minsk: MET Ltd. 2001, 222–228 (In Russian).
20. St-Pierre, R. G. Growing saskatoons. A manual for orchardists. Department of Horticulture Sciences, University of Saskatchewan. Saskatoon: SK, 1997, 338 p.

21. Poyarkova, A. I. Genus 730. Juneberry – *Amelanchier* Medik. Flora of the USSR. In 30 volumes. Moscow-Leningrad: USSR Academy of Sciences Publishing House, 1939, Vol. 9. 408–413 (In Russian).

22. Aldasoro, J. J., Aedo, C., Navarro, C. Phylogenetic and phytogeographical relationships in Maloideae (Rosaceae) based on morphological and anatomical characters. Blumea. 2005.Vol. 50, №1, 3–32.

23. Janick, J., Cummins, J. N., Brown, S. K., Hemmat, M. Apples. Fruit Breed. Vol. 1: Tree and Tropical Fruits [Eds. J. Janick and, J. N. Moore]. N. Y.: John Wiley and Sons, 1996, 1–78.

24. Campbell, C. S., Evans, R. C., Morgan, D. R. et al. Phylogeny of subtribe Pyrinae (formerly the Maloideae, Rosaceae): Limited resolution of a complex evolutionary history. Plant systematics and evolution. 2007, Vol. 266, №1–2. 119–145.

25. Dickson, E. E., Arumuganathan, K., Kresovich, S. et al. Nuclear DNA content variation within the Rosaceae. American Journal of Botany. 1992, Vol. 79, №9, 1081–1086.

26. Dickinson, T. A., Lo, E. Y. Y., Talent, N. Polyploidy, reproductive biology, and Rosaceae: understanding evolution and making classifications. Plant Systematics and Evolution. 2007, Vol. 266, №1–2. 59–78.

27. Evans, R. C., Alice, L. A., Campbell, C. S. et al. The granule-bound starch synthase (GBSSI) gene in the Rosaceae: Multiple loci and phylogenetic utility. Molecular Phylogenetics and Evolution. 2000, Vol. 17, №3, 388–400.

28. Evans, R. C., Campbell, C. S. The origin of the apple subfamily (Maloideae; Rosaceae) is clarified by DNA sequence data from duplicated GBSSI genes. American Journal of Botany. 2002, Vol. 89, №9, 1478–1484.

29. Evans, R. C., Dickinson, T. A. Floral ontogeny and morphology in Gillenia ("Spiraeoideae") and subfamily Maloideae, C. Weber (Rosaceae). International journal plant science. 2005, Vol. 166, №3, 427–447.

30. Phipps, J. B., Robertson, K. R., Smith, P. G. and Rohrer, J. R. A checklist of the subfamily Maloideae (Rosaceae). Canadian Journal of Botany. 1990, Vol. 68, №10, 2209–2269.

31. Potter, D., Eriksson, T., Evans, R. C. et al. Phylogeny and classification of Rosaceae. Plant systematics and evolution. 2007, Vol. 266, №1–2. 5–43.

32. Rehder, A. New species, varieties and combinations from the herbarium and the collections of the Arnold Arboretum. Journal of the Arnold Arboretum. 1920, Vol. 1, №4, 254–263.

33. Velasco, R., Zharkikh, A., Affourtit, J. et al. The genome of the domesticated apple (*Malus* × *domestica* Borkh.). Nature Genetics. 2010, Vol. 42, №10, 833–839.

34. Weber, C. The genus *Chaenomeles* (Rosaceae). Journal of the Arnold Arboretum. 1964, Vol. 45 №2, 161–205 and №3, 302–345.

35. Medicus, F. C. *Amelanchier*. Philosophische Botanik: mit kritischen Bemerkungen. Von den mannigfaltigen Umhüllungen der Saamen. Mannheim, 1789, Vol. 1, 135.

36. Matthias de l'Obel. *Amelanchier*. Kruydtboeck oft beschrÿuinghe van allerleye ghewassen, kruyderen, hesteren, ende gheboomten. T'Antwerpen: By Christoffel Plantyn, 1581, 223.

37. Tournefort, J. P. Genus II. *Mefpilus* Neflier. Institutiones rei herbariæ. Parisiis: E Typographia Regia, 1700–1703, Vol. 1, 641–642.

38. Linnaei C. *Chionanthus*. Species plantarum, exhibentes plantas rite cognitas, ad genera relatas, cum differentiis specificis, nominibus trivialibus, synonymis selectis,

locis natalibus, secundum systema sexuale digestas. Holmiae: Laurentii Salvii, 1753, Vol. 1. 479–480.

39. Annenkov, N. I. *Amelanchier* Med. Botanical Dictionary. St. Petersburg: Imperial. Academy of Sciences, 1878, 27–28 (In Russian).

40. Kaden, N. N., Terentyeva, N. M. Etymological Dictionary of the Latin names for plants found in the agrobiological station MSU "Chashnikovo" vicinity. Moscow: Lomonosov Moscow State University Press, 1975, 20, 169–170 (In Russian).

41. Jones, G. N. American species of *Amelanchier*. Illinois biological monographs. 1946, Vol. 20, №2, 136 p.

42. Kobiv, Y. Dictionary of Ukrainian scientific and vernacular names for plant. Kyiv: Naukova Dumka, 2004, 800 p. (in Ukrainian).

43. Chernik, V. V., Dzhus, M. A., Sautkina, T. A. et al. Systematics of higher plants. Angiosperms. Dicotyledonous class. Minsk: Belarus State University Press, 2010, 203–212 (In Russian).

44. An update of the Angiosperm Phylogeny Group classification for the orders and families of flowering plants: APG III. The Angiosperm Phylogeny Group. Botanical Journal of the Linnean Society. 2009, Vol. 161. 105–121.

45. Engler A. Syllabus der Pflanzenfamilien. Eine Übersicht Über das gesamte Pflanzen-system mit Berücksichtigung der Medicinal- und Nutzpflanzen nebst einer Übersieht über die Florenreiehe und Florengebiete der Erde zum Gebrauch bei Vorlesungen und Studien über specielle und medicinisch-pharraaceutische Botanik. – Berlin: Verlag von Gebrüder Borntraeger, 1903, 233 p.

46. Zielinski, Q. B., Thompson, M. M. Speciation in *Pyrus*: Chromosome number and meiotic behavior. Botanical gazette. 1967, Vol. 128, №2, 109–112.

47. Gladkova, V. N. Pink, or rose family order (*Rosales*). Plant life. In 6 vol. Moscow. 1980, Vol. 5/2. 175–189 (In Russian).

48. The Plant List by the Royal Botanic Gardens Kew and Missouri Botanical. 2013, URL: http://www.theplantlist.org/tpl1.1/search?q=Amelanchier (Accessed 12 May 2014).

49. Timonin, A. K., Sokolov, D. D. and Shipunov, A. B. Ordo *Rosales* – Rosales order. Botany. In 4 volumes. Moscow: Publishing Center "Academy," 2009, Vol. 4. Systematics of higher plants. Book 2. [Ed. A.K. Timonin]. 239–241.

50. Phipps, J. B. *Mespilus canescens*, a new *Rosaceous* endemic from Arkansas. Systematic botany. 1990, Vol. 15, №1, 26–32.

51. Opalko, A. I., Kucher, N. M., Opalko O.A. and Chernenko, A. D. Phylogeny and phytogeography pome fruits. Autochthonous and alien plants. The collection of proceedings of the National Dendrological park "Sofiyivka" of NAS of Ukraine. 2012, Vol. 8. 35–44. (in Ukrainian).

52. Opalko, A. I., Andrienko, O. D. and Opalko O.A. *Amelanchier* Medik. at the NDP "Sofiyivka" of the NAS of Ukraine. News Biosphere Reserve "Askania Nova." Vol. 14, Special Issue. 2012, 194–198. (in Ukrainian).

53. International Code of Nomenclature for algae, fungi, and plants (Melbourne Code) adopted by the Eighteenth International Botanical Congress Melbourne, Australia, July 2011, Section 2. Names of families and subfamilies, tribes and subtribes. Chapter III. Nomenclature of taxa according to their rank. Article 19. URL: http://www. iapt-taxon.org/nomen/main.php?page=art19 (Accessed 12 May 2014).

54. Haston, E., Richardson, J. E., Stevens, P. F. et al. The Linear Angiosperm Phylogeny Group (LAPG) III: a linear sequence of the families in APG III. Botanical Journal of the Linnean Society. 2009, Vol. 161, №2, 128–131.

55. Landry, P. Le concept d'espece et la taxonomie du genre *Amelanchier* (Rosacees). Bulletin de la Société Botanique de France. 1975, Vol. 122, №5–6. 243–252.

56. Catalogue of Life: 2014 Annual checklist. Catalogue by Royal Botanical Gardens Kew. URL: http://www.catalogoflife.org/annual-checklist/2014/search/all/key/Amelanchier/match/1 (Accessed 8 May 2014).

57. Campbell, C. S., Wright, W. A. Apomixis, hybridization, and taxonomic complexity in eastern North American *Amelanchier (Rosaceae)*. Folia Geobotanica and Phytotaxonomica. 1996, Vol. 31, №3, 345–354.

58. Wolfe, J. A., Wehr, W. C. Rosaceous *Chamaebatiaria*-like foliage from the Paleogene of western North America. Aliso. 1988, Vol. 12, №1, 177–200.

59. Axelrod, D. I. Evolution of the Madro-tertiary geoflora. Botanical Review. 1958, Vol. 24, №7, 433–509.

60. Takhtajan, A. L. Systema et phylogenia Magnoliophytorum. Moscow-Leningrad: Nauka, 1966, 612 p (In Russian).

61. Sax, K. The origin and relationships of *Pomoideae*. Journal of the *Arnold arboretum.* 1931, Vol. 12, №1, 3–22.

62. Schuster, M., Büttner, R. Chromosome numbers in the *Malus* wild species collection of the genebank Dresden-Pillnitz. Genetic resources and crop evolution. 1995, Vol. 42, №4, 353–361.

63. Darlington, C. D., Moffett, A. A. Primary and secondary chromosome balance in *Pyrus*. Journal of Genetics. 1930, Vol. 22, №2, 129–151.

64. Talent, N., Dickinson, T. A. Polyploidy in *Crataegus* and *Mespilus (Rosaceae, Maloideae)*: evolutionary inferences from flow cytometry of nuclear DNA amounts. Canadian journal of botany. 2005, Vol. 83, №. 10. 1268–1304.

65. *Amelanchier* (Shadbush). Maps. URL: http://eol.org/pages/29970/maps (Accessed 19 May 2014).

66. Takhtajan, A. L. Floristic regions of the world. University of California Press, 1986, 544 p.

67. Tzvelev, N. N. Genus Juneberry – *Amelanchier* Medik. Flora of Eastern Europe. Volume X. [Ed. N. N. Tsvelev]. St. Petersburg: World and Family, 2001, 552–555 (In Russian).

68. Wu, Z., Raven, P. H., Hong, D. et al. *Amelanchier* Medikus. Flora of China. Vol. 9: Pittosporaceae through Connaraceae. Missouri Botanical Garden Press. 2003, 190.

69. Country statistics. European network on invasive alien species (NOBANIS). Gateway to information on alien and invasive species in North and Central Europe. URL: http://www.nobanis.org/Search.asp (Accessed 20 May 2014).

70. Mosyakin, S. L., Fedoronchuk, M. M. Vascular plants of Ukraine: a nomenclatural checklist. [Ed. S. L. Mosyakin]. Kiev: National Academy of Sciences of Ukraine 1999, 286.

71. Dobrochaeva, D. N., Kotov, M. I., Prokudin, Y. N. et al. Genus. Saskatoon (Juneberry) – *Amelanchier* Medik. Determinant of higher plants of Ukraine. Kyiv: Naukova Dumka, 1987, 159–160 (In Russian).

72. Bilyk, O. V., Vegera, L. V., Dzhym, M. M. et al. Catalogue of the Dendrological Park "Sofiyivka" Plants. Uman. 2000, 94. (in Ukrainian).

PART III

ECOLOGICAL PECULIARITIES OF THE FOOTHILLS OF THE NORTHEN CAUCASUS: CYTOGENETIC ANOMALIES OF THE LOCAL HUMAN POPULATION

CHAPTER 12

SOURCES OF FRESH AND MINERAL WATER IN NORTH OSSETIA—ALANIA

MARGARITA E. DZODZIKOVA

CONTENTS

ABSTRACT

According to the Federal State Institution "Water Centre" Republic of North Ossetia-Alania, Territorial Division of Water Resources of the Republic of North Ossetia-Alania and the results of our field research conducted a comprehensive analysis of moisture in the Republic of North Ossetia Alania. Revealed that the freshwater sufficient to ensure the needs of the population of and commercial facilities in the Republic of household drinking water. Impressive reserves of drinking and mineral water Alagir area (46.4% of all stocks in the country), namely the territories relevant to the reserve, and this despite the fact that the territory of North Ossetian State Nature Reserve is only 12.86% of the area of North Ossetia.

12.1 INTRODUCTION

The Republic of North Ossetia-Alania is situated on the northern depression of the Main Caucasian ridge and is part of the Central Caucasus and Eastern Ciscaucasia between 42° 38′ – 43° 50′ northern latitude and 43° 25′ – 44° 57′ east longitude. Its length from north to south is 125 km and from west to east is 120 km. Its area is 7971 km², of which 4121 sq. km are on the plane. The population of the Republic, according to the State statistics Committee of Russia, is 706.1 thousand people (data as of 2013), and the capital city of Vladikavkaz – 350 thousand people. The Republic of North Ossetia-Alania is one of the most densely populated regions of the Russian Federation and takes on this indicator 5th place (after Moscow, St. Petersburg, Moscow region, and Republic of Ingushetia) [1]. The real density of population in places of residence of the main part is over 140 people/km². More than half – 56% of the population lives in Vladikavkaz.

North Ossetia is on the same parallel with Bulgaria, Central Italy and southern France. To the north it borders with the Stavropol region, in the west with the Kabardino-Balkar Republic, on the east with the Chechen and Ingush republics, south of the border goes along the watershed of the Main Caucasian range and borders with the Republic of South Ossetia and with Georgia. The main Caucasian ridge delays the entry to the Republic of North Ossetia-Alania warm and moist air masses from the Black Sea. The impact on the thermal regime of the water balance of the basin of the Caspian Sea, due to its great distance, slightly. The weak impact of the seas caused continental climate with moderately hot and long summer. The vegetation period lasts from may to October inclusive [2].

The mountainous part of North Ossetia-Alania consists of five ridges stretching from north-west to south-east parallel to each other: Woody, Pasturable, Rocky, Lateral and Main. To the north lie the North-Ossetian sloping plain. Its height above sea level decreases in the direction from south to north from 700 up to 400 m. The northern part of the Republic is the Terek-Kuma lowland, which is separated from the southern part of the Sunzha and Terek ranges. There are large differences in climate and moisture mountain and lowland parts of North Ossetia-Alania. In the north of the Republic of features of the continental climate are manifested most strongly. Here is the biggest absolute annual amplitude of temperatures (76°C), absolute lowest winter temperature (−34°C) and maximum summer (+42°C). The area is characterized by a small number of annual rainfall, frequent dry winds and drought. However, the low temperature in the area is rare in the winter, as a rule, is soft, the summer is hot and long [3].

The mild climate typical for the North-Ossetian sloping plain. Average temperature of January is −4.5°C, in July +20.1°C. Rainfall for the year 600–700 mm. In the mountains cool summer, long and cold winter, less amplitude of fluctuations of temperatures, abundant rainfall [4].

For historical, orographic, soil and climatic conditions of the territory of North Ossetia-Alania is divided into two distinct areas: the mountainous and flat. In turn, these districts are divided into sub-districts. Today plain and Ossetian artesian pool was formed in after Jurassic period (approximately 25–30 thousand years ago) on the basis of Vladikavkaz depression (deflection). Boulder-pebble deposits fluvioglacial origin was covered by sediments of the Terek river and its many tributaries, river bed gradually moved [5].

The purpose of the study was the analysis of security of the Republic of North Ossetia-Alania drinking and mineral (medicinal and table) water and water for domestic purposes.

12.2 MATERIALS AND METHODOLOGY

Apart from literary data [6] were used operational reports of the Centre of water resources of the Republic of North Ossetia-Alania ("Center for the study, use and protection of water resources of the Republic of North Ossetia-Alania – URL: http://comready.ru/company/4534140) and the results of their own field research.

12.3   RESULTS AND DISCUSSION

Within Ossetian artesian basin in the thickness of Quaternary deposits, lies a powerful horizon of underground waters. The power of this horizon occurs mainly in the southern part of the plain, at the expense infiltrate from river beds of the waters. Nutrition also participate rain falling within this plain. The unloading of the horizon is in the northern part of her within tracts Becan and Tuaca.

The depth of groundwater in the area of the cities of Vladikavkaz and Alagir is from 70 to 110 m in the area, Beslan – 50 m, to the north from the village Humalag reduced to zero. The capacity of the aquifer in the southern part of the plain of about 100 m, and in the district, Beslan exceed 180 m. Consumption spring water in the discharge zone is estimated at 20–25 m^3/sec. (Becan 15 m^3/sec.; Tuaca 5–10 m^3/sec). Deposits and lots of fresh underground waters in territory RNO-Alania with approved reserves are shown in Table 12.1.

The number of freshwater springs and their total output in the fourth hydrogeological area are shown in Table 12.2.

Arrangement of freshwater springs on the territory of North Ossetia-Alania Republic is represented on the map in Figure 12.1.

List of natural mineral springs of North Ossetia-Alania shown in Table 12.3.

Deposits of mineral waters used are shown in Table 12.4.

Arrangement of sources of mineral water North Ossetia-Alania Republic is represented on the map (Figure 12.2).

The complex analysis of data on security with water resources of all settlements RNO-Alania showed that expected operational resources of underground fresh waters are sufficient for providing both current, and perspective requirement of the population, economic and technical objects for economic drinking water that doesn't contradict earlier published data [7, 8].

The total amount of expected operational resources on the republic is estimated at 803 million m^3/year. Needs for economic drinking water make about 240 million m^3/year. The specified requirement is satisfied due to water selection on fields of underground fresh waters with explored

TABLE 12.1 Fields of Fresh Underground Waters in the Territory RNO-Alania

Water utility	Name fields and sectors of fresh groundwater	The reserves, thousand m³/day	Year of approval stocks
HDW	Ordzhonikidzevsky (total) including sites:	527.60	1985
IW	Redant 1	210.0	1985
	Balta	140.0	1985
	Redant 2	13.8	1985
	Dlinnodolinsky	14.8	1985
	Chmiysky	80.0	1985
	Yuzgny	70.0	1985
HDW	Alagirskoe	25.0	1986
HDW	Tamiskskoe	4.1	1972
HDW	Gizeldonskaya	30.8	1989
	Abstraction borehole	25.0	1989
	3 Capturing and springs	5.8	1989
HDW	Ardon	50.0	1973

TABLE 12.1 Continued

Water utility	Name fields and sectors of fresh groundwater	The reserves, thousand m³/day	Year of approval stocks
HDW + IW	Beslan (total) including sites:	73.7	2008
	Municipal Unitary Enterprise "Production management of water and sanitation"	20.00	2008
	Open Joint Stock Company "Salute"	5.99	2008
	Fayur-Union	8.5	2008
	Limited Liability Company "Vladikavkazsky Food Plant of North Ossetia Consumer Union"	0.061	2008
	Closed Joint Stock Company "Ariana"	0.085	2008
	Company Limited Responsibility "Forward-S"	0.060	2008
HDW	Michurinskoe	4.446	2007
IP	Iraf (total) including sites:	0.98	1981
	Tagar-totors	0.37	1981
IP	Hoskharanrag-Habal	0.12	1981
	Gachina	0.22	1981
	Regah	0.27	1981
WFI	Mikhailovskoe	9.6	1987

TABLE 12.1 Continued

Water utility	Name fields and sectors of fresh groundwater	The reserves, thousand m³/day	Year of approval stocks
WFI	Humallag-Zilginskoe (total)	19.2	1987
	including sites:		
	Humalagsky	6.4	1987
	Zilginsky	12.8	1987
WFI	Levoberegnoye	154.1	1979
WFI	Veselovskoye	80.2	1979
WFI	Kizlyarskoye	25.0	1974
WFI	Kievskoye	61.4	1987
WFI	Vinogradnenskoe	112.2	1970
HDW	North Chermensky	0.0009	2008
HDW	Koban	0.0105	2008
HDW	*Lots 1 and 2 in the Suburb area RSO-A*	28.1	2007
HDW + IW	Mozdok	0.1406	2008

Note. Direct deposits are water font, italics – areas of these deposits. HDW – *household and drinking water purposes*, WFI – water for irrigation (watering), IW – water for industrial purposes, IP – water for irrigation of pastures.

TABLE 12.2 Freshwater Springs and Their Total Production Rate

Administrative region	Number springs	Non springs	Total flow rate, l/s	Non-springs used for water supply
Alagirsky	83	31–92, 111, 116–118, 119–124, 129–136, 147	837	30, 32, 34, 54, 56, 92, 111
Ardonsky	8	103–110	3733	—
Digorsky	2	29–30	45	—
Irafsky	30	1–28	1014	1, 2, 14
Kirovsky	4	99–102	7	—
Pravoberegny	6	93–98	7	93,95,96
Prigorodny	37	112–115, 125–128, 137–160, 170–174	527	112, 113, 125, 126, 173
Vladikavkaz	9	161–169	852	161, 163, 166, 168
Total	179	—	7022	

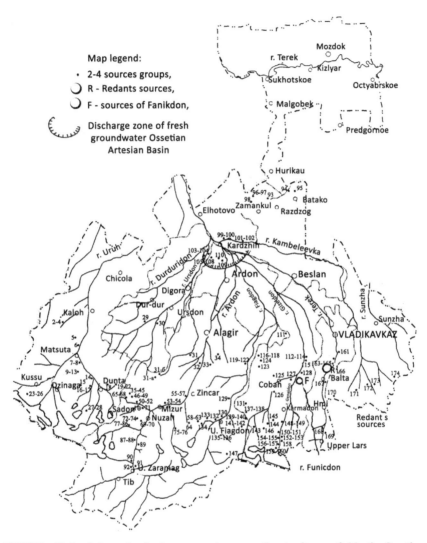

FIGURE 12.1 Schematic freshwater springs on the territory of North Ossetia-Alania. Scale 1:570–000 (Symbols: (i) 2–4 – group of springs, (ii) P – Redantskie springs, (iii) Φ – spring Fanykdon, (iv) Discharge zone of fresh groundwater basin Ossetian atrezianskogo).

TABLE 12.3 Natural Mineral Waters North Ossetia-Alania

Names of mineral springs	Type of water salt composition	Mineralization (M g/L)	Output (Q l/s)
Kartasuar	Carbonic	1.2–2.8	0.3–2.5
Zgil	Bicarbonate-calcium		
Kalak	Sodium		
Kamskho	Sodium		
Dvuhgolovi	Sodium		
Lisri	Sodium		
Halatsa	Carbonate-calcium-sodium	1.8–2.0	0.1–0.5
Abana	Carbonic hydro-carbonate-sodium-iron-magnesium	1.8–2.0	0.3–0.5
Tib 1	Carbonic hydro-carbonate-calcium-magnesium	1.0–1.6	0.1–0.5
Bubu	Carbonate-hydro-carbonate calcium	1.2–1.8	0.2–0.4
Kaliat	Carbonate-hydro-carbonate calcium	0.9–1.4	0.1–0.2
Kudzahta	Carbonate-hydro-carbonate calcium	1,4–1,6	0.5-l.5
Zaramag	Carbonate-hydro-carbonate calcium	5–6.5	0.5–1.0
Lyakau	Carbonate-hydro-carbonate calcium	3.2–3.8	0.1–0.2
Tapankau	Carbonate-hydro-carbonate calcium	2.8–3.1	0.05–00.1
Gurkumta	Carbonate-hydro-carbonate calcium	1.8–2.8	0.01–0.05
Narski	Carbonate-hydro-carbonate calcium	2–2.2	0,01–0.05
Hasievsky (Zrug)	Carbonate-hydro-carbonate calcium, magnesium	9.4	0.6–1.2
Ginat (Zrug)	Carbonate-hydro-carbonate calcium, magnesium	6.8–8.2	0.01–0.02

TABLE 12.3 Continued

Names of mineral springs	Type of water salt composition	Mineralization (M g/L)	Output (Q l/s)
Zacka	Carbonate-hydro-carbonate calcium	5–5.5	60.1–0.2
Zintsar	Chloride-sulfate	7.6–8.8	0.01–0.02
Kalotikau (Hilak)	Carbonate-chloride-bicarbonate-sodium, iron, boron, silicon	5–6.0	0.5–0.6
Suar 1	Sodium chloride	11.7	0,005
Zilahar	Sodium chloride	14.6	0.005
Tamisk	Hydrogen sulfide-sulfide-calcium-magnesium	0.5–2.5	7.5–20.0
Humesidon	Carbonate-bicarbonate-chloride-sodium	1.5	0.05
Tanadon	Hydro-bromo-iodine	4.5	0.3
Koltasuar	Carbonate-bicarbonate-chloride-sodium	2.7	0.01–0.03
Haznidon	Sodium chloride	4.6–6.1	0.01
Karidon	Chloride-sodium-sulfate-calcium	8–12.0	8–9.0
Skottat	Chloride-sodium-sulfate-calcium	6.5–6.8	0.05
Masota	Carbonate-bicarbonate-chloride-calcium	1.2–1.3	0.1
Upper Karmadon	Carbonate-chloride-sodium	0.6–8.5 $t = 10$–60 C^0	6.0
Unalsky	Sodium chloride	2.1	0.1
Suargom	Sulphate-bicarbonate	1.9	0.2
Chmi	Sulphate-bicarbonate	2.5	0.05–0.1
Kesatikau	Carbonate-bicarbonate	4.5–4.8	0.1–0.6
Abaytikau	Carbonate-bicarbonate	2.8–3.2	0.2–0.3
Dzinaga	Carbonate-bicarbonate	1.8–2.4	0.05–0.06

TABLE 12.4   Deposits Used Mineral Water

Number map Figure 12.2	Name field (plot)	Inventories m³/day	Use of water
1	Lower-Karmadon	2200	water treatment
2	Upper Karmadon	178	water treatment
3	Tamisskoe	492	water treatment
4	Zaramagskiye	55	water treatment
5a	Tibskoe – Tib-1	97	water treatment
5b	Tibskoe – Tib-2	84	water treatment, industrial bottling
6	Redantskoe	704	water treatment
7	Korinskoe	630	water treatment
8	Zamankulskoe, wells 2–3 and 8–3	34	industrial bottling
9	Zamankulskoe new wells in 1991	250	industrial bottling
10	Razdolnoe plot, well 1-M	1081	water treatment
11	Plot Biragzang (well 1 BT)	484	water treatment
12	Tsemzavod plot, well 3-T	1080	industrial bottling
13	Plot Tib source 7–56, "Fatima"	250	industrial bottling

reserves (about 63% of the general water selection) and on sites of water intakes with unconfirmed stocks.

In the territory of North Ossetia operational stocks on 29 fields of drinking water of 1678.94 thousand m³/days are reconnoitered and approved, from them 13 fields are intended for economic and drinking water supply. In total for economic and drinking water supply it is reconnoitered stocks in number of 1086.08 thousand m³/days, for production water supply – 129 thousand m³/days. Other stocks are intended for an irrigation of lands and flood of pastures.

Besides the reconnoitered fields in the republic for economic and drinking water supply and other purposes the borehole water intakes located on sites with not explored reserves are used. The general water selection in the republic in 2011 made 503.35 thousand m³/days or 183.7 million m³/year

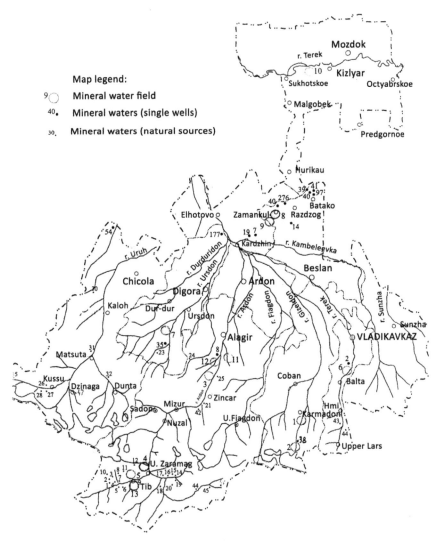

FIGURE 12.2 Layout sources deposits mineral waters North Ossetia-Alania. Scale 1:570–000 (Symbols: 1 – mineral water, 2 – single wells, which revealed mineral water, 3 – natural mineral waters).

that makes about 23% of the total amount of operational resources of the territory RNO-Alania.

The North Ossetian National Natural Park is located in the Alagirsky area RNO-Alania on a northern slope of Big Caucasian ridge within heights of 650–4646 m above sea level. From the North to the south the reserve in whole or in part includes the ridges of the Caucasus: Foothill, Woody, Pasturable, Rocky, Lateral and Main. To NONNP territory, 592 rivers, with a general extent of 831.5 km are related, 76 glaciers with a total area about 37 sq.km are registered. The Tseysky glacier, the largest in the reserve, has the area of 9.7 sq.km and length of 8.6 km. It comes to an end at the height of 2–300 m, in a forest zone. From glaciers the reserve rivers – Ardon (biggest), Arkhondon, Baddon, Bugultydon, Sadon, Fiagdon, Tsazhiudon, and others originate. There are also some lakes – Tsazhiutsad and others. There is a small mineral lake at the settlement Zgil [9].

In the territory of NONNP there is a large number a little studied, but springs used by the population (Figures 12.3 and 12.4).

List of mineral springs territories Alagirsky region studied with healing properties is shown in Table 12.5.

List of mineral springs territories Alagirsky region with poorly known healing properties is shown in Table 12.6.

FIGURE 12.3   Springs drinking water on the left – on the left bank of the Ardon, before entering the first tunnel Transkam (about town Chyramad); Right – on the left bank of the river Fiagdon, just above her left tributary of the river Tsazhiudon (Photo Dzodzikovoy M.E.).

FIGURE 12.4   Tsazhiudon Left River – a tributary of the left Fiagdon. Right – a spring of fresh water, 5 meters left of the river Tsazhiudon (Photo Dzodzikovoy M.E.).

Thus, from 179 springs of fresh water registered in the territory RNO-Alania (see. Table 12.2), the most part – 83 springs, that is 46.4% are on territories of the Alagirsky area, goes further Suburban – 37 (20.7%), Irafsky – 30 (16.8%), Vladikavkaz – 9 (5.02%), the Ardonsky area – 8 (4.5%), Right-bank – 6 (3.4%), Kirovsky – 4 (2.2%) and least of all in the Digor area – 2 (1.1%).

As officially used, it is registered 39 natural (see the Table 12.3) sources of mineral water, from them 25 (64.1%) fall on the territory of the Alagirsky area (see the Table 12.2). In the territory of NONNP, a security zone and border sites the chemistry of a surface water, dynamics of chemical seasonal changes, influence of waters of various genesis on the frequency of emergence of the induced tumors is comprehensively studied, radiometric measurements of the ruslovykh deposits along coast of the rivers and vicinities of mineral sources [10–12] are conducted. At the same time on lands of the Reserve there are more than 70 mineral sources (see the Table 12.5) from which it is studied and only 10%, curative properties of 45 registered sources (are used see the Table 12.6) are insufficiently studied [13–15].

TABLE 12.5 Mineral Springs Territories Alagirsky Region (by Dontsov V.I. and Tsogoev V.B. [6])

Name and chemical characteristics of the source	Belonging to the River Basin	Territorial identity, the name of the gorge	Mineralization of water, Q g/L
Healing mineral springs			
"Tib-1," carbonate-bicarbonate-sodium-calcium medical-table water	Ardon	Alagirskoe	4.16
"Tib-2," bicarbonate, magnesium-calcium, medical-table water	Ardon	Alagirskoe	1.38
"Zaramag" chloride-bicarbonate-sodium, medical-table water	Ardon	Alagirskoe	5.98
"Tsey" sodium chloride, medical-table water	Ardon	Alagirskoe	5.52
"Hilak" medical-table, boric sredneuglekislaya, ferruginous water of low mineralization	Ardon	Alagirskoe	1.79
"Biragzang" medical-table water. GOST 13273–88	Ardon	Alagirskoe	2.1
"Sadon" medical-table, sulfate-sodium bicarbonate water	Ardon	Alagirskoe	0.71
"Tamisk" medical-table water	Ardon	Alagirskoe	1.08

TABLE 12.6 Mineral Springs Territories Alagirsky Region (by Dontsov V.I. and Tsogoev V.B. [6])

Name and chemical characteristics of the source	Belonging to the River Basin	Territorial identity, the name of the gorge	Mineralization of water, Q g /l
	Mineral water with healing properties insufficiently studied		
Kalotikaudon[1]	Fiagdon	Fiagdonskaya	
Hanikomdon[2]	Fiagdon	Fiagdonskaya	
Zaramagskiye mineral water[3]	Ardon	Alagirskoe	3.5–11.2
Nara (Hasievskie)[4]	Ardon	Alagirskoe	0.5–0.7
Zakkinskie[5]	Ardon	Alagirskoe	0.5–0.75
Gurkumta	Ardon	Alagirskoe	
Tapankau	Ardon	Alagirskoe	
Lyakau	Ardon	Alagirskoe	
Ginat	Ardon	Alagirskoe	
Kartasuar	Mamyshondon	Mamyshonskoe	2.0
Zgil (Kubaladzhy Suar)	Mamyshondon	Mamyshonskoe	1.1–1.2
Kalak	Mamyshondon	Mamyshonskoe	1.1–1.5
Kalak 2	Mamyshondon	Mamyshonskoe	1.6–2.3
Kamskol	Mamyshondon	Mamyshonskoe	1.5–2.8
Kamskho 2	Mamyshondon	Mamyshonskoe	1.6–3.0
Dvuhgolovy	Mamyshondon	Mamyshonskoe	1.6–2.5
Lisri	Mamyshondon	Mamyshonskoe	0.8–3.6

TABLE 12.6 Continued

| | Mineral water with healing properties insufficiently studied | | |
Name and chemical characteristics of the source	Belonging to the River Basin	Territorial identity, the name of the gorge	Mineralization of water, Q g /l
Halatsa contains boric acid (H₃BO₃) and 150 mg/L, and silicon (Si) 23.4 mg/L	Mamyshondon	Mamyshonskoe	
Abana (2 sources)[6]	Mamyshondon	Mamyshonskoe	
Tibskies – 1, 2, 4	Mamyshondon	Mamyshonskoe	8.7-6.4
Bubu	Mamyshondon	Mamyshonskoe	1.2–1.8
Kaliat, carbonic sodium-calcium bicarbonate.	Mamyshondon	Mamyshonskoe	
Kudzahta	Mamyshondon	Mamyshonskoe	1.6.0
Zintsar, sulfate-sodium chloride.	Ardon	Alagirskoe	
Unalsky	Ardon	Alagirskoe	2.1

Note. [1] has 2 outputs the river Bugultadon, [2] follows the northern outskirts of the village Dallagkau [3] has 18 wells, [4] has 6 sources, [5] has 6 sources, [6] carbonated mineral water.

12.4 CONCLUSIONS

The comprehensive analysis of data on security with water of all settlements of the Republic Northern Ossetia-Alania showed:

1. Expected operational resources of fresh waters are sufficient for ensuring requirement of the population and economic objects of the republic in economic drinking water.

2. It is revealed that most of all stocks of drinking and mineral water it is registered in the Alagirsky area, namely in the territories concerning North Ossetian National Natural Park and this with the fact that the territory of the reserve makes only 12.86% of all square RNO-Alania.

3. Profound studying of a chemical composition of water and its therapeutic effect of the North Ossetian National Natural Park could expand recreational opportunities of these territories significantly.

KEYWORDS

- Alagirsky region
- artesian basin
- freshwater springs
- North Ossetian National Natural Park
- underground waters

REFERENCES

1. Official portal of the Republic of North Ossetia-Alania. Administration of the Head of the Republic of North Ossetia-Alania and the Government of the Republic of North Ossetia-Alania. About Republic. Geographical Position. Mode of access: http://www.rso-a.ru/index.php/o-respublike-severnaya-osetiya-alaniya.html. Natural resources: Population (May 26, 2014) (In Russian).

2. Grigorovich, S. F. The mountains and plains of North Ossetia: Satellite tourist, local historian and sightseers. 2nd edition, revised, Ordzhonikidze North Ossetian Book Publishers, 1960, 128 p (In Russian).

3. Beroev, B. M. North Osetii. Moscow. Physical Education and Sports. 1984, 144 p (In Russian).

4. Amirkhanyan, A. M. North Ossetian State Reserve. Ordzhonikidze. 1989, 104 p (In Russian).

5. Kovalev, P. V. Patterns of development of glaciation in the Greater Caucasus. Abstract for the degree of Doctor of Science. Moscow. 1966, 95 p (In Russian).

6. Dontcov, V. I., Tsogoev, V. B. Natural Resources of the Republic of North Ossetia-Alania. Water resources. Vladikavkaz. Projects Press. 2001, 367 p (In Russian).

7. Phallagova, D. M. Mineral waters of North Ossetia and their chemical characteristics. Vladikavkaz. Ir. 1992, 207 p (In Russian).

8. Dzodzikova, M. E. Mineral springs of North Ossetian State Natural Reserve. Proceedings of the scientific practical conference "The role of protected areas in the sustainable development of North Ossetia." Vladikavkaz. 2011, 35–41 (In Russian).

9. Dzodzikova, M. E. Water resources of North Ossetia Reserve, problems and improvement of the ecological situation. Proceedings of the 10th International Congress "Environmental and children." Anapa. Spa Association "Change". 2013, 303–305 (In Russian).

10. Dzodzikova, M. E., A. Pogosyan. Rivers and glaciers of the North Ossetian Nature Reserve. Collection of scientific papers on the 75th anniversary of, B. M. Beroeva "Mountain regions: XXI Century". Vladikavkaz. North Ossetian State University, K. L. Khetagurov. 2011, 175–179 (In Russian).

11. Dzodzikova, M. E., Pavlova, I. G., Gabaraeva, V. M. Effect of treatment of various origins on the incidence of mammary tumors in rats induced by methyl-nitroso-urea. Proceedings of the 7th International Conference "Sustainable development of mountain areas in the context of global change". Vladikavkaz. North-Caucasian Mining and Metallurgical Institute (State Technical University). 2010, 124–125 (In Russian).

12. Dzodzikova, M. E., Tedeyev, T. G. Radiometric measurements of channel deposits in the valley of the rivers and Zymagondon Mamyshondon. Proceedings of the I-th International scientific conference "Regional Development in the 21st Century". Vladikavkaz. North Ossetian State University, K. L. Khetagurov. 2013, 170–172 (In Russian).

13. Dzodzikova, M. E., Gridnev, E. A., A. Pogosyan Water chemistry of the North Ossetian Nature Reserve. Collection of scientific papers on the 75th anniversary of Professor, B. M. Beroeva "Mountain regions: XXI Century". Vladikavkaz. North Ossetian State University, K. L. Khetagurov. 2011, 173–175 (In Russian).

14. Dzodzikova, M. E., Gridnev, E. A., A. Pogosyan Dynamics of changes in water chemistry of some areas of the North Ossetian State Natural Reserve. Bulletin of the International Academy of Ecology and Life Safety. St Petersburg. 2013, T. 18. №4, 56–58 (In Russian).

15. Dzodzikova, M. E., Kabolov, Z. H., Kasabieva, E. E. Water resources of North Ossetia Reserve. Proceedings of the All-Russian scientific-practical conference "Historical and cultural heritage of the peoples of the South of Russia: state and prospects for conservation and development." Grozny. Academy of Sciences of the Chechen Republic. 2009, 143–145 (In Russian).

CHAPTER 13

INTRODUCTION OF CLOVER SPECIES (*TRIFOLIUM* L.) IN THE NORTH CAUCASUS

SARRA A. BEKUZAROVA, LIDIA B. SOKOLOVA, and
IRINA T. SAMOVA

CONTENTS

ABSTRACT

In contrast environmental conditions studied wild species of clover on different heights of mountains (600, 800, 1300, 1600 and 2000 m above sea level), comparing them with the recognized cultivars of red clover cultivars Daryal. Based on 45-years-olds of have been established the biological characteristics and prospects their as a starting material for breeding. Created valuable source material and formed new adaptive to mountain conditions cultivars Farn, Alan, Iriston 1, and Iriston 7.

13.1   INTRODUCTION

In the evolution of the family arose legumes and biological diversity of its species growing on the slopes of the North Caucasus. Gene banks plant communities remain the most adapted properties that determine their resistance to adverse factors. Adapted to stress conditions species have valuable economic traits acquired during evolution. This is primarily their ability to accumulate in the soil of organic nitrogen oxide owing to the presence of nodule bacteria on the roots.

Following the principles of preparation of the starting material in the creation of cultivars method of introduction, initial assessment of the samples was carried out in wild places growing at altitudes of 600–2000 m above sea level.

In the evolution of plants played a major role and the introduction – as old as the ancient agriculture. Introduction of plants humanity began to deal with the transition time from collection to growing plants.

Modern diversity of cultivated plants was the result of the introduction of plants for thousands of years passing [1, 2].

However, given the long period of the method of introduction, it should be noted that her search continues, and it is not an independent science. This important research industry is at the crossroads of botanical knowledge and practices of cultivation of plants. In most cases, success is determined by the introduction of soil and climatic conditions where the plant is introduced into the culture. Certain elements are common introduction directly to plant breeding [3].

The role of plant introduction at the present stage of its development is quite versatile. This is primarily determined by the botanical science and experimental studies of agricultural science, in particular using the selection methods. This is one of the methods for the study of plants is natural habitats (ex situ), which in recent years has given special priority in the program of conservation of plant diversity.

Introduction of plants has its own methods: selection of introduced species in phytocoenosis (plant community), introductory test, and determination of the degree of adaptation of introduced species. The process itself consists of several stages, the main ones are: search of introduction, primary and secondary introductory test [4].

Introductions objects are all vegetable organisms of our planet.

History of the development of perennial grasses, legumes in particular, shows a direct link phylogeny and the introduction of high-yielding cultivars on delivery.

In particular, according to some researchers [5–7] sowing clover in Russia began to develop in 1766 on the initiative of the Free Economic Society. As a result of the introduction of long bean plants were obtained new high-yield cultivars.

In the North Caucasus field grass cultivation began in the 80s of the XIX century on the initiative of Ardasenova [6]. In North Ossetia, the main work on the study, collection and introduction of wild species began in the 30s under the leadership of Vavilov [8]. Great contribution to the development of sowing clover in North Ossetia have [6].

In previously conducted studies did not fully address the factors of longevity, hardiness, high regrow capacity of plants, disease resistance, seed production.

13.1.1 SOIL AND CLIMATIC CONDITIONS OF THE NORTH CAUCASUS

North Caucasus is characterized by a large variety of environmental conditions. On soil and climatic conditions of the area is divided into five zones extending parallel to the Greater Caucasus Range. They are arranged in a sequence from north to south:

1. Steppe zone on chestnut soils, insufficient moisture;
2. Steppe zone on chernozem soils, unstable moistening;
3. Forest-steppe zone of sufficient moisture on leached chernozem;
4. Preforest zone of excess moisture in sod-podzolic soils;
5. Mountainous area on a mountain meadow soils.

Species of clever are grown in the second, third and fourth zones. The fifth mountain zone is used for overseeding clover meadows and pastures.

Hillsides, delaying the flow of air masses, are forcing them climb up, thereby the temperature of the air mass decreases rapidly, and moisture saturation increases, what leads to precipitation. The distribution of annual precipitation is extremely uneven. It ranges from 250 to 2500 mm. The smallest amount of precipitation in the north-east and east. For long-term data, the average January temperature in the whole Ciscaucasia ranges +2–5°C, in the mountain zone to a height of 2000 m the average temperature is +5–9°C. The average July temperature in West Ciscaucasia is +23–24°C, East Ciscaucasia +25–29°C, in the mountainous area, it reaches only +12–15°C.

However, in the mountainous area observed an unusual pattern: in summer climb to 100 m accompanied by a decrease in average temperature 0.5–0.6°C, while in winter more often the opposite is true.

According to the annual sum and monthly average rainfall in the course of Ciscaucasia is important not only vertical zones, but also relief, exposure, and a number of other factors that make a great variety of areas on the annual amount of precipitation. As at the storage temperature inversion precipitation is observed here.

Mountainous area is 1000–1500 mm of rain per year. With a height of rainfall increases, and at an altitude of 2000 m can reach 2500 mm per year. In the lower zone of the mountains, from 1000 to 2000 m, rainfall is less, but still enough for germination of clover. Because of the abundance of rain in the highlands during the winter accumulate a large amount of snow that causes avalanches. Winter formed a steady snow cover height on average 50–75 cm Snow held from mid October until the first half of May. Length of the frost-free period at an altitude of 2000 m is 100 days.

The vertical zonation of mountain chernozems replaced by meadow-steppe soils that stretch marks between 2000–2700 meters above sea level and is a transitional link to the mountain meadow.

Elevations spread these soils undergo significant variations depending on the degree of exposure, topography, steep slopes and other natural conditions of the area.

Environmental analysis, taking into account environmental factors different mountain heights enables reveal mechanisms that determine the dependence of phenotypic diversity of legumes on environmental factors, to establish their influence on some signs.

13.2   MATERIAL AND METHODOLOGY

Our studies were conducted on different soils: leached chernozem (altitude 600 m), meadow-chernozem (800–900 m), leached sod in varying degrees ashed (900–1000 m), dark brown (1300–1600 m), and mountain meadow (2000 m).

To elucidate the mechanisms of adaptation of introduced plants to extreme mountain conditions, and the selection of promising plants to produce cultivars and their practical use in these conditions evaluated material model ecological genetic experiment in which samples are tested in different contrast conditions at altitudes of 600, 800, 1300, 1600 and 2000 m.

On different mountain altitudes every 200 m horizontally measured air temperature at a height of 1–2 m from the surface of the soil and soil-ground temperature at a depth of 0.1–1 m. In positive air temperatures and root layer carried along seeding promising cultivars of clover for seed – seed plots.

13.3   RESULTS AND DISCUSSION

In today's complex environmental conditions, the productivity of mountain pastures and hayfields unstable and low and fully depends on many factors: weather, anthropogenic, zoogenic, man-made and other As a result, most of the valuable forage and medicinal herbs falls, reduced their biomass and abundance, and the low-eaten and poisonous plants predominate. Besides, destroyed grass, sloping land degradation, erosion intensified.

Such a state of natural grassland requires the use of a number of measures to increase productivity and save valuable gene pool feed, food and medicinal herbs. One of those activities that contribute to the increase of forage natural lands is overseeding of herbs with high adaptive properties adapted to data mining conditions. Therefore, on the basis of extant species must perform the evaluation, selection and reproduction with the aim of re-introduction in degraded pastures and biodiversity conservation.

Study species of legumes, especially clover, mountain grasslands and pastures, the definition of the size of their habitat, adaptive traits, economic and biological evaluation of the vertical zones and reproduction of individuals are a major problem in conservation genetic phytocenoses mountain and on their basis, the creation of new cultivars with high adaptive properties.

In order to create a starting material for the formation of new cultivars studied wild species of clover, isolated from natural ecosystems of the North

Caucasus mountain areas at altitudes of 800–2000 m above sea level. These include species of clover: *Trifolium alpestre* L., *Trifolium ambiguum* Bieb., Trifolium canescens Willd., *Trifolium hybridum* L., *Trifolium pratense* L., *Trifolium repens* L., and *Trifolium trichocephalum* Bieb.

During the growing season phenological observations were carried out for the development of plants. Take into account the economic and biological features of each of the studied species within 4 to 5 years of life.

All indicators compared to the zoned grade of a clover Daryal who is created by a method of artificial hybridization on the basis of the introduced samples from the Southern Sakhalin (a fatherly form) and Yugoslavia (a maternal form). A characteristic feature is its regionalized cultivars longevity, stable data and fodder seed productivity, and speed of regrowth after cutting, high winter hardiness, disease resistance. All samples were studied in the collection nursery on every mountain top. Area plots was within 2–5 m².

For years of researches (45 years) more than 200 wild-growing samples from which the red clover is presented seventy by forms from different mountain belts were studied. The others are presented by types of a clover: *Trifolium pratense* L., *Trifolium ambiguum* Bieb., *Trifolium hybridum* L., *Trifolium trichocephalum* Bieb., *Trifolium alpestre* L., *Trifolium canescens* Willd. и *Trifolium repens* L. Wild-growing populations were characterized by high winter hardiness, good fodder advantages and exceeded the zoned grade on a crop of green material for 20–60%, on winter hardiness for 3–20% and foliage on 2–6% (Table 13.1).

In our studies of wild clover populations collected in North Ossetia in different places of growth, were also characterized by high levels of the protein content (19.7–21.2%), carotene (3.45–4.50 mg/100 g wet wt), low in fiber (17.2–21.6%). According to these indicators, they were significantly superior to standard – grade Daryal (Table 13.2).

In field experiments in the foothill area at an altitude of 600 m was determined the most important morphological, biological and seed-growing properties of a type of Trifolium pratense. According to our research, the taproot of wild forms is longer and the root collar is submerged in the soil at a shallower depth. The difference in the interstices is noticeable, especially in the first mowing. The height of the stems varies depending on the phase of plant development. Clover cultivar Daryal stooling

TABLE 13.1 Characteristics of Wild Populations of Red Clover for Some Economically Valuable Features

Variety	Hardiness, %	Height, cm	Number of leaves, %	Green weight of 1 plant	
				gr	% of the standard
Daryal – the standard	80	62	57	479	100.0
Dargavsky	100	58	63	618	129.0
Zamankulsky	83	52	60	575	120.0
Urukhsky	98	60	63	725	152.0
Tseysky	90	52	59	610	127.3
Sapitsky	90	56	57	761	158.8

TABLE 13.2 Nutrient Content and Number of Leaves of Red Clover in the Flowering Stage

Place of gathering, altitude	Protein is in dry matter, %	Carotene mg, 100 g of crude material	The fiber in absolutely dry matter, %
Daryal – the standard, 600 m	17.43	2.45	23.18
village Zamankul, 500 m	21.24	3.60	20.16
village Urukh, 600 m	20.23	4.50	19.15
Sapitsky box (Suburb of Vladikavkaz), 800 m	19.70	4.63	21.58
village Dargavs, 1560 m	20.20	3.45	17.18

phase reaches 16–17 cm, 5–7 cm above the wild forms. During flowering stem length is the same in both forms. In the stage flowering number of flowering heads (of flowers) of cultivars was higher on 50% than of the "savages". Growing wild forms have the advantage in number of leaves. However, in the first year of life in some samples number of leaves 5–6% lower, than at a standard grade Daryal.

All wild-growing forms possess higher winter hardiness, exceeding on this indicator the zoned cultivar Daryal. The maximum yield of green mass in the third year of life was observed in similar types of clover and hair of head.

Cultivar on the third year of vegetation reduces their economic and biological signs, yielding parametric wild-growing species. In the course of studying the features of wild plants clover was also revealed that they are much less affected cultural diseases. Most species have a high resistance: similar, grizzled and hair of head whose lesion score was not more than 1–3%.

It was also established that the maximum development of "savages" to reach 4–5 years of life, while red clover cultivars to this period completely disappear.

Topical issue in breeding is to create cultivars, along with having good feeding advantages and high seed productivity. This figure is especially necessary for exotic species used in hayfields and pastures. We determined that the amount of the produced seeds is fully dependent on climatic factors. The correlation coefficient (r) with a probability of 99.9% is 0.7–0.9.

Revealed that the seed production of the species studied has significant differences (Table 13.3).

Table 13.3 data indicate that fluctuations in the number of seeds formed rather high, especially in the species *Trifolium alpestre*, *Trifolium*

TABLE 13.3 Seed Production Clover on the Third Year of Cultivation

Species	Average of heads, pieces	Average of flowers, pieces	Formed seeds, % (from-to)	Coefficient of variation V, %
сорт Дарьял— стандарт	64	86	32–42	15,8
Trifolium pratense	76	108	25–45	16.4
Trifolium ambiguum	57	70	28–56	18.1
Trifolium canescens	32	62	17–26	13.2
Trifolium hybridum	72	89	21–43	17.6
Trifolium repens	68	72	45–50	12.6
Trifolium alpestre	26	117	15–41	20.1
Trifolium trichocephalum	64	76	26–48	18.4

ambiguum and Trifolium trichocephalum (18–20%). By most different kinds of heads Trifolium pratense, Trifolium hybridum, Trifolium repens. Less than other types of generative organs form clover Trifolium ambiguum and Trifolium repens, because of their biological characteristic form from 2 to 8 seeds in one ovary.

We found that with increasing mountain height is reduced leg length, number of leaves and increased contamination of inflorescences. At an altitude of 2000 m clover plants less sick anthracnose, lower leg length, but the number of internodes increased from 5–6 to 7–9.

It is known [9, 10] that wild-growing samples are characterized by big diversity of population on economic and biological signs, including on the phenological. This makes it possible to conduct screening in populations in the direction of reducing the growing season (ripening), and in the late-side. In the study of the influence of anthropogenic factors have been identified that isolated areas seed production is much higher than under the influence of anthropogenic factors (Table 13.4).

Unlike other types of clover more undergoes a change of environment.

Our supervision over feature of blossoming and formation of seeds (antekologiya) confirms that under the influence of stressful factors the blossoming cycle is broken, the mass of each plant decreases, the quantity of puny seeds increases. However, existence of puny seeds testifies not only to lack of pollinators at the time of blossoming, but also to influence of weather conditions.

TABLE 13.4 Seed Production Clover Species in Plant Communities of Mountain Meadows

Species	Weight is 1000 pieces of seeds		Number of leaves, %	
	anthropogenous influence	the isolated site	anthropogenous influence	the isolated site
Trifolium pratense	1.42	1.86	36.5	44.5
Trifolium ambiguum	1.68	1.92	31.4	64.8
Trifolium hybridum	0.72	0.88	36.0	66.8
Trifolium alpestre	1.28	2.12	32.9	41.5
Trifolium canescens	1.62	2.22	26.4	52.1
Trifolium repens	0.48	0.62	12.3	45.0

The given results of researches allow concluding that the introduced plants resume development and seed efficiency. The studied, selected and multiplied plants fill up a biodiversity of the degraded site when.

At sufficient genetic richness of resources of introduced wild types and existence of ekologo-geographical localization of various populations on genetic structure in the nature the choice of the best plants from them often was the key to success of further selection work.

According to the content of irreplaceable amino acids wild-growing samples for the fifth year of life didn't reduce indicators. The quantity them decreased according to a phase of development of plants (stooling phase – budding – flowering).

In Table 13.5 average data on amount of irreplaceable amino acids are provided in green material of wild-growing samples of a clover meadow depending on a phase of their development, a place of growth and year of vegetation.

The data show that the greatest amount of amino acids was found in the phase stooling. In the future, the development of plants, the amino acid content decreased.

In conditions of foothill zone on the fourth or fifth year of vegetation "savage" of amino acids in the green mass did not decrease, but rather increased. In the flowering stage, this figure was higher than in the budding stage. In addition, the wild form, resettled from the mountain (1600 m) in the foothills (600 m) area, marked increase in the content of amino acids with plant age (correlation coefficient – 0.72 in the budding phase and 0.68 in the flowering stage).

Valuable feature for breeding of wild species of clover is their productive longevity. According to our research, crops of clover in the third year of vegetation decreased, and the fourth – all cease to exist, while the wild species in this period reached its greatest development.

On cultural soils (in collectible nurseries) several wild plants change their signs. However, signs remain stable high foliage, high dry matter yield and longevity. At an altitude of 600 m above sea level (Experimental industrial enterprise "Mikhailovskoe") wild forms of the 2nd year of life inferior to standard yield of green mass by 32–63% and surpassed it by the number of leaves on a 9–27% yield and dry matter – on 1,2–5,2%. The same samples, at an altitude of 800 m (Tarskaya hollow) also

TABLE 13.5 Amount of Essential Amino Acids, Expressed in Grams per 100 g of Absolute Dry Matter, in the Green Mass of Wild Clover Grown Up in the Foothill Zone – 600 m and in the Mountain Zone – 1600 m a.s.l

Development phase	Foothill zone					Mountain zone				
	Years of life									
	1	2	3	4	5	1	2	3	4	5
Wild-growing										
Stooling phase	9.60	6.13	8.46	9.12	9.28	9.78	9.26	10.02	10.48	11.06
Budding	7.82	4.72	7.12	7.84	7.18	8.22	7.52	8.40	9.45	9.82
Flowering	8.36	4.80	7.00	8.32	7.94	7.26	8.22	8.38	9.02	9.42
Variety Daryal										
Stooling phase	9.35	6.00	6.46	–	–	8.42	7.06	7.12	–	–
Budding	9.58	5.00	5.98	–	–	7.26	5.86	6.86	–	–
Flowering	7.81	4.59	4.82	–	–	6.92	5.46	5.65	–	–

yielded cultural variety in yield of green mass and surpasses its foliage on 2,2–21,8% and dry matter – at 1.5–4%.

Completely opposite indicators on yield of green mass had samples at 2000 meters above sea level, on the cultural soil sown in nurseries collection. Here are some "savages" significantly (80–93%) were superior cultivars and native sample (Table 13.6).

Comparing grade Vladikavkaz on 3 altitudes (Table 13.7), it should be noted that the yield per unit area at an altitude of 2000 m is 2.5 times lower than in the foothill area at an altitude of 800 m studied wild forms, on the contrary, with the rise heights exceed the standard and native not only forage crop, but also on the grounds of foliage, dry matter basis.

In this connection, in the evaluation samples of wild considered an important indicator of isoflavone content. Chemotaxonomic study genus *Trifolium*, showed that phytoestrogens are 100 kinds (from 300). Identified formononetin, biochanin A, gene stein as glucosides and their quantitative content. Determined that the maximum amount of isoflavones is at the base of the root after overwintering. [6]

We also found that the content of estrogenic isoflavones in the green mass of the samples studied depended on the variety, gathering places on the vertical zonation of mountain areas (Table 13.8).

From Table 13.8 it can be concluded that wild specimens growing in places have a higher content of biochanin A, formononetin, kumesteron and genistein than regionalized grade.

Carried out and the overall biochemical assessment of collection of wild plants in nurseries (Table 13.9).

Table 13.9 shows that wild samples collected from various heights under the cultivated soil had a higher dry matter content, crude protein and amino acids. Contents estrogenic substances under these conditions (600 m above sea level) on 5.7–23.8 mg higher than cultivar Daryal.

Consequently, wild specimens are valuable economic and biological characteristics: a high number of leaves and dry matter content, the maximum amount of nutrients and productive longevity, high winter hardiness and disease resistance. Wild specimens – a valuable source material for breeding.

The vertical zonation pattern Mountains picked clover plant development. With increasing height highlands increased number of leaves

TABLE 13.6 Fodder Yields (kg/m²) Samples of Wild Clover At An Altitude of 2000 m in the Fourth Year of Life

Variety, sample	Crop of the green			
	weight for a total of two mowings, kg/m²	% of the standard	% for the better grade	% of a native (aboriginal) mean
Daryal, standard	2.02	100.0	77.1	86.7
Vladikavkazsky, best grade	2.61	129.0	100.0	112.0
Wild-growing, native	2.33	115.0	89.2	100.0
Wild-growing, the village Sioni village (it is built at the height of 1850 m)	3.90	193.2	149.0	167.4
Wild-growing, the village Hidikus (it is built at the height of 1800 m)	3.64	180.1	139.0	156.2

TABLE 13.7 Comparative Indicators Economically Valuable Wild çclover Samples at Altitudes

Variety, sample	Altitude								
	600 m			800 m			2000 m		
	green mass, kg/ m²	dry matter, %	number of leaves, %	green mass, kg/ m²	dry matter, %	number of leaves, %	green mass, kg/ m²	dry matter, %	number of leaves, %
Vladikavkazsky (standard)	2.5	14.7	40.0	2.7	16.3	54.6	1.09	17.1	56.4
Wild populations									
Sanibansky	1.9	15.1	44.5	2.3	18.3	62.1	1.36	18.7	58.3
Alagirsky	1.2	15.9	49.0	2.1	20.3	76.4	1.56	20.1	60.2
Dargavsky	0.9	19.9	67.0	2.7	21.2	56.8	1.62	19.0	59.0

TABLE 13.8 Contents Estrogenic Isoflavones in the Green Mass of Wild Populations of Red Clover (% on dry matter in 1 g)

Sample	Gathering place, altitude	Phase of development	Biochanin A + formononetin	Coumestans	Genistein
Grade Daryal	Tarsky hollow, 800 m	blossoming beginning	0.34	0.26	0.25
Grade Daryal	Kabardino-Sunzhensky spine, 800 m	full blossoming	0.18	0.17	–
Wild-growing	Kabardino-Sunzhensky spine, 800 m	full blossoming	0.09	0.15	0.09
Wild-growing	Dargavs, 1760 m	full blossoming	0.085	0.12	0.085
Wild-growing	Dargavs, 1760 m	budding	0.23	0.01	0.32

TABLE 13.9 Biochemical Characterization of Samples Clover in Collection Nursery on Altitude of 600 m a.s.l.

Sample	Origin, altitude	Dry matter, %	Protein, %	Amount of amino acids, g/100 g	Sum estrogenic substances, mg/100 g
Daryal – standard	North Ossetia	14.7	19.5	19.41	20.7
Wild-growing	Tarsky hollow, 800 m	15.3	21.4	20.62	33.8
Wild-growing	Alagirsky gorge, 900 m	17.7	19.0	19.78	26.4
Wild-growing	Unal, 1700 m	14.5	18.6	20.63	44.5
Wild-growing	Dargavs, 1760 m	17.4	21.8	21.1	32.3

of plants, seed yield, resistance to anthracnose, and the content of essential amino acids, conversely, reducing the length of the branches, and the content of estrogenic isoflavones.

First implemented in the North Caucasus of introduction experiment involving more than 70 samples of wild species collected from different mountain heights. The regularities of the development of red clover plants, depending on the altitude level of seed origin.

The result of breeding mountain grasslands and pastures is created promising cultivars, Alan, Iriston 1, and Iriston 7. The main advantages of this class: productive longevity (5–6 years), high seed production in all years of life (0.2–0.3 t/ha), winter hardiness (97–98%). This grade has not yet spread widely since received an insufficient number of seeds.

13.4 CONCLUSION

Evolutionary adaptation method phytocenoses wild in the mountains, the selection of the initial forms of the various groups, recurrent selection on years of life and at different stages of development provide a comprehensive assessment of the population with the directed formation of desirable features.

Based on the study of biological and economic features of the wild species, the formation conditions of their generative organs, productivity and quality, depending on environmental factors, the use of effective methods for the selection of exotic species, was created starting material for grades grassland in mountain conditions.

Evaluation of wild clover species in natural ecosystems for seed production allowed carrying out the selection of the most productive forms. Maximize productivity have species *Trifolium ambiguum* Bieb., *Trifolium canescens* Willd., and *Trifolium hybridum* L., which were much higher than in the isolated area zoned grades cultivar Daryal on various parameters within 10–50%.

Environmental assessment of samples at different heights and the selection of promising plants, their pollination and reproduction reduce the breeding process, produce seeds hard-hybrid population, which enable the creation of perennial adapted to mountain conditions highly plastic cultivars.

KEYWORDS

- adaptation
- breeding
- ecology
- nodule bacteria
- phytocoenosis

REFERENCES

1. Sventitskii, I. I., Bashilov, A. M. Theory of biological evolution and ecology the origins of agricultural science. Advances in science and technology of agriculture. 2002, №12, 21–23.
2. Sokolova, L.B, Patoshina, A. N. The influence of environmental conditions on the formation of the firstborn of flowering/Biodiversity Caucasus. Abstracts. Nalchik. 2001, 155–157.
3. Novoselova, A. S., Novoselov, M. Yu., Bekuzarova, S. A. et al. Adaptive selection and variety of new generation for different soil and climatic conditions of Russia. Adaptive fodder: problems and solutions. Collection of scientific papers of Institute of Forages. Moscow. 2002, 271–278.
4. Mirkin, B. M., Naumova, L. G., Khaziakhmetov, R. M. Managing the agroecosystems: ecological aspect. Biology Bulletin Reviews. 2001, Vol. 121. №3, 227–240.
5. Habibov, A. D. Some results of the introduction of species Trifolium, L. in mountain botanical garden. Mining resources of introduction of crop. Makhachkala. 1996, 46–49.
6. Bekuzarova, S. A. Breeding of red clover. Vladikavkaz. Publisher Gorsky Agrarian State University, 2006, 175 p.
7. Hamsutdinov, Z. I., Ionis, Y. I., Pilipko, S. V. Ecological-evolutionary principles breeding arid forage plants. Grassland. 1997, №11, 18–25.
8. Ecological breeding and seed clover. The results of 25 years of research of the creative association "Clover." Moscow. All-Russian Scientific Research Institute named VR feed Williams. 2012, 288 p.
9. Bekuzarova, S. A., Samova, I. T., Kotaeva, M. A. Invention "Method of selection of the initial samples clover." Patent №2464778. Published 27.10–2012. MPK A01H/04.
10. Bekuzarova, S. A., Tsopanova, F. T., Kotaeva, M. A. et al. Invention "Method of forming grassland cultivars of legumes" Patent №236615. Published 10.09.2009. Bull. №25, MPK A01H1/04.

CHAPTER 14

DETOXIFICATION OF SOILS CONTAMINATED WITH HEAVY METALS

GALINA P. KHUBAEVA, SARRA A. BEKUZAROVA, and
KURMAN E. SOKAEV

CONTENTS

ABSTRACT

The chapter presents the results of four years of experiments of soil contamination with heavy metals. Explored ways detoxification of polluted soils, as a result of artificial pollution so as and the contamination of soil with heavy metals from the atmosphere from hazardous industries.

14.1 INTRODUCTION

The soil cover together with its microcosm performs the functions of a universal absorber, destroyer and converter of various contaminants. Despite the protective properties of the soil, there are limits beyond which led to irreversible processes. Therefore, of particular importance is detoxification of soils, for example, restoration of disturbed lands technologically.

Almost all regions are marked with the level of cropland soil pollution by heavy metals. Agrochemical Service of the Russian Federation is monitoring the content in the soil agrochemical parameters, as well as heavy metals, pesticide residues, radionuclides, and has accumulated some data on various soil-climatic zones [1, 2]. Near cities and highways of soil contaminated with lead, zinc, copper, nickel, cadmium. Heavy metals are found in crop production even at relatively low levels of their content in the soil [3].

Entering the soil heavy metals are exposed to various types of transformation, depending on soil properties and biological characteristics of plants. The main factors affecting the mobility of heavy metals in the soil, their transformation and availability of plants considered to be the solubility of heavy metals, the pH of the soil environment, the content of organic matter in the soil particle size distribution and cation exchange capacity, the type and the level of heavy metal contamination of soil, species and biological particular crops. For soil contaminated with heavy metals, methods that reduce their translocation into the plant, based on transfer of cations of heavy metals in the form of poorly available to plants or movable connection with the subsequent leaching of [3]. The most common methods are based on the transfer of metal cations in sedentary when using large doses of organic fertilizers, lime, phosphorites and claying, as well as the use of zeolites [3].

But the translation of heavy metals in mobile connections before they will leach into the underlying horizons, plants can have time to accumulate them in large enough quantities.

A major role in the migration and sorption of heavy metals plays the organic composition of the soil. Organic matter increases the absorption capacity, buffering capacity of the soil, help to reduce the toxic effect of heavy metals, reduce the concentration of salts in the soil solution, reduce the phytotoxicity of multivalent heavy metals and prevent their penetration into plants [4]. The duration of action of making high doses of organic fertilizers is shown on light soils with low absorptive capacity. When light soil remediation as effective reception claying sometimes used, for example, making clays containing aluminum silicates of the montmorillonite type [5]. This method is expensive, difficult to achieve technologically. In recent years, wide-spread use of natural sorbents, such as zeolites. Zeolites exhibit the highest efficiency in highly contaminated soils, reducing the mobility of heavy metals. Actions zeolite increases with manure or various non-traditional fertilizers [5].

Many researchers [6–8] offer such effective technique that reduces the mobility of heavy metals, such as liming of acid soils. The need for liming depends on the structure and chemical composition of the soil in each zone and not productive in relation to neutral and slightly alkaline soil, acidic soils on the mobility of heavy metals above that increases the supply of them in plants.

While there is no consensus on rational method of transformation of heavy metals in the right direction, and there is no way to assess the degree of detoxication of soils for production of environmentally friendly products. Reliable specific recommendations to reduce the availability of heavy metals from contaminated soils into plants so far not been sufficiently developed.

Taking into account the above, the aim of our research was to study the translocation of heavy metals in soil-plant system and methods of detoxification for environmentally friendly products.

14.2 MATERIAL AND METHODOLOGY

Multivariate microfield experience in the study of the translocation of heavy metals in soil-plant system and methods of detoxification was conducted at the experimental field of the North Caucasus Research Institute of mountain and foothill agriculture in 2002–2004, according to the Methodological Instructions of the Central Institute of agrochemical service of agriculture (currently VNIIA) [9]. The experiment was conducted on the ground, which is a low-power leached chernozem in the gravels.

Experience pawned four replications in cellophane vessels without a bottom, size $40 \times 40 \times 30$ cm, the surface area of the vessel – 0.16 м².

To study the basic variants of the experiment at the same polluted background was created artificially by a background application to the soil of heavy metals: $CuSO_4 \cdot 5H_2O$; $ZnSO_4 \cdot 7H_2O$; $Pb(C_2H_3O_2)_2 \cdot 3H_2O$; $CdSO_4 \cdot 8H_2O$; $Ni(NO_3)_2 \cdot 6H_2O$ based on a pure element Cu – 150, Zn – 300, Pb – 100, Cd – 5 и Ni – 100 mg/kg of soil.

Background mineral fertilizers (nitrogen, phosphorus, potassium), agromelioranty, the semi-rerotting manure and heavy metals introduced into the soil separately. The soil was mixed thoroughly. The vessel is placed on the subsurface layer of soil to a depth of 25 cm. Bottom 5 cm were filled with a layer of subsurface area, another 20 cm – the test soil. The upper edge of the vessel was allowed to protrude above the soil. A 5 cm soil during the filling of vessels compacted compaction to avoid shrinkage during the growing season.

In the experience was cultivated potato varieties Vladikavkaz. Planting potatoes produced in the second decade of April.

Years of experiment varied considerably due to meteorological conditions. Rainfall in 2002 was 837 mm, in 2003–722, and in 2004–996 mm.

Perhaps results of experiments are include emissions of nearby steel mills in different years, which contaminate soil and plants [10].

Also provided is a method for evaluating phytoindication soil contamination with heavy metals of industrial origin [10, 11].

14.3   RESULTS AND DISCUSSION

Studies have shown that during the potato vegetation occurred some changes in the content of heavy metals in soil. From the data in Table 14.1 show that the content of heavy metals in the initial soil insignificant. Mineral fertilizers had no significant effect on the amount of metals in the soil. The content of heavy metals in the soil in the control and background options decreased during the growing season of potato about the same.

Artificially created in the soil during the laying of the experience of heavy metal pollution pattern (Cu – 150, Zn – 300, Pb – 100, Cd – 5 and Ni – 100 mg/kg of soil, based on the pure metal) during the potato vegetation

TABLE 14.1 Dynamics of Heavy Metals (HM) in the Soil Under Potatoes (on Average Over 3 years), mg/kg

Treatments	Before planting potatoes					Flowering stage					After cleaning				
	Cu	Zn	Cd	Pb	Ni	Cu	Zn	Cd	Pb	Ni	Cu	Zn	Cd	Pb	Ni
Control	12.9	25.4	0.27	26.2	3.2	14.6	24.7	0.22	24.2	3.0	10.1	20.8	0.23	23.0	1.8
$N_{30}P_{30}K_{30}$ – Background	12.9	25.4	0.27	26.2	3.2	14.7	25.6	0.19	24.6	3.0	10.3	22.2	0.19	24.5	1.5
Artificially created contaminated background															
Background + HM (Cu, Zb, Cd, Pb, Ni)	150	300	5	100	100	106.6	251.2	4.76	97.7	89.0	96.1	236.3	4.6	89.8	78.4
Background + HM + lime (6 t/ha)	150	300	5	100	100	112.3	237.2	4.84	95.7	91.7	99.7	228.4	4.2	89.3	77.9
Background + HM + manure (20 t/ha)	150	300	5	100	100	107.5	240.9	4.82	95.8	89.6	94.3	216.8	4.6	90.4	77.0
Background + HM + lime (6 t/ha) + manure (20 t/ha)	150	300	5	100	100	112.2	198.5	4.85	95.1	87.3	97.5	188.9	4.7	88.3	78.2
Background + HM + irlit 1 (2 t/ha)	150	300	5	100	100	98.7	227.3	4.78	98.5	89.5	76.2	206.6	4.7	94.7	72.0
Background + HM + irlit 7 (2 t/ha)	150	300	5	100	100	98.3	222.6	4.80	92.1	85.3	80.0	210.5	4.6	88.2	67.2
Background + HM + irlits 1+7(1+1 t/ha)	150	300	5	100	100	100.1	224.0	4.77	96.5	88.6	76.2	205.7	4.5	92.6	63.0
$N_{60}P_{60}K_{60}$ + HM	150	300	5	100	100	110.3	249.5	4.75	96.8	86.1	95.5	232.6	4.3	83.9	77.1
MAC						100	150	1	100	100					

Note. MAC – maximum allowable concentrations.

has undergone significant changes, notably and gradually decreasing towards the end of the growing season for potatoes all embodiments of the experiment (Table 14.1), which is apparently due, on the one hand, the accumulation of these metals in leaves and tubers of potatoes during the growing season and, on the other hand, the leaching of their in the lower horizons of the soil.

Application of lime and manure at separate their introduction did not have a significant impact on the content of Cu, Cd and Ni in the soil with respect to the embodiment of Background ($N_{30}P_{30}K_{30}$) + HM, but they are markedly reduced content of Zn and Pb, especially when sharing their introduction. Influence irlits 1 and 7 to reduce the heavy metal content was more pronounced, especially in the versions with the introduction irlit 7. Apparently the high acidity of this natural material (pH 3.8 and hydrolytic acidity 10.8 mEq/100 g) dissolved the harder metals, which contributed to more intense their leaching from the root layer of soil, and resulted in a significant decrease in their content in the topsoil to the flowering stage and harvesting compared to the artificially contaminated with a background in the tab experience.

Studying the behavior of heavy metals in the soil-plant system, determine the size of the income and the removal of them from the harvests of crops is crucial in the development of ways to detoxify the soil and produce environmentally friendly crop production.

The results of studies on the content of heavy metals in potato tubers are shown in Table 14.2.

Should be said about a large difference in the accumulation of heavy metals in the tubers in different years and in different variants of the experiment. Weighted average metal content in the control varied over the years in the range: Cu – 2.2–5.2; Zn – 1.8–5.2; Cd – 0.03–0.04; Pb – 0.1–3.5; Ni – 0.48–1.65 mg/kg. Such variation in partly due to years of research have developed in different weather conditions, and possibly contact with soil and plants of heavy metals from nearby steel plants at different times during the processing of different-quality ore.

Roughly the same pattern is observed in the form $N_{30}P_{30}K_{30}$. Application of mineral fertilizers had no significant effect on the heavy metal content in the tubers. The data show that the control variant (initial soil) concentration of the heavy metals (Pb, Ni) in potato tubers in some years exceeds

TABLE 14.2 Effect of Agrochemicals on the Content of Heavy Metals (HM) in Potato Tubers (Average Over 3 years), mg/kg

Agrochemicals	Cu	Zn	Cd	Pb	Ni
Control	4.13	3.50	0.03	2.06	1.20
$N_{30}P_{30}K_{30}$ – Background	4.23	3.63	0.04	2.17	0.99
Background + HM	5.93	5.27	0.06	4.03	2.52
Background + HM + lime	3.57	4.40	0.035	2.43	1.73
Background + HM + manure	3.70	4.57	0.045	3.03	1.79
Background + HM + lime + manure	3.23	3.67	0.035	1.77	1.56
Background + HM + irlit 1	3.95	4.30	0.02	1.70	2.14
Background + HM + irlit 7	4.40	4.20	0.03	2.80	2.43
Background + HM + irlits 1+7	4.10	3.80	0.02	2.60	2.25
$N_{60}P_{60}K_{60}$ + HM	5.83	5.63	0.065	4.07	3.63
MAC	5	10	0.03	0.5	0.5

Note. MAC – maximum allowable concentrations.

standards of maximum allowable concentrations (MAC). In 2004 Cu concentration in the tubers exceeded the MAC at 1.04 times, Cd in 2002–1.33 times, Pb in 2002 and 2003–5.2 and 7.0 times, respectively, Ni in 2003 and 2004–2.9 and 3.3 times, respectively. This is despite the fact that experience to bookmark content of these metals in the soil was significantly lower MAC.

This supports the view that the MAC of heavy metals in the soil does not always guarantee a clean crop production and talks about the need to find and develop measures to detoxify the soil to produce uncontaminated by heavy metals products.

The maximum accumulation of heavy metals in potato tubers in all the years of research going on in the options Background + HM and $N_{60}P_{60}K_{60}$ + HM, that is when you make a single, respectively, and a double dose of fertilizer and the establishment in the soil under potatoes contaminated by heavy metals from which the background without ameliorants plants absorb much more heavy metals than in options where melioranty were made. Moreover, a significant product contamination occurred following heavy metals: Cu, Cd, Pb and Ni with maximum concentration limit excess at 1.41–1.37; 1.99–2.16; 7.03–10.34 and 8.0–8.1 times, respectively, the elements and options (Table 14.2). Such contaminated products used for food purposes is not recommended. Contaminated with heavy metals potato tubers is best to use at the appropriate processing for the production of starch or alcohol.

Entering into the soil of agroameliorants contributed to the reduction of heavy metals in potato plants. The best action in this regard has provided liming (Table 14.2). Thus, soil limestone powder at a dose of 6 t/ha significantly reduced the concentration of heavy metals in potato tubers in all the years of research. This is our conclusion with respect liming role coincides with the opinion of other authors [7, 10–12], who argue that the basic method of reducing the mobility of most heavy metals in acid soils and their uptake by plants is liming in the combined effects on the soil. On the one hand, the metals in the form of carbonates and hydroxides of low solubility, the other – lime in acidic soils considerably increases the microbial mass and micro-organisms, in turn, are capable of absorbing and retaining many metals [12]. As a result of this somewhat increases and the absorption capacity of the cationic soil that inhibits the delivery of heavy metals in plants.

In our study on the accumulation of heavy metals in potato tops action semi-rerotting manure was also effective (Table 14.3), but slightly inferior calcification (Table 14.2).

On average over 3 years inhibitory effect of manure on heavy metals in plants exceeded one $N_{30}P_{30}K_{30}$ + HM on Cu – 1.6; Zn – 1.2; Cd – 1.5;

TABLE 14.3 Effect of Agrochemicals on the Content of Heavy Metals (HM) in the Tops of Potatoes (Average Over 3 years), mg/kg

Agrochemicals	Cu	Zn	Cd	Pb	Ni
Control	5.83	13.46	0.06	7.43	3.73
$N_{30}P_{30}K_{30}$ – Background	6.40	18.23	0.09	8.33	3.96
Background + HM	40.16	45.03	0.22	11.9	9.03
Background + HM + lime (6 t/ha)	30.73	35.73	0.14	8.67	7.60
Background + HM + manure (20 t/ha)	34.36	36.06	0.16	9.46	7.37
Background + HM + lime + manure	32.73	36.07	0.13	9.30	7.90
Background + HM + irlit 1 (2 t/ha)	44.60	49.40	0.20	9.30	8.95
Background + HM + irlit 7 (2 t/ha)	44.6	53.75	0.18	9.50	12.55
Background + HM + irlits 1+7 (1+1 t/ha)	39.60	40.90	0.20	8.65	8.35
$N_{60}P_{60}K_{60}$ + HM	40.26	46.50	0.24	12.23	9.43

Pb – 1.3 and Ni – 1.4 times. Joint application of lime and manure on the background $N_{30}P_{30}K_{30}$ + HM somewhat improved their positive effect on the restriction of heavy metals in plants.

Application irlits had a positive effect on reducing the heavy metal content in potato tubers, slightly inferior lime and manure, and in some cases even exceeding them. For example, in 2003 in potato tubers on the options using irlits was found less than Cu and Zn, than options with the introduction of lime and manure. Greater impact on reducing the accumulation of heavy metals by plants provided irlit 1 Action irlit 7 was weaker, probably due to the very high acidity, dissolving acting on acid-soluble forms of metals, which facilitates their uptake and accumulation by plants [13].

Between the aerial part and root system of plants is a constant exchange of substances. Both synthetic lab – leaf and root – are mutually dependent on each other's work, using the "semi," formed in each of them, to continue synthesis. Therefore, in determining the content of nutrients, including trace elements and heavy metals in the plant, it is important to know in this particular case, their content is not only cash crops (tubers), but also its by-products (tops).

In this regard, laboratory tests were performed to determine the content of heavy metals in potato tops in the most critical phase (flowering stage), depending on the anthropogenic load and application of various agro-chemicals and ameliorants (Table 14.3). On versions of the experiment with the introduction of heavy metals pollution in creating background concentration of heavy metals in the tops greatly increased compared with the control and the background for this. In this case, the excess of the average content was on the Cu – 6.3 times, Zn – 2.5, Cd – 2.4, Pb – 1.4 and Ni – 2.3 times compared with the fertilized background ($N_{30}P_{30}K_{30}$).

Liming, application of manure and irlits significantly reduced the concentration of heavy metals in the tops. Most of all, the content of heavy metals decreased soil application of lime, followed by the application of manure both separately and combined application of lime. Application irlits also be effective, but slightly inferior to the action of lime and manure.

Removal of heavy metals from the main crop and by-products of crops is an important indicator of the biological cycle of the environment. The magnitude of the removal of non-constant and is determined by soil and climatic conditions, size of the yield, the metal content in the soil, their availability to plants, etc.

The calculations to determine the economic removal of heavy metals from the harvest of agricultural products (potato tubers + tops) are presented in Table 14.4.

Adding $N_{30}P_{30}K_{30}$ markedly increased the economic take-out all the studied metals in all the years of research. On average over 3 years exceeding the removal of heavy metals in this version compared to the control was: Cu – 0.05 kg/ha, Zn – 0.12, Pb – 0.05 and Ni – 0.01 kg/ha. On artificially contaminated soil background (option 3) hardware removal increased significantly and on average over 3 years exceeded the background option for copper in 4.1 times, and zinc – 2.3 times, for cadmium – 2.7 times, for lead – 1.4 times and nickel – 3.0 times. While the share of foliage as a percentage of hardware removal amounted to Cu – 86.0–87.9; Zn – 79.6–94.0; Cd – 80.0–85.7; Pb – 50.0–80.7 and Ni – 66.7–77.8%.

Liming had no noticeable effect on the removal of heavy metals from the main crop and by-products. Apparently this is due to the fact that the introduction of lime to a certain degree facilitates transfer of heavy metals in a less accessible to the plants the compound which limits absorption by plants, and the concentration of metals in plants is lower and hence the stem downward. Adding lime contributes to a noticeable increase in yield as the primary (tubers) and incidental (tops) production of potatoes. Apparently, such a combined effect of lime in this case mutually balanced and therefore the difference in the removal of heavy metals between the variants Background + HM and Background + HM + lime, was negligible (Table 14.4). Application of manure both separately and together with lime helped to improve the economic removal of heavy metals in all the years of research, which is associated with a significant increase in yield on these options.

The maximum removal of all hardware elements marked on the form $N_{60}P_{60}K_{60}$ + HM and fluctuated over the years in the range: Cu – 0.77–1.01 kg/ha, Zn – 0.86–1.12, Cd – 0.006–0.008, Pb – 0.11–0.51 and Ni – 0.06–0.41 kg/ha.

It should be added that in 2004, there was the smallest removal of lead in all variants of the experiment: an order of magnitude lower than in other years, that, in all probability, due to the large amount of rainfall and

TABLE 14.4 Removal of Heavy Metals (HM) From the Soil With a Crop of Potatoes, Depending on the Application of Agrochemicals

Agrochemicals	Hardware removal, kg/ha					Removal from the tops, % of hardware removal				
	Cu	Zn	Cd	Pb	Ni	Cu	Zn	Cd	Pb	Ni
Control	0.10	0.16	0.001	0.10	0.04	58.9	78.3	72.3	83.0	65.9
$N_{30}P_{30}K_{30}$ – Background	0.14	0.28	0.001	0.15	0.05	61.4	80.2	63.4	81.2	42.0
Background + HM	0.58	0.61	0.005	0.23	0.12	87.3	87.4	82.8	69.3	73.2
Background + HM + lime	0.54	0.62	0.003	0.21	0.12	90.3	87.4	83.4	74.2	80.8
Background + HM + manure	0.73	0.78	0.005	0.27	0.14	90.9	88.4	84.5	75.0	80.7
Background + HM + lime + manure	0.73	0.80	0.004	0.26	0.16	91.7	89.7	84.1	87.1	82.7
Background + HM + irlit 1	0.72	0.73	0.003	0.17	0.15	92.4	93.9	90.9	75.0	81.2
Background + HM + irlit 7	0.67	0.79	0.003	0.19	0.18	91.1	92.6	88.2	62.4	83.1
Background + HM + irlits 1+7	0.64	0.67	0.003	0.18	0.14	91.4	92.7	90.9	54.7	81.5
$N_{60}P_{60}K_{60}$ + HM	0.88	0.97	0.006	0.36	0.20	88.7	89.1	80.6	72.5	74.0

waterlogged soil, which led to greater dissolution and leaching of this element in the underlying soil.

Application irlits also significantly increased hardware removal of heavy metals compared to the control and background options in connection with a higher yield, but noticeably inferior variants with manure.

Determining soil toxicity by method bioassay [10] was evaluated by the nitrogen-fixing ability of root nodules legumes – clover, astragalus, sweetclover, alfalfa, sainfoin grown in soils contaminated by heavy metals Pb, As, Hg, F, Zn. The degree of toxicity of contaminated soil was determined by the ability of the nodules of legumes to the synthesis of leghemoglobin in conditions of maximum humidity. In the case of a pink or red coloring of more than 50% nodules soil condition was evaluated as satisfactory, in the case coloring 20–50% of nodulesthe soil condition was environmentally risky, and staining less than 20% of nodules talked about environmental disaster. Toxicity of soil is also determined with the accumulation of heavy metals in the parts of plants under the soil surface at the flowering stage of ragweed, clover, alfalfa, sainfoin.

In different places was investigated the degree of soil pollution with heavy metals – Cd, Zn, Pb at doses which exceed the maximum allowable concentrations [11]: near highway Rostov-Vladikavkaz, near the plant "Electrozinc" and on the pilot site of the North Caucasus Research Institute of Mountain and Foothill Agriculture. In the area of greatest contamination (plant "Electrozinc") of ragweed in the flowering stage cadmium content exceeded the maximum allowable concentration of 1.5 times, zinc – 37 times, lead – by 2.2 times. In other cultures studied at the flowering stage is also observed excess of MPC selected heavy metals. However, ragweed plants sorbed heavy metals in much larger quantities than the other studied culture.

Comparison of sorption capacities of different plants in the same phases of development, but under different environmental conditions can detect species and culture with the highest bioindicatelly opportunities. To quantify the ability of ambrosia to the accumulation of heavy metals in the above-ground mass in comparison with other crops with similar sorption properties (clover, alfalfa, sainfoin), experiments were carried out on the territory of the metallurgical plant, near the road and in the agricultural areas (Table 14.5).

TABLE 14.5 Content of Heavy Metals in Plants At Different Stages of Their Development of Crops, mg/kg of Dry Matter

Crop	Stem growth	Budding	Flowering
Plant "Electrozinc," Cadmium (Cd)			
Ragweed	3.42	4.32	4.52
Clover	2.12	2.86	3.42
Alfalfa	2.26	2.78	3.24
Sainfoin	1.86	2.06	2.78
Highway Rostov – Vladikavkaz, Cadmium (Cd)			
Ragweed	2.08	3.65	4.11
Clover	1.78	3.04	3.86
Alfalfa	1.96	3.96	4.02
Sainfoin	1.65	2.92	3.58
Experimental field plot NCRIMFA, Cadmium (Cd)			
Ragweed	0.89	1.1	1.2
Clover	0.58	0.38	1.01
Alfalfa	0.50	0.50	0.74
Sainfoin	0.42	0.80	0.49
Maximum allowable concentrations (MAC)	3.0	3.0	3.0
Plant "Electrozinc," Zinc (Zn)			
Ragweed	314,02	325,89	968,6
Clover	78,12	86,46	114,3
Alfalfa	84,32	92,18	124,62
Sainfoin	64,44	72,02	88,14
Highway Rostov – Vladikavkaz, Zinc (Zn)			
Ragweed	236.73	280.25	620.0
Clover	48.48	50.24	56.18
Alfalfa	56.26	68.18	76.16
Sainfoin	54.12	0.42	6.16
Experimental field plot NCRIMFA, Zinc (Zn)			
Ragweed	44.47	139.12	156.18
Clover	2.86	23.92	31.62
Alfalfa	2.13	24.46	50.64
Sainfoin	2.03	31.2	34.46
Maximum allowable concentrations (MAC)	26.1	26.1	26.1

TABLE 14.5 Continued

Crop	Stem growth	Budding	Flowering
Plant "Electrozinc," Lead (Pb)			
Ragweed	4.30	7.98	11.2
Clover	1.12	3.12	6.46
Alfalfa	1.62	3.58	6.12
Sainfoin	0.86	2.18	2.92
Highway Rostov – Vladikavkaz, Lead (Pb)			
Ragweed	3.24	7.64	8.22
Clover	2.32	4.86	5.48
Alfalfa	2.68	3.14	3.68
Sainfoin	1.98	2.08	2.36
Experimental field plot NCRIMFA, Lead (Pb)			
Ragweed	2.18	6.04	8.12
Clover	0.86	3.08	4.12
Alfalfa	1.12	2.15	5.16
Sainfoin	1.76	3.12	5.0
Maximum allowable concentrations (MAC)	5.0	5.0	5.0

Considering the feature of vascular plants to concentrate heavy metals at the beginning of the growing season in a minimum amount, with a gradual increase in their content to the flowering stage, bioindicatelly evaluating several plant species were carried out in different phases of development (stem growth, budding, flowering).

14.4 CONCLUSIONS

Application agromeliorantes and organic fertilizers: limestone powder, local zeolite-clay – irlits, semi-rerotting manure on soil contaminated by heavy metals is a highly welcome detoxification of soils and reduce inputs of heavy metals in the soil of the potato plant.

Since the tops of potato absorbs from the soil and accumulate in the mass significant quantities of heavy metals that exceed their content in the tubers (up to 12 times or more) and is not used for commercial purposes, then collect it from the subsequent disposal can be considered one of the ways to detoxify the soil from heavy metals.

Developed and effective methods for determining soil toxicity bioassay method with contaminated soil with heavy metals as a result of the introduction of offsets into the atmosphere from a nearby factory or highway. The content of heavy metals in the soil was determined using the methods phytoindication. The degree of toxicity of contaminated soil was also determined by the ability of the nodules of legumes at different stages of plant development to the synthesis of leghemoglobin in conditions of maximum humidity.

KEYWORDS

- cadmium
- fertilizers
- irlit
- lead
- lime
- manure
- potatoes
- zinc

REFERENCES

1. Sokayev, K. E. The problem of man-made pollution agricultural soils. K. E. Sokayev. International Academy of Ecology and Life Protection. St. Petersburg, 2007, T. 13. №3, 102–106.
2. Ovcharenko, M. M. On the development of agrochemical service. M. M. Ovcharenko. Chemicals in Agriculture. 1994, №3, 5–8.
3. Sokayev, K. E. Agroecological monitoring of soil and fertilizer efficiency in the foothills of the Central Caucasian. K. E. Sokayev. Vladikavkaz. Publishing and printing company to them. B. Gassieva. 2010, 287 p.
4. Black, N. A. Abatement of phytotoxicity of heavy metals. N. A. Black, M. M. Ovcharenko, L. L. Popovicheva. Agrochemicals. 1995, №9, 101–107.

5. Ovcharenko, M. M. Factors of soil fertility and product contamination with heavy metals. M. M. Ovcharenko, V. V. Babkin, N. A. Kirpichnikov. Agrochemical Gazette. 1998, №3, 31–34.

6. Aliyev Sh. A. Agromeliorants as a means of greening agriculture. Agrochemical Gazette. 2001, №6, 26–28.

7. Shilnikov, I. A. Problems liming. I. A. Shilnikov, N. A. Kirpichnikov, L. P. Udalova et al. Chemicals in Agriculture. 1996, №5, 18–21.

8. Guidelines for the conduct of research on the topic: "To study the translocation of heavy metals in soil, agricultural products, and to develop methods of detoxification of soil to produce clean products." Moscow, 1993, 18 p.

9. Ogluzdin, A. S. Sapropel as meliorant soils contaminated with heavy metals/A. S. Ogluzdin, Y. V. Alekseev, N. I. Vyalushkina. Chemicals in Agriculture. 1996, №4, 5–7.

10. Zaalishvili, V. B. Invention: Method of estimation of technogenic pollution by heavy metals. Zaalishvili, V. B., Bekuzarova, S. A., Komzha, A. L., Kozaeva, O. P. Patent №2485477, published on 20.06.2013, the IPC G01 N3/48.

11. Zaalishvili, V. B. The invention: A method for determining soil toxicity. Zaalishvili, V. B., Bekuzarova, S. A., Komzha, A. L., Bekmurzov, A. D. Patent №2490630, published on 20.08.2013. IPC G01N33/24.

12. Shilnikov, I. A. Migration of cadmium, zinc, lead and strontium from the root layer of sod-podzolic soils. L. A. Shilnikov, M. M. Ovcharenko, M. V. Nikiforov, N. I. Akanova. Agrochemical Gazette. 1998, №5–6. 43–44.

13. Sokayev, K. E. Ecology Environment of Vladikavkaz and its suburbs. K. E. Sokayev, G. P. Khubaeva. Vladikavkaz. Publisher Olympus-Business. 2014–206 p.

CHAPTER 15

GENETIC HEALTH OF THE HUMAN POPULATION AS A REFLECTION OF THE ENVIRONMENT: CYTOGENETIC ANALYSIS

LIDIA V. CHOPIKASHVILI, TATIANA I. TSIDAEVA,
SERGEY V. SKUPNEVSKY, ELENA G. PUKHAEVA,
LARISSA A. BOBYLEVA, and FATIMA K. RURUA

CONTENTS

ABSTRACT

The paper considers demographic and cytogenetic aspects of monitoring of population living in the conditions of the North Caucasus foothills of Republic of North Ossetia-Alania. Among the examined men and women of reproductive age 51.5% of examined persons had frequency of chromosome aberrations from 1 to 3%, 40.9% had it from 3 to 4%, 7.6% had it above 4%. Causes of high variations in the mutations accumulation of Cd-production workers (from 4.9%±1.76 to 12%±2.65) are discussed. Cytogenetic analysis of women with obstetric anamnestic record, which level of chromosome aberrations varied from 2.5%±1.27 to 9.0%±2.34, was conducted. The reality of mutation load in the population of the given region correlation with demographic situation dynamics in 1996–2000 and 2008–2012 is shown.

15.1   INTRODUCTION

Wide spectrum of problems arising in connection with the effect of pollutants on living systems requires the monitoring of gene pool of humans as integral part of the biosphere. The mankind and all living organisms exist until genetic possibilities of organism correspond to the parameters of the environment; their inconsistency leads to the extinction. Consequently, medical and genetic state of the population of regions with industrially tense ecological situation may be a model when estimating complex effect of environmental pollutants on human gene pool and contribute to the elaboration of population health forecast and management principles.

Metallurgy giants "Elektrotsink" and "Pobedit" are the main suppliers of gene toxicants to the environment of the Republic of North Ossetia-Alania (RNO-A). The wastes of metallurgy (liquid, hard and in the form of slag) represent a high danger, their mass reaches 3.2 million tons, they include 4604 tons of lead, 9289.6 tons of zinc, 14.74 tones of cadmium etc. At a distance of one kilometer from the factory total indicator of the pollution by

4 heavy metals (Pb, Cd, Zn, Cu) amounts from 400 to 2000 mg/kg of soil, which corresponds to the category of extreme pollution [1].

It is necessary to take into account that Vladikavkaz (capital of RNO-A) has unique geographic position: it is surrounded by mountain chain, industrial emissions are not dispersed, but precipitate and pollute soil, water, air, territory of the city, accumulate in the plants.

Anxiety of the current situation is aggravated by that heavy metals are then transmitted by the food chain and reach human organism with the food. Beside of heavy metals, medical products, microbes and viruses (infections) also represent a danger as they are biological factors evoking mutagenesis [2].

Taking into account peculiarities of ecological situation in RNO-A, the aim of the study consisted in the elaboration of genetic monitoring system for the evaluation of possible consequences of complex effect of environmental factors on human organism. The realization of elements of this system is performed, from one side, by means of retrospective analysis of histories of pregnancies and deliveries including occurrence, epidemiological structure and dynamics of congenital malformations of development (CMD), registration of spontaneous abortions (SA), stillborn (SB) in comparative aspect: I period from 1996 to 2000 and II period from 2008 to 2012. From other side, we studied genetic health of the population, including cytogenetic examination of metallurgy workers, men and women of reproductive age, not connected with unhealthy production, women with obstetric anamnestic record (OAR), which had children with CMD, SA and MP were observed.

15.2   MATERIALS AND METHODOLOGY

The study of man-caused pollutants (Pb (II) ions) content in plant objects of Vladikavkaz industrial zone was conducted by means of atomic absorptive spectrometry. Plants were sampled in the locations exposed to maximum anthropogenic influence, namely close to "Elektrotsink" and "Pobedit" factories, as well as near frequent avenues of the city. Aboveground parts of plants were cut out, dried, crumbled up to the size of 3–5 mm and carefully mixed. Analytic samples with the weight of 100–200 mg were prepared from the samples described above by means of quartering and then they were incinerated in the muffle furnace under the temperature of 350–500°C with addition of ammonium sulfate in order to transmit Pb (II) ions into low-volatile forms. The obtained residuum was then dissolved in 1% solution of nitric acid. The analysis of lead content was performed on the atomic absorptive spectrometer "Kvant-Z-ETA" in the regime of thermal-electric atomization. The content of the metal was determined using state standard sample.

In our study of genetic health of population of RNO-A we used the method of human lymphocytes cultivation [3]. Blood samples were taken from cubital vein. Blood was cultivated in nutrient medium composed of: RPMI-1640 medium (6 mL), cattle serum (1.5 mL) and biological stimulator phytohemagglutinin "PanEco" (0.1 mL). The mixture was put in sterile flasks; 0.5 mL of heparinized blood was added. Flasks were stored under the temperature of 37°C. Fixation of cultures was performed after 48 hours of cultivations. 2 hours before the end of cultivation we introduced colchicine in the concentration 0.5 µg/mL. Further processing was conducted according to the common methodology: hypotonization by 0.55% solution of KCl, fixation by the mixture of ethanol and ice acetic acid in the proportion 3:1, dripping of cell suspension on previously cooled and degreased object-plates. Then the preparations were colored by Gimza dye according to Romanovsky. The analysis of preparations was performed using Eunoval microscope under the magnification 10 × 100. Metaphase plates sampling was performed following common requirements: integrity of metaphase plates, clearness of color, absence of chromosomes superposition, medium degree of condensation, isolation of metaphase plates from each other [4].

In order to study the state of gene pool of RNO-A population during last five years 238 individuals were examined, above 35700 metaphases were analyzed (about 150 for each individual). 112 individuals among all examined individuals constituted the group for revealing of spontaneous mutation level. We studied also genotype of 27 employees of metallurgy production (12 employees of factory management, 15 workers of Cd-workshop), 15 individuals constituted reference group. From each of the explored workers were analyzed by 130–150 metaphases. Sample workers cadmium workshop was small (15 people) from each of the surveyed we analyzed metaphases 300–500.

In order to study the dynamics of demographic processes we applied retrospective method of pregnancy and delivery course analysis. Totally 97582 case histories of pregnancy and delivery were analyzed.

15.3   RESULTS AND DISCUSSION

15.3.1   THE STUDY OF BIVALENT LEAD IONS CONTENT IN PLANT OBJECTS OF VLADIKAVKAZ

Mushrooms and conifers showed the highest and the lowest accumulation capacity of Pb (II) ions, respectively. This study showed that the mushrooms *Marasmius oreades (Bolton) Fr.*, *Laetiporus sulfureus (Bull)*, *Psathyrella candolleana (Fr)* – contain from 0.66 to 10.44 mg/kg of Pb; they are followed by herbaceous plants, in particular by clover (*Trifolium pratense L.*) – 2.31–3.05 mg/kg; conifers close the series: spruce (*Picea pungens Engelm*) – 0.022–0.34 mg/kg; Scots pine *(Pinus sylvestris L.)* – 0.25–0.92 mg/kg; thuja (*Thuja occidentalis L.)* – 1.3–1.57 mg/kg.

15.3.2   CYTOGENETIC RESEARCH OF THE HEALTH OF WORKERS, CONTACTING WITH HEAVY METALS (Co, Mo, Cd) IN RNO-ALANIA

We focused a particular attention on the workers employed in cadmium production (Cd-production), because the preliminary examination of molybdenum and cobalt production workers showed that effect in these groups was expressed in "soft" form relatively to gene-toxic effect of cadmium. In average 4.8% of chromosome aberrations (ChA) was observed of molybdenum (Mo) and cobalt (Co) production workers.

Cytogenetic study of workers was performed when taking into account age (18–39 years), record of service and profession. Investigated individuals were divided into 3 groups: group I was control one (15 individuals) consisting of Vladikavkaz inhabitants not connected with unhealthy production (9 men and 6 women), group II (12 individuals) consisted of employers of factory management not contacting directly with cadmium (6 men and 6 women), group III consisted of workers of Cd-workshop (15 individuals from 45 workshop workers, 9 men and 6 women). Record of service of workers varied from 6 months to 24 years.

The analysis of obtained data showed that in the group I (spontaneous mutagenesis) average ChA percentage amounted 2.4. In the group II ChA level was higher, than in group I and varied within the limits from 2.9 to 3.8%. In the group of workers variations were wider: from 4.9 to 12%. It is necessary to notice that no direct correlation between the level of chromosomes damages and the duration of record of service was observed in cadmium as well as in molybdenum and cobalt production. In Cd-workshop 13.3% of examined workers with record of service from 12 to 14 years had ChA level close to the level of group II namely from 4.4 to 4.9%. At the same time the amount of chromosomes anomalies of 26.7% of workers with record of service 6–7 to 12 months varied within the limits from 8 to 11% which corresponds to the level of damages typical for the group of workers with record of service above 15 years (60% of examined individuals) (Figure 15.1). As the sample size of cadmium workshop workers was small (15 individuals), we analyzed from 300 to 500 metaphases plates for each examined individual.

Cytogenetic analysis allowed to reveal the character of chromosome damages: aberrations of chromosome type (double fragments, dicentrics) and of chromatid type (singular fragments), first ones dominated more than 2 times relatively to the last ones. This fact is evidence of that heavy metals have the highest negative effect during pre-synthetic phase of mitosis, G_1 [5, 6].

The presence of specific damages in the spectrum of chromosome aberrations of Cd-production workers, characteristic only for this sample group, which do not occur at individuals contacting with cobalt and molybdenum, evokes a particular interest. Endoreduplications of chromosomes,

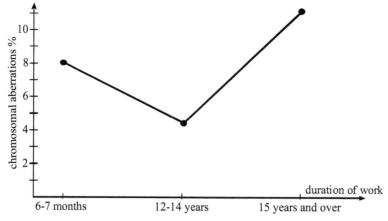

FIGURE 15.1 Individual peculiarities of chromosome aberrations in dependence from record of service of cadmium workshop workers.

"fluffy" and sprayed chromosomes belong to such chromosome anomalies. The amount of sprayed chromosomes was 0.7% from the total spectrum of aberrations, the number of endoreduplications (intranuclear polyploidy) amounted 0.8% [7, 8]. High frequency of cell polyploidy usually is detected in tumorous cell populations. One of mechanisms in polyploids formation is the excluding of mitosis stage from cellular cycle and direct transition from stage G_2 to G_0 and then to the stage G_1 [9, 10].

Occurrence of "fluffy" chromosomes is related, according to our hypothesis, with the breaking of deoxynucleoproteid integrity, probably with methylation, acetylation processes, which underlie possible anomaly of epigenetic genome organization and can display phenotypically such a way [11]. Cytogenetic analysis of Cd-production workers allowed to detect direct correlation between ChA level in workers genome and their profession: ChA level of machine operatives and cathode operators amounted in average from 9 to 12%; electricians and loaders had lower ChA level, namely 5.7% in average.

When analyzing obtained results it is possible to derive the conclusion that Co, Mo and Cd have mutagenic effect. Cd showed to be the most dangerous gene-toxicant in our study. It is necessary to notice that the peculiarities of genetic constitution of individual play considerable role in the effect of heavy metals. It is known that the combination of polymorph genes and p53 gene, taking part in the functioning of protection systems determines the individual character of human cells response to the effect of gene-toxic factors [12]. In our experiments the level of heredity damage among examined individuals did not always depend from the duration of worker contact with heavy metals (record of service) and in some cases it was even close to control values under considerable record of service and vice versa. This fact bases the recommendation of preliminary cytogenetic (in vitro) examination of workers before their acceptance to the work in unhealthy production.

Modern studies in the field of molecular biology can explain the reason of broad variation of our data about the accumulation of mutations in organisms of examined workers (small record of service with high ChA level and high record of service with ChA level witch does not exceed significantly the control). Baranov [13] notices that the reaction of genome to different exogenous (ecological) factors is determined in a considerable measure by functional peculiarities of genes of metabolism and genes of DNA reparation system. It is known now that all human genes have molecular differences as a result of mutations accumulation, which leads to the polymorphism of gene pool of factory workers and the population in the regions. Mutant genes lead to the synthesis of proteins with modified functions [14, 15]. Consequently each human is genetically unique and has his unique biochemical portrait. This explains the fact of presence of individuals with genotypes, which allow them to work in metallurgy factory.

15.3.3 CYTOGENETIC STUDY OF REPUBLIC NORTH OSSETIA-ALANIA POPULATION HEALTH

During the period of 2008–2012, in order to study the state of population gene pool we examined 112 RNO-A inhabitants, men и women of reproductive age (18–39 years). At least 150 metaphase plates were analyzed of each inhabitant (totally about 16800 metaphases) [16].

The analysis of obtained results showed that ChA level varied within broad range: 51.5% of examined individuals had ChA level within the limits from 1% to 3% of aberrant cells; 40.9% had ChA level from 3 to 4% and 7.6% had above 4% of ChA (Figure 15.2).

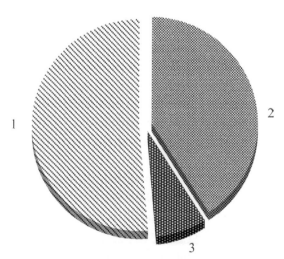

FIGURE 15.2 Level of spontaneous mutations in the gene pool of Vladikavkaz city inhabitants (sectors 1, 2 and 3 mean <3%, 3–4% and 4–6% of ChA, respectively).

Our earlier study performed in 1997–2000 showed similar result: more than 50% of examined individuals had elevated amount of genetic anomalies (the ChA level of examined individuals exceeded control data almost 2 times) [16]. As potential parents, they can reproduce descendants with deviations, which can be the reason of the organization of prophylactic measures aimed to the recreation of the population gene pool.

15.3.4 ANALYSIS OF CYTOGENETIC STUDIES OF WOMEN WITH OBSTETRIC ANAMNESTIC RECORD (OAR)

About 84 women were examined: 48 healthy (control group) and 36 with OAR. More than 12–600 metaphases were analyzed. Group of women with OAR consisted from those who had SA, children with CMD. The study showed that genotypes of 35.3% of examined women correspond to norm (2.5–3.0% of ChA), in 33.2% ChA level exceeded 4%, and in 31.5% it varied from 5% and upper, four examined women had 9% of aberrant cells (Figure 15.3).

When analyzing spectrum of chromosome anomalies, we noticed that the portion of chromosome type aberrations detected at donors with

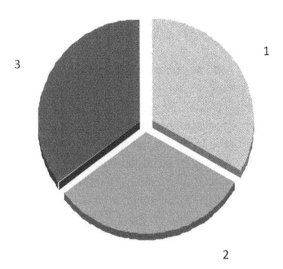

FIGURE 15.3 Cytogenetic study of women with OAR (sectors 1, 2 and 3 mean <3%, 3–4% and 4–6% of ChA, respectively).

reproduction disturbances amounts 52.5%, whereas in the group of healthy inhabitants it does not exceed 36%.

When comparing our data obtained during last years with earlier publications [17], we noticed the absence of positive dynamics in the improvement of genetic health of RNO-A population. In the group of healthy donors the amount of individuals with normal ChA level (from 1 to 3%) decreased from 57.4 to 51.5%, at the same time the amount of individuals with elevated ChA level (above 4%) increased by 13%. In the group of women with OAR we observe the decrease of amount of individuals with normal genotype from 39.8 to 35.3% and the increase of amount of individuals with ChA level above 5% from 26.5 to 31.5%. The presence of negative dynamics of gene-toxic processes is apparently conditioned from one side by the increase of sources of environmental pollution: traffic emissions, o medical products, foodstuff containing mutagens. From other side the generation, which ancestors lived also in ecologically unfavorable environment, reached childbearing age.

Thus, our cytogenetic studies (state of genome of metallurgy production workers, women with OAR, healthy individuals of reproductive age) testify to the serious mutation load accumulated in the population of RNO-A.

15.3.5 DYNAMICS OF DEMOGRAPHIC SITUATION IN THE REPUBLIC NORTH OSSETIA-ALANIA

Reality of mutation load in the population of the given region can be illustrated by the biological quality of new-born children, high level of chromosome aberrations (ChA) detected at the individuals of reproductive age, considerable portion of sterile marriages (above 15%). According to the Altukhov [17], who, by means of generalizing scientific data about European population during 30 years, showed that at least 50% of primary gene pool is not reproduced in the next generation, represents a particular interest within the frame of the topic under consideration. Genetic component of this process amounts in average 20–30%, up to 20% of humans in reproductive age do not marry, about 10% of marriages are sterile.

The materials of retrospective analysis of pregnancies and deliveries that we performed are presented at Figures 15.4–15.6. They show the dynamics of demographic situation in the region during 1st (1996–2000)

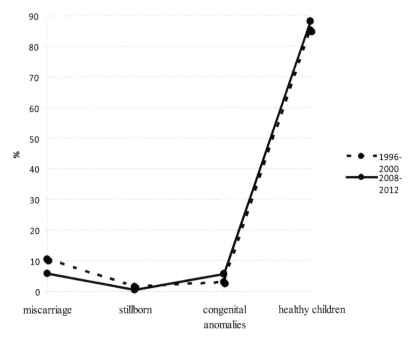

FIGURE 15.4 Dynamics of miscarriage, SB (stillborn) and CA (congenital anomalies) for two periods: 1996–2000 and 2008–2012 Dynamics of (miscarriage), SB (stillborn) and CA (congenital anomalies) for two periods: 1996–2000 and 2008–2012.

and 2nd (2008–2012) periods of study. They indicate that in the 1st period in average 7660 children were born annually, whereas in the 2nd period – 10300 children per year.

Mean frequency of spontaneous abortions per year (period I) amounted 10.53% from the total number of planned pregnancies; in the period II this value decreased 1.8 times and amounted 5.98% from the total number of planned pregnancies. The frequency of stillbirths in the period I amounted in average 1.40%; in the period II it decreased more than 3 times and amounted 0.42%. The frequency of birth of children with CMD in the period I amounted 2.84% or 31.71 per 1000 new-born children; in the period II it increased 2 times and amounted 5.61% or 59.62 per 1000 new-born children (see Figure 15.4). The expressed correlation can be detected: the higher are SA and still birth frequencies, the lower is the probability of birth of children with CMD and vice versa: the less SA and still birth are observed, the higher is the frequency of children with CMD birth. This relation is logical: the accumulation of considerable amount of ChA in the

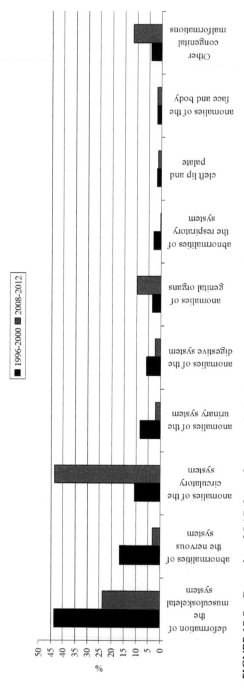

FIGURE 15.5 Dynamics of OAR forms for two periods: 1996–2000 and 2008–2012.

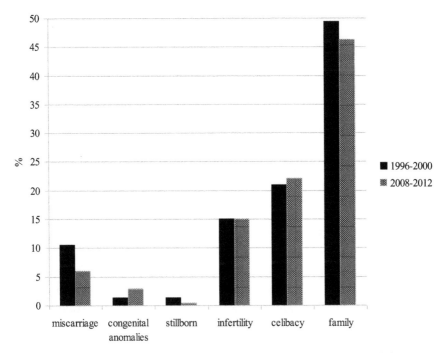

FIGURE 15.6   Comparison of annual biological losses spectrum for two periods: 1996–2000 and 2008–2012.

genome of parents and their embryos could not provide faultless embryogenesis, which is displayed in the frequencies of SA, still birth and birth of children with CMD. As a result of natural selection gene pool "cleaning" occurs. It is necessary to remember in this connection the warning of N.P. As a result of natural selection gene pool "cleaning" occurs, which allows to the nature to correct of its own errors.

It seems to be interesting to compare the dynamics of CMD forms (spectrum) in the 1st (1996–2000) and 2nd (2008–2012) period (see Figure 15.5).

The analysis of CMD spectrum in the 2nd period showed considerable changes relatively to the first period: the anomalies of the blood circulation system came to the first place according to the frequency of occurrence (43.34%, which is 4 times higher than in the 1st period); bone and muscles pathologies got the second place (23.86%). Anomalies of the nervous system decreased 5 times from 16.74 to 3.24%. The frequency of congenital anomalies of genitals increased almost 3 times: from 3.62 to 9.88%. The frequency of anomalies of digestive apparatus decreased 2 times: from 5.90 to 2.29%. Frequency of anomalies of other organs and systems (anomalies of face and body, crack of lip and palate, respiration apparatus) did not change considerably between these periods (less than 2%).

The data that we presented on the frequency of SA, still birth and CMD are incontestable fact testifying high mutation load accumulated in the population of RNO-A, which argues once more the importance of the research of the complex effect of mutagens of chemical and physical nature on the population gene pool.

15.4   CONCLUSIONS

1. Environmental pollution underlies the accumulation of mutation load in the examined by us of Republic North Ossetia-Alania (48.5% of population have the frequency of chromosome aberrations from 3.1 to 7.6%). Anomalies of chromosome complex underlie biological losses – spontaneous abortions, stillbirth and congenital malformations of development.

2. The analysis of demographic situation during the periods from 1996 to 2000 and from 2008 to 2012 showed, that reproductive losses in the 1st period are represented by spontaneous abortions – 10.53%, still birth – 1.40%, sterility of young families – 15%. It is necessary to notice that 21% of women were not married at al. At least half of individuals with congenital anomalies (2.84% of population) will not take part in the reproduction of new generation (this is ~ 1.42%). In the second period of study (2008–2012) the pattern of biological losses is the following: spontaneous abortions – 5.98%; still birth – 0.42%; sterility – 15.00%; not married – 22.10%; with congenital malformations of development – 5.61% (~2.82% will not take part in the reproduction). As a result 46.31% of population will not take part in the reproduction of new generation.

3. Cytogenetic analysis of cadmium production workers revealed correlation between the level of chromosome aberrations of workers and their profession. Genotoxic effect of heavy metals on human organism not always depends on the duration of contact with them. Peculiarities of genetic constitution of individual play an

important role. This fact bases the recommendation of preliminary cytogenetic examination (in vitro) of workers, which allows detecting individuals with genotype resistant to the given type of mutagens.

4. In the group of women with obstetric anamnestic record 35.3% had genotype within the limits of norm (2.5–3.0% of chromosome aberrations), 33.2% had the level of chromosome aberrations above 4% and 31.5% had it from 5% and upper, for of examined women had 9% of aberrant cells, which is related to the increase of sources of environmental pollutants as traffic and industrial emissions, medical products, foodstuff containing mutagens. From other side the generation, which ancestors lived also in ecologically unfavorable environment, reached childbearing age.

5. Retrospective analysis of pregnancy and delivery case histories showed the presence of the following correlation: the higher is the frequency of spontaneous abortions and still birth, the lower is the probability of birth of children with congenital anomalies and vice versa: the less spontaneous abortions and still birth are observed, the higher is the frequency of birth of children with congenital anomalies. This way by means of natural selection the gene pool "cleaning" takes place.

KEYWORDS

- chromosome aberrations
- congenital malformations
- heavy metals
- spontaneous abortions
- still birth

REFERENCES

1. Alborov, I. D., Kharebov, G. Z., Gasinov, S. A., et al. Effect of non-ferrous metallurgy waste products on the ecology of the region. Herald of the International Academy of Ecology, Man and Nature (Journal MANEB). 2013, Vol. 18, №4, 9–13 (In Russian).
2. Chshiyeva, F. T., Chopikashvili, L. V., Dzhagayeva, Z. K. et al. Analysis of clastogenesis level of children with gastroduodenitic diseases. Bulletin of experimental biology and medicine. 2010, Vol. 149, No 4, 415–417.

3. Hungerford, P. A. Leukocytes cultured from small inoculate of whole blood and the preparation of metaphase chromosomes by treatment with hypotonic KCl. Stain Technology. 1965, Vol. 40. 333–338.

4. Bochkov, N. P. Clinic genetics. Moscow. Meditsina pub. 1997, 288 p (In Russian).

5. Inglot, P., Lewinska, A., Potocki, L. et al. Cadmium-induced changes in genomic DNA-methylation status increase aneuploidy events in a pig Robertsonian translocation model. Mutation Research. 2012.Vol. 747, №2, 182–189.

6. Yilmaz, D., Aydemir, N. C., Vatan, O. Influence of naringin on cadmium-induced genomic damage in human lymphocytes in vitro. *Toxicology* and *Industrial Health*. 2012, Vol. 28, №2, 114–121.

7. Kovaks, J. B., Sadah., Hoene, E. Binucleate cells in a human renal cell carcinoma with 34 chromosomes. Cancer Genet. Cytogenet. 1988, Vol. 31, №2, 211–215.

8. Allen, J. W., Collins, B. W., Setzer, R. W. Spermatid micronucleus analysis of aging effects in hamsters. Mutation Res. 1996, Vol. 316, №5–6. 261–266.

9. Fergusson, L. R., Whiteside, G., Holdaway, K. M. et al. Application of florescence in situ hybridization to study the relationship between cytotoxicity, chromosome aberrations, and changes in chromosome number after treatment with the topoisomerase II inhibitor amsacrine. Environmental and Molecular Mutagenesis. 1996, Vol. 27. 255–262.

10. Gateva, S., Jovtchev, G., Stergios, M. Citotoxic and clastogenic activity of CdCL2 in human lymphocytes from different donors. Environmental Toxicology and Pharmacology. 2013, Vol. 36, №1, 223–230.

11. Pendina, A. A., Grinkevich, V. V., Kuznetsov, V. V., Baranov, V. R. Methylation of DNA as universal mechanism of regulation of genes activity. Ecological genetics. 2004, Vol. II, №1, 27–28 (In Russian).

12. Abilev, S. K. Polymorphism of genes as indicator of sensitivity of humans to environmental factors. Proceedings of united plenum of scientific councils of Russian Federation on the human ecology and environmental hygiene. [Eds Yu.A. Rakhmanin, N. F. Izmerov] Moscow. 2010, 22 (In Russian).

13. Baranov, V. R. Ecological genetics and predictive medicine. Ecological genetics. 2003, Vol. I. 22–29 (In Russian).

14. Halasova, E., Matakova, T., Musak, L., et al. Evaluating chromosomal damage in workers exposed to hexavalent chromium and the modulating role of polymorphisms of DNA repair genes. International Archives of Occupational and Environmental Health. 2012, Vol. 85, №5, 473–481.

15. Stephan, J., Pressl, S. Chromosomal aberrations in peripheral lymphocytes from healthy subjects as detected in first cell division. Mutation Research. 1999, Vol. 446, №2, 231–237.

16. Bochkov, N. P., Chebotaryov, A. N., Katosova, L. D. et al. Database for the analysis of quantitative characteristics of chromosome aberrations frequency in the culture of lymphocytes of peripheral human blood.. Genetics. 2001, Vol. 37, No 4. 549–557 (In Russian).

17. Chopikashvili, L. V., Bobyleva, L. A., Tsallagova, L. V., et al. Evaluation of genetic health of the RNO-A population in the conditions of high anthropogenic press. Vladikavkaz medical and biological bulletin. 2002, Vol. I, Issues 1 and 2, 21–23 (In Russian).

18. Altukhov, Yu. P. Genetic processes in populations. Moscow. Akademkniga pub. 2003, 431 p (In Russian).

PART IV

PHENOGENETIC STUDIES OF CULTIVATED PLANTS AND BIOLOGICAL PROPERTIES OF THE SEEDS

CHAPTER 16

ECOLOGICAL AND BIOLOGICAL STUDY OF COLLECTION OF THE GENUS *HORDEUM L.*

NINA A. BOME, NIKOLAY V. TETYANNIKOV,
ALEXANDER YA. BOME, and OLGA N. KOVALYOVA

CONTENTS

ABSTRACT

The chapter presents the results of study of barley samples of different ecological and geographic origin from world collection of N.I. Vavilov Research Institute of Plant Industry in the conditions of southern Tyumen region. The evaluation of samples is performed using a set of characters valuable for selection in comparison with approved for cultivation in the Tyumen region varieties (standards). The data on field germination capacity of seeds, plant probability of survival, plant height variability, ear length, resistance to lodging, grain production are presented.

Considerable differences in meteorological characteristics among vegetation seasons allowed to detect some peculiarities of variability of characters. The span of variation of characters under study can characterize in a certain extend barley ecological plasticity under complicated soil and climatic conditions.

16.1   INTRODUCTION

During recent years climatic conditions on the Earth become less favorable because of climate changes and atmospheric pollution. On the territory of Russia the conditions of plant cultivation in many cases are rather complicated (low or elevated air temperature, deficit of water, oxygen, excess of salts in the soil, insufficient biotic diversity of agrocenoses), which leads to dramatic decrease of yields and even to the crop destruction. Consequently the issue of plant adaptive properties increase becomes more and more important, special attention is focused on the creation of highly resistant varieties. This work is conducted in different directions including the revealing of genes determining plant resistance. A set of transgenic plants resistant to viral infections, herbicides and insects is already obtained. Works aimed to create plants with characters providing resistance to abiotic factors (drought, salinity, oxidative stress) are also conducting [1].

Tyumen region is characterized by high potential of lend resources and it is the zone of risky agriculture, what is conditioned by hard natural climatic conditions and low biological potential [2].

Variety plays a considerable role in the obtaining of stable yields, improvement of production quality, increasing of economic effectiveness of leading cereals including barley. However, when choosing varieties for concrete conditions an objective evaluation of plant resistance to unfavorable environmental factors according to a set of economically valuable characters is necessary. In order to study single plants or their groups the method of their transmission from different parts of species areal, contrasting by ecological conditions in the homogenous environment of experimental plot, garden or greenhouse is applied [3].

It is evident from the above that the issue of plasticity, adaptation and resistance of the initial material to the environmental factors remain actual.

The aim of the present study is the ecological and biological research of the barley collection fund according to the set of selection-valuable characters under conditions of southern Tyumen region.

The following problems were set within this aim: to perform geographical analysis and botanic description of barley samples from the collection of genetic resources of N.I. Vavilov Research Institute of Plant Industry (VIR); to study growth and development peculiarities of samples of different ecological and geographic origin according phenotypic display of valuable characters (field germination capacity of seeds, plant probability of survival during vegetation season, plant height, resistance to lodging, elements of grain production); to detect basing on the complex evaluation plant forms with high adaptive and productive properties.

16.2   MATERIALS AND METHODOLOGY

The ecological and biological study of barley collection fund according to the complex of selection-valuable characters was conducted in 2012–2013 in Tyumen research station of VIR and in the laboratory of biotechnological and microbiological research of Tyumen State University cathedra of botany, biotechnology and landscape architecture.

Field study was performed at the experimental site of "Kuchak lake" biological station of Tyumen State University, situated in sub-boreal forest zone of Tyumen region (at the border of northern part of Tyumen district and southern part of Nizhnetavdinsk district). The soil of the site is ameliorated, sod-podzolic, sandy loamy. The reaction of medium in salt soil extract was 6.6, for example, alkalescent. Humus content was 3.67%. Soil analysis was performed in the Laboratory of Ecotoxicology of Tobolsk complex research station of Russian Academy of Sciences Ural branch. Sampling was performed according to all-Russian State Standard 28168–89.

Sowing of barley samples was performed manually using marker. The area of each sample plot is 0.5 m^2 (length of one row was about 1 m, row-spacing was 15 cm and the number of rows in a plot was 3. Sowing depth was 5 cm, distance between plots in one tier was 30 cm and the distance between tiers was 60 cm. 200 seeds were sowed in each plot.

The foundation of collection nursery, sowing, observations and calculations were performed according to directions of VIR on the research and conservation of world barley collection [4]. The description of morphological characters, biological properties, yield capacity and yield structure of samples was performed according to International COMECON classifier of *Hordeum* L. genus [5].

For each character averages, standard errors of means and coefficients of variation were calculated. Mathematical processing of experimental data was performed according to the methods described by Dospekhov and Lakin [6, 7].

Vegetation season (May–August) of 2012 was characterized by very hard weather conditions, namely by strong moisture deficit (28.9–63.6% of normal) on the background of elevated air temperatures daily means (2.4–4.1°C above normal). Temperature regime in 2013 was close to long-term averages, water deficit was observed in June and August (about 62% of normal). Generally the conditions for barley growth and development were more favorable than in the previous year.

16.3 RESULTS AND DISCUSSION

Field testing included 333 samples of different ecological and geographic provenance: from Peru, Tadjikistan, Kazakhstan, Germany, Ethiopia, Czechoslovakia, USA, France, Ukraine, Poland, the Netherlands, Hungary, Belgium, Great Britain, Syria, Iraq, Brazil, Egypt.

Collection fund is presented by two subspecies: double-row barley (*Hordeum distichon* L.) and multi-row barley (*Hordeum vulgare* L.). Varieties recommended for cultivation in Tyumen region (Acha variety from Novosibirsk region and Kedr variety from Krasnoyarsk region) were applied as standard. Standards were situated at the beginning and at the end of the collection, as well as after every 20 numbers.

Collection material was characterized by high diversity, as samples belonged to 59 botanical varieties. The determination of the varieties (var. in abbreviated form) of cultivated barley is based on the morphological characters: filmy or naked seeds, ear density (friable, dense, very dense), width of glumaceous scales (narrow or broad), presence and character of ear aristas (barbate, awnless and furcaty forms, serrated or smooth aristas), color of flower and spicate scales (yellow, violet, black, orange) [8, 9].

The most numerous varieties in the collection: *pallidum* (faded), *pyramidatum* (pyramidal), *nutans* (drooped).

Var. *pallidum* (faded) belongs to subspecies of multi-row barley. Its ear is friable, elongate, yellow, barbate. The aristas are denticulated, long, 1.5–2 times longer than ear. Crops are spring, semi-winter and winter. This variety is widely spread in all cultivation zones.

Var. *pyramidatum* (pyramidal). It also belongs to multi-row, filmy barley. It is characterized by dense ear, yellow color, barbate. Spicate scales are narrow.

Var. *nutans* (drooped). Ears are yellow, friable. Side flowers are staminal, steril. Spicate scales are narrow. The aristas are denticulated, long, 1.5–2 times longer than ear. Spring forms dominate, semi-winter forms are rare, winter forms are isolated. It is cultivated in all continents. This is the most widely spread variety of double-row barley [8, 9].

In the collection the most numerous were diversities: *nigripallidum* (black pallidum), *nigrum* (black), *coeleste* (celestial), *ricotense*.

Var. *coeleste* (celestial) – is the variety of multi-row naked barley. It is characterized by friable, barbate ear of yellow color.

Var. *nigripallidum* (black pallidum). Ear in the period of full maturity are black or gray, friable. Spicate scales are narrow. Aristas are yellow, serrated, long, 1.5–2 times exceeding the length of ear. колоса в 1,5–2 раза. Spring, winter and semi-winter forms are presented. It is cultivated in all continents. Most often it occurs as admixture.

Var. *nigrum* (black) belongs to multi-row barley subspecies. It is characterized by friable, barbate ear of black or gray color. Aristas are also black.

Var. *ricotense*. Ears are yellow, friable. Aristas are smooth or very slightly serrated in the upper part only, long, 1.5–2 times exceeding the ear length. Spring, semi-winter and winter forms are cultivated. It occurs as bigger or smaller admixture in all zones of barley cultivation.

Var. *erectum* (straightly standing). It belongs to double-row barley subspecies. Ears are yellow, broad, and dense. Side flowers are anthers. Spicate scales are narrow. Aristas are mostly parallel to the ear, serrated, long, 1.5–2 times exceeding the ear length. It is spring form. It is spread in all European countries, Asia (Transcaucasia, Iran, Central Asia, Siberia, AND Japan), America (USA) [8, 9].

The following varieties are represented by smaller amount of samples: *schimperianum, himalaense* (Himalayan), *hypatherum* (semi-barbate), *sinicum* (Chinese), *parallelum* (parallel), *trifurcatum, duplinigrum* (twice black), *erectum* (straightly standing), and others.

Thus, during two years (2012 and 2013) a large heterogeneous set of barley collection samples characterized by broad geographic, botanical and selection diversity was studied according to the complex of selection-valuable characters.

The results of the study of genus *Hordeum* L. Collection showed considerable differences between samples according to quantitative characters.

The effect of both, genotype, as well as meteorological factors on the seeds field germination capacity and plant survival during the vegetation season was found when evaluating the collection in 2012 and 2013. The barley samples from Peru C.I.11055 (k-30649) and Germany HVS 81583/72 (k-25160) showed maximum field germination capacity. The variation of the character in the first year and second of study was within the limits from 23 to 94% and from 0.5 to 92.5%, respectively, with the average value of the collection 69.7% and 54.9%, respectively. Decreased daily mean air and soil temperature in 2013 did not contribute to the active germination of seeds, which was reflected in the average germination values of the collection.

The conditional classification of samples according to the field germination capacity allowed to reveal the differences within the collection according to germination capacity of seeds in the changing environmental conditions. During the period of study of this character the group with average field germination capacity (51–70%) dominated, but the amount of samples within this group and other ones varied from year to year. In particular, in the conditions of 2013 the quantity of the group with low field germination capacity (below 50%) considerably increased. At the same time the group with very high germination capacity (above 90%) considerably decreased relatively to 2012 (Figure 16.1).

It is possible to evaluate the reaction of plant organism on the environmental conditions at the initial stages of its development by field germination capacity and later by its biological stability, for example, by the

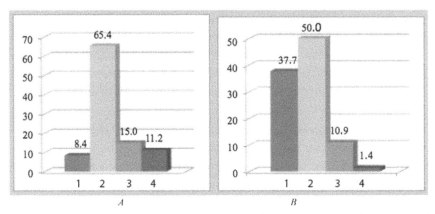

FIGURE 16.1   Distribution of barley samples according to field germination capacity of seeds, % [Note: 1 – low (<50%); 2 – medium (51–70%); 3 – high (71–90%); 4 – very high (>90%), *A*– 2012, *B* – 2013].

indicator of its probability of survival during the period of plant growth and development [10].

Considerable part of collection samples was characterized by high resistance to environmental factors. In spite of very complicated weather conditions of vegetation season 2012, many barley samples showed good adaptive properties. The all-collection average portion of plants survived until the end of vegetation season relatively to sprouts was 81.9–89.7%.

The lodging of cereal crops is rather often phenomena. It can occur in different phases of plant growth and development. According to Tretyakov and Yakovlev [11], the probability of this henomena exists since the moment of exit in tube and lasts until the full maturity of grains.

The disposition of cereals to be lodged limits their production potential, leads to the considerable change of metabolic processes in plants, accelerated development of fungi diseases, decreasing of grain quality and complicates the crop harvesting. The high level of lodging can be related with soil and climatic conditions, as well as with breakdown in cultivation technology: increasing of doses of mineral fertilizers both by the area and in the plow-layer, inconvenience with the norm of seeds sowing [12, 13].

Such investigators as M.I. Rudenko, V.P. Vorontsova, V.A. Kumakov, and M.N. Chaylakhyan in their publications express the idea that the higher is the stem of cultivated cereal plant the higher is the probability to be lodged. That is why an important role in the solution of the issue of resistance to lodging belongs to short-stem varieties [13].

Genotypes studied under uncontrolled meteorological conditions differed in the phenotypic display of the character in dependence from the relation between precipitation and air temperature during the period of active plant vegetation. It is important to study the limits of the variability within a certain genotype and to reveal morphological type of plants with the height optimal for the given soil and climatic conditions.

In the vegetation season 2012 minimal and maximal heights of barley plants in the collection in the phase of earing amounted 45.8 cm (samples: Local, k-14952 from Tajikistan) and 84,0 cm (samples: C. I. 11070, k-30662 from Peru), respectively, with the average value of 63.8 cm.

In 2013, minimal height of 49,7 cm and maximal height of 103,6 cm were observed in the samples EB 1427 (AHOR4576/7) (k-21976) and Dans Sainte Croix (k-25782), respectively. Both samples originate from Germany. The average height within the collection amounted 70.8 cm.

In 2012, the group of middle low plants dominated (50.5% of samples). In 2013 the highest amount of low plants was observed (52.3%). We relate the above-mentioned differences with unfavorable water supply conditions in the first case and with more favorable ones in the second one (Figure 16.2).

Ear formation in the conditions of 2012 passed during hot and dry weather, which evidently affected their length. The variation of the character was from 5.5 cm in the sample C. I. 10997 (k-30631) from Peru to 22.6 cm (Kamyshinskiy variety 23, sample k-30244 from Volgograd region), with average value of 16.6 cm.

In some cases the inhibition and depression of growth processes were so strong that the ear formation did not occur or, after ear was formed, its drying was observed.

No expressed dependence between plant height and ear length was detected. Only some samples had big ears and relatively tall plants. It is possible to cite as examples such samples as (k-15223) originating from Egypt, C. I. 11017, (k-30637) and C. I. 11029, (k-30643) from Peru. The height of these plants exceeded 70 cm, and the ear length exceeded 20 cm.

A

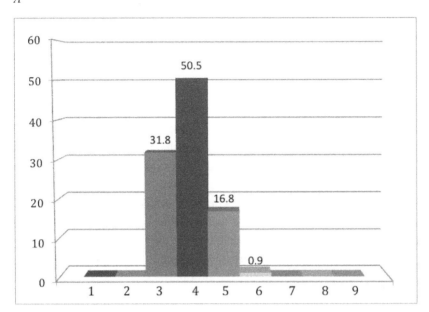

B

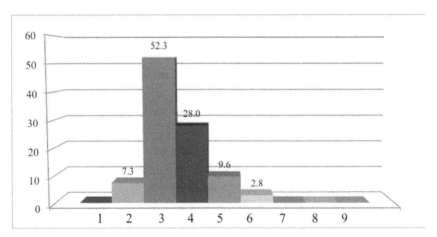

FIGURE 16.2   Distribution of barley samples in the collection by the plant height in the earing phase, % [Note: 1 – dwarf (<41 cm); 2 – very low growing (41–60 cm); 3 – low growing (61–70 cm); 4 – medium low growing (71–80 cm); 5 – medium growing (81–95 cm); 6 – medium tall growing (96–110 cm); 7 – tall growing (111–125 cm); 8 – very tall growing; 126–140 cm) 9 – extremely tall growing (>96–140 cm), *A* – 2012, *B* – 2013].

The length of ear in the conditions of 2013 varied from 1.85 cm (samples: Hadmerslebener 46459, k-23492, Germany) up to 28.0 cm in the sample Vaie (k-22774, USA) with the average value of 19.48 cm. In the vegetation season 2013 we also did not detect expressed dependence between plant height and ear length.

Most of samples can be characterized as resistant to lodging in the conditions of 2012 and 2013. The group with very high level of resistance was the most numerous; it included 51.2 and 84.7% of samples in 2012 and 2013, respectively. These samples got the highest grade – 9 points. No dependence between plant height and resistance to lodging was found.

The amount of plants reaching the moment of full maturity of grains and, consequently, productive stems depends both on individual peculiarities of varieties and environmental factors. The capacity of plants to form bushing out sprouts has a considerable effect on the formation of certain stocking density of stems in crops. The amount of productive stems per 1 m^2 is the indicator determining the productivity of selection sample or variety. This character depends both on genotype and meteorological conditions of the given vegetation season.

In the conditions of 2012 barley samples were characterized by low capacity to form lateral sprouts. Consequently total and productive bushiness was low. In the given conditions it was 244 productive stems per 1 m^2 in average within the collection. The variation of this character among samples was high, namely from 48 to 532 stems/m^2.

When classifying samples by the amount of productive stems according to International CMEA classifier of *Hordeum* L. genus [5] three groups were found differing according the measure of expression of this character.

The group with very low (<200–300) amount of productive stems was the most numerous (80.0% of samples); 19% of samples were characterized by low stem density per unit area (301–500).

Only one variety represented the third group and was characterized by medium value of the character, namely Astana 200 from Kazakhstan (532 stems/m^2). It is possible to notice that other varieties obtained from Research-and-production center of A.I. Barayev grain farm had better characteristics of the productive stems density and bushiness than a number of other collection samples.

The following varieties were also attributed to the best ones: Arna (k-738, Kazakhstan), Целинный 30 (k-003, Kazakhstan), Tselinny 5 (k-003, Kazakhstan), Karabalykskiy 150 (k-30149, Kazakhstan), Kedr (standard, Russia, Krasnoyarsk Territory). The domination of samples from Kazakhstan allows to derive the conclusion about their high ecological plasticity relatively to extreme conditions of 2012, which were expressed in moisture deficit and elevated air temperature.

In the conditions of 2013 the amount of productive stems per 1 m^2 varied within the range from 6 in the samples from Germany (k-25792 and Parseteer Giattgranning) to 568 (samples k-26620 and Mestnyj from Ethiopia).

Barley samples were classified into three groups according to International classifier. The group with medium characteristics was the scanty. Only 4 (1.8%) samples were attributed to it: (L-2048/63/2 Lageiewnik k-22176, Gitte k-25170, Meta k-25682, Local k-26620) from Poland, Germany, the Netherlands and Ethiopia.

The amount of productive stems in standard varieties Acha and Kedr amounted in average 394 and 378 per 1 m^2, which corresponds to the group with small amount.

The environment is characterized by unstability of its factors and all-inclusiveness of its effect on plant organism during the periods of its growth and development [14, 15]. In response to the given environmental conditions cultivated plants must have a broad adaptive potential in order to give high, high-quality and stable yield.

The combination of productivity with high resistance to unfavorable environmental factors in the same variety is one of the main selection tasks [16, 17].

Variety remains not only the tool of productivity increasing, but it is also the factor, which absence prevents to imply the advances of science and technique. In the agriculture variety represents a biological system, which could not be replaced by anything [18].

In the conditions of 2012 the mass of grains from a plot expressed in g/m^2 varied within the range from 11.7 in the sample C. I. 10997 (k-30631) to 218.1 in the sample C. I. 11070 (k-30662) from Peru with the mean value of the whole set under study 95,9 g/m^2.

Among the samples evaluated in 2013 the sample Jackson N1 C.I.7045 (k-24964) from USA was characterized by the lowest grain mass of 4.2 g/m². The maximum value (398.8 g/m²) of grain mass was observed in the sample Arni 1 (k-25783) from Germany. The mean collection value amounted 165.8 g/m².

According to A.A. Zhuchenko [19], E.A. Tyumentseva et al. [20], E.I. Koshkin [21] the high role in adaptive selection belongs to samples, characterized by stability of selection-valuable characters display in changing environmental conditions.

It is important for selection practice to find in collection fund the sources of valuable characters, which can be valuable initial material when applying the methods of hybridization, experimental mutagenesis, polyploidy.

Basing on the results of field and laboratory evaluation of barley samples we proposed sources of some selection-valuable characters (Table 16.1).

It is known that favorable environment is this one providing stable functioning of natural ecosystems in various natural and natural-anthropogenic conditions [22].

The agriculture in Tyumen region is directed to the increase of areas under cereal crops, namely spring wheat, barley, oats [23]. In the present climatic conditions harsh for agriculture the varieties of cultivated plants with broad adaptive potential become especially valuable. In order to create selection material under permanently changing environmental conditions the adaptive selection has predominated importance [19, 24, 25].

It is known that there is no direct connection between high productivity and elevated adaptivity of variety; even vice versa, the higher is potential It is potential productivity of the variety the lower is resistance of its yield under unfavorable conditions. Consequently the revealing of plant forms combining high productive and adaptive properties is now one of the most important issues. The issue of cultivated plants biodiversity conservation is considered as not less important. In order to provide the security of the country from the point of view of food, biological resources, ecology, in order to provide new raw materials for the industry secure conservation, enrichment, research and careful use of plant diversity is important. At the same time plant diversity is endangered by "genetic erosion" [26–28].

The Department of Gray Breads of N.I. Vavilov All-Russian Research Institute of Plant Industry owns one of unique collections of world-wide

TABLE 16.1   Sources of Selection-Valuable Characters of Barley

Character	Samples
Field germination capacity of seeds $n = 12$	C. I. 11007(k-30009, Peru); C.I. 11055 (k-30649, Peru); C.I. 11056 (k-30650, Peru); C.I.11126 (k-30687, Peru); C.I.11118 (k-30682, Peru); Kamyshinskiy 23 (k-30244, Volgograd region); Karabalykskiy-150 (k-30149, Kazakhstan); k-15519 (Kazakhstan); Tselinny 30 (k-003, Kazakhstan); Arna (k-738, Kazakhstan); O-334/71 (k-22210, Czechoslovakia); Kharkovskiy 410 (k-23460, Ukraine); HVS 81583/72 (k-25160, Germany)
Probability of plant survival $n = 7$	Local (k-14933, Tajikistan); Local (k-14944, Tajikistan); Local (k-14967, Tajikistan); C.I. 11118 (k-30682, Peru); C.I. 11056 (k-30650, Peru); Karabalykskiy −150 (k-30149, Kazakhstan); Arna (k-738, Kazakhstan)
Plant height, resistance to lodging, the length of the ear $n = 18$	Local (k-14951, Tajikistan); Local (k-14963, Tajikistan); Karabalykskiy 15 (k-30149, Kazakhstan); Kamyshinskiy 23 (k-30244, Volgograd region); C. I. 10975 (k-30624, Peru); C. I. 11095 (k-30671, Peru); Z-1218/72 (k-22229, Czechoslovakia); Sladkovicovo k-152–8-6 (k-22791, Czechoslovakia); Cree C.I. 15256 (k-22335, США); Vale (k-22774, США); Belle (k-23562, Germany); Athiopien-AB. 1225/47 (k-21986, Germany); Wuirtnds Ivana (k-22133, Germany); Lada (k-25677, Germany); WGA 72–7(k-23873, Ethiopia); DZo 2–770 (k-22019, Ethiopia); Loia (k-21969, France); Cosmos 34 (k-25977, Poland)
Amount of productive stems per 1 m² $n = 5$	Astana 2000 (k-696, Kazakhstan); L-2048/63/2 Lageiewnik (k-22176, Poland); Gitte (k-25170, Germany); Meta (k-25682, the Netherlands); Local (k-26620, Ethiopia)
Grain mass per unit area (g/m²) $n = 5$	Gitte (k-25170, Germany); Haarer Isdania (k-25746, Germany); Arni 1 (k-25784, Germany); Local (k-26584, Ethiopia); Meta (k-25682, the Netherlands)

diversity of barley, oats, rye (the actual name is Department of genetic resources of oats, rye, barley). During several decades professor A.Ya. Trofimovskaya (1903–1991) was the head of the Department of Gray Breads. During this time, she founded the principles and developed approaches to the research of the total diversity of barley and oats. As a result of long-term work with the world-wide gene pool of barley A.Ya. Trofimovskaya with collaborators created one of the most extensive collections. The collection contains forms of different geographic provenance belonging to 150 botanical varieties of barley, representing practically the full world genetic diversity of this culture with the widest range of the variability of the most important, characters including selection ones [29].

The formation of barley collection fund, the study and revealing of sources of selection-valuable characters for soil and climatic conditions of Tyumen region represent the interest for selection and genetic programs.

16.4   CONCLUSIONS

1. Results of the research of *Hordeum* L. Collection represented by 333 samples from 19 countries of the world belonging to 59 botanical varieties showed considerable differences between samples in the display of quantitative characters.

2. The effect of genotype and meteorological factors of the years 2012–2013 on field germination capacity of seeds and plant survival probability during the vegetation season was found. High field germination capacity of seeds was provided by barley samples from Peru – C.I.11055 (k-30649) and from Germany – HVS 81583/72 (k-25160). Considerable part of collection samples was characterized by high resistance to the environmental factors.

3. When classifying the collection by plant height 5 groups were detected: very low, low, moderately low, medium and moderately high. The predominated morphological type was represented by low plants (61–70 cm) in 2012 and in 2013 moderately low plants (71–80 cm), respectively.

4. No expressed dependence was detected between plant height, ear length and resistance to lodging. Most of samples (51.2% in 2012 and 84.7% in 2013) were characterized by high (9 points) stem stability.

5. Diversity of the material studied, different reaction of samples on changing environmental conditions provide general biological stability and high productivity of barley (up to 218.1 g/m^2 in 2012 and up to 398.78 g/m^2 in 2013) in the conditions of southern Tyumen region. Maximum grain production was provided by such samples as Gitte

(k-25170, Germany), Haarer Isdania (k-25746, Germany), Arni 1 (k-25783, Germany), Meta (k-25682, the Netherlands), and Local (k-26584, Ethiopia).

KEYWORDS

- barley
- cultivar
- genetic resources
- sample
- Tyumen region

REFERENCES

1. Yakushkina, N. I. Plant physiology. Manual for university students. N. I. Yakushkina, E.Yu. Bakhtenko. Moscow. Humanitarian Publishing Center "VLADOS," 2005, 463 p (In Russian).
2. Sapega, V. A. Bioclimatic potential of soil and climatic zones of Northern Trans-Ural region/V. A. Sapega, G.Sh. Tursumbekova, N. N. Zhuravleva, S. V. Sapega.. Bulletin of Tymen State University. 2011, №4, 41–44 (In Russian).
3. Magomedmirzayev, M. N. Introduction in quantitative morphogenetics. M. N. Magomedmirzayev. Moscow. Nauka. 1990, 212 p (In Russian).
4. Methodical guidelines on the study and conservation of world-wide collection of barley and oats (4ᵗʰ edition, completed and remade). [Ed. I.G Loskutov]. St. Peterburg. VIR. 2012, 63 p (In Russian).
5. International CMEA classifier of *Hordeum, L.* genus. Leningrad, 1983, 55 p (In Russian).
6. Dospekhov, B. A. Methods of field experiment (with the basics of statistical processing of results of research)/B. A. Dospekhov. Moscow. Kolos, 1979, 416 p (In Russian).
7. Lakin, G. F. Biometry. Moscow. Publisher "Vysshaya Shkola." 1990, 295 p (In Russian).
8. Cultivated flora of USSR: vol. II, part 2. Barley. Lukyanova, M. V., Trofimovskaya, A.Ya., Gudkova, G. N. et al. Leningrad. Agropromizdat Leningrad Branch, 1990, 421 p (In Russian).
9. Gryaznov, A. A. Qualifier of within-species taxons of cultivated (sowing) barley. Educational visual aid, Kostanay: Kosanayskiy Printing House. 2007, 107 p (In Russian).

10. Guzhov, Yu.L. Selection and seed-farming of cultivated plants. Eds. Yu.L. Guzhov, A. Fux, P. Valichek. Moscow. Mir, 2003, 539 p (In Russian).
11. Tretyakov, N. N. Effect of variable lodging intensity on the yield formation and sowing qualities of spring barley seeds. Eds. N. N. Tretyakov, A. F. Yakovlev. Biological Basis of Increasing the Productivity of Crops. Proceedings of Timiryazev Agricultural Academy. Moscow: Timiryazev Agricultural Academy Publishers. 1984, 54–58 (In Russian).
12. Nettevich, E. D. Short stem character and barley selection for the resistance to lodging. Eds. Nettevich, E. D., Sergeyev, A. B. Selection of cereals and leguminous plants for the Nonchernozem zone of Russia. 1974, Issue 32, part 1. 66–69 (In Russian).
13. Bome, N. A. Resistance of cultivated plants to unfavorable environmental factors. Eds. Bome, N. A., Bome, A.Ya, Belozerova, A. A. Tyumen. Tyumen State University Publisher, 2007, 192 p (In Russian).
14. Kosulina, L. G. Physiology of plant resistance to unfavorable environmental factors. Eds. L. G., Kosulina, E. K. Lutsenko, V. A. Axenova. Rostov-on-Don. Rostov State University Publisher, 2011, 236 p (In Russian).
15. Zverev, A. T. Basic lows of ecology/A. T. Zverev Moscow. Publishing House "Paganel.' 2009, 171 p (In Russian).
16. Kadyrov, M. A. Some aspects of selection of varieties with broad agro-ecological adaptation. Eds. M. A. Kadyrov, R. I. Grib, F. N. Baturo. Selection and seed-farming, 1984, №7, 8–11 (In Russian).
17. Koval, R. R. Complex selection of valuable genotypes at the provocative background in autogamous cultivated plants. R. R. Koval. Agricultural biology, 1985, №3, 3–13 (In Russian).
18. Efremova, V. V. Changing of variety composition of winter field agrocoenosis: anniversary issue on 75-years of Krasnodar State University. Eds. V. V. Efremova, Yu.T. Aistova, N. I. Terpugova. Krasnodar. Agro-Ecological Monitoring in the Krasnodar Territory Agriculture, 1997, 468 p (In Russian).
19. Zhuchenko, A.A. Adaptive Potential of Cultivated Plants: Ecological and Genetic Fundamentals. Monography. Kishinev. Shniitsa, 1988, 766 p (In Russian).
20. Tyumentseva, E. A. Reaction of Winter Wheat on Hydrothermic Factors of Wintering and Vegetation Concerning Biological Stability of Plants in the Conditions of South of Tyumen Region of Russia. Eds. E. A. Tyumentseva, N.An. Bome, A.Ya. Bome. Ecological Consequences of Increasing Crop Productivity. Plant Breeding and Biotic Diversity. Chapter 13. Eds, A. I. Opalko, L. I. Weisfeld, S. A. Bekuzarova, N. A. Bome, GE. Zaikov. Toronto-New Jersey. Apple Academic Press Inc. 116–123.
21. Koshkin, E. I. Physiology of Agricultural Crops Stability: Manual. Moscow. Drofa, 2010, 638 p (In Russian).
22. Federal low from 10.01.2002 №7-F3 "About the protection of environment" http://www.rg.ru/2002/01/12/oxranasredy-dok.html (In Russian).
23. Information on the state of the agroindustrial complex of Tyumen region in 2011–2013. November 14, 2013, Official site of the Ministry of Agriculture of Russian Federation. http://www.mcx.ru/documents/document/v7_show/25618.htm (In Russian).

24. Bome, N. A. Selection of Cultures and Methods of Varieties Creation in the Extreme Conditions of Northern Trans-Ural Region. Abstract of DSc. thesis. St. Petersburg: N. I. Vavilov All-Russian Research Institute of Plant Industry, 1996, 46 p (In Russian).
25. Bome, N. A. Resistance of Cultivated Plants to Unfavorable Environmental Factors. Monography. Eds. N. A. Bome, A.Ya. Bome, A. A. Belozerova. Tyumen. Tyumen State University Publishers, 2007, 191 p (In Russian).
26. Alexanyan, R. M. Strategy of World Genetic Pools Interaction in the Condition of Globalizatio. Proceedings on Applied Botany, Genetics and Selection. 2007, Vol. 164, 11–33 (In Russian).
27. Convention on Biodiversity. The Interim secretariat for the CBD, Geneva, Executive Center, 1992.
28. Bome, N. A. Applied and Theoretical Aspects of Cultivated Plants Genetic Pool Formation. Eds. Bome, N. A., Bome, A.Ya., Kolokolova, N. N. Fruit and Berry Growing in Russia: Proceedings State Research. All-Russian Selection and Technological Institute of Horticulture and Nursery of Russian Agricultural Academy. Moscow. 2012, Vol. 34(1), 106–113 (In Russian).
29. Loskutov, I. G., Trofimovskaya, A. Ya. The Development of Works of the Department of Grey Breads. Proceedings on Applied Botany, Genetics and Selection. 2009, Vol. 165, 8–12 (In Russian).

CHAPTER 17

REACTION OF COLLECTION SAMPLES OF BARLEY (*HORDEUM* L.) AND OATS (*AVENA* L.) ON CHLORIDE SALINIZATION

NINA A. BOME and ALEXANDER YA. BOME

CONTENTS

ABSTRACT

The results of study of 53 samples of barley and 36 samples of oat from the collection of N.I. Vavilov Research Institute of Plant Industry, Bol'shaya Morskaya St., St. Petersburg, 190000, Russia according to their reaction on salinity stress in the laboratory experiments are presented. The paper discusses the expediency of the application of the ratio of shoot and root mass, beside of the characters of seeds laboratory germination capacity, for the evaluation of samples. All samples were classified into three groups according to the salt resistance basing on the results of the extensive study taking into account the display of characters. Salt resistant samples can be used as initial material for adaptive selection.

17.1   INTRODUCTION

Salinization of soils is one of important abiotic factors leading to the productivity decrease of cereals, including oat. Under widespread and permanent increase of the salt lands area the objective evaluation of plants according to the salt resistance gets high importance. In the forest-steppe of Tyumen region there are 602.1 thousand ha of solonetzic soils and 158.7 thousand ha of solonetzic soils [1]. Different measure of salinization negatively affects the soils fertility and imposes special requirements to the cultivated plant species [2].

Salinization resistance conditions the need of adaptation to three independent factors at the same time: increase of osmotic pressure, toxic effect of ions and oxidative stress [3]. When using salt soils in the agricultural production is necessary to select cultures and species with the highest salt resistance and to recommend them for the cultivation in different soil and climatic zones [4].

The evaluation of plants salt resistance in the natural conditions of cultivation can give the most complete and authentic view of it. However, this is a long and laborious process. Hence, salt resistance of plants is usually determined in precisely controlled conditions of vegetation or laboratory experiment.

17.2   MATERIALS AND METHODOLOGY

The determination of barley and oat salt resistance was performed in the laboratory of biotechnological and microbiological studies of the Cathedra of Botany, Biotechnology and Landscape Architecture of Tyumen State University according to methods described by V.V. Polevoy [5]. Samples of barley were represented by two subspecies: double-row (*Hordeum distichon* L.) and multi-row (*Hordeum vulgare* L.); oat samples belonged to three species: *Avena sativa* L., *Avena strigosa* Schreb. and *Avena abyssinica* Hocht.

Seeds were laid in Petri dishes, preliminary heated under 175°C during one hour in drying box, on the filter paper in 0,98% solution of NaCl salt (experiment) and in distilled water (control). Before putting seeds were treated during 10 minutes by 1% solution of $KMnO_4$ in order to prevent the fungi development. Sample size was 50 seeds in each dish, experiment was repeated 4 times. Seeds germination was performed in the thermostat TPS-2 under constant temperature of 22°C.

Laboratory seeds germination capacity was determined 7 days after the start of the experiment. The following parameters were measured in order to characterize germs according to quantitative characters: amount of germ roots, length and mass of roots and shoots.

The mathematical processing of experimental data was performed according to standard methods [6].

17.3   RESULTS AND DISCUSSION

When evaluating initial material by standard laboratory methods, such indicators as germination energy and laboratory germination capacity of seeds are common criteria of salt resistance. However, in the scientific literature it is noticed that the change of plant germination characteristics under salinization often has only a weak correlation with salt resistance level of plants [7]. Consequently it is impossible to derive the conclusion about the rate of plants reaction on salinization basing solely on the indicators characterizing the capacity of seeds to germinate, as these data are not informative relatively to the germs development [8].

Therefore, in our experiments, in order to obtain more reliable and objective results we performed the estimation of quantitative characters of primary root system and aboveground organs, namely amount of germ roots, length of roots and shoots, as well as their mass.

When studying morphometric parameters of germs, the inhibition of growth processes since the moment of seeds germination was observed in the experiment relatively to control (Table 17.1).

The results obtained revealed considerable difference in the display of chloride salinization effect on three biological characteristics of barley and oat (amount of germ roots, length of roots and length of shoots). Salinization had harder inhibiting effect on the development of oat germs. The decrease

TABLE 17.1   Variability of Morphometric Parameters of Germs When Germinating Seeds at the Provocative Background With Salinization (Averaged Data)

Character	*Hordeum* L. (n=53)		*Avena* L. (n=36)	
	Control	**Experiment (NaCl)**	**Control**	**Experiment (NaCl)**
Amount of roots, pieces	5.6±0.09	5.3±0.13*	4.8±0.36	3.8±0.21*
Root length, mm	73.1±3.92	52.8±2.78*	97.5±2.24	54.3±1.69*
Shoot length, mm	81.3±4.08	61.7±3.47*	97.6±2.21	44.9±1.18*

*differences between control and experiment are significant with P<0.05.

of morphological characters on the provocative background amounted in barley from 5.4% (amount of germ roots) to 27.8% (roots length) and in oat from 20.8% (amount of germ roots) to 54.0% (shoot length).

The amount of germ roots showed the lowest sensitivity to salinization in the cultures under study. In barley and oat the inhibiting effect was more expressed in root length and shoot length, respectively.

In spite of general regularity expressed in the inhibition of root system and aboveground part of germs, a considerable difference among the samples under consideration in the reaction on salt stress was observed.

When analyzing morphometric parameters of each oat sample the decrease of the character (in %) in the variant with NaCl was determined, which allowed to select samples with favorable display of characters both in control, as well as under salinization. The following samples were attributed to this group: Local (k-14677, Spain), Klock 1 (k-13573, Sweden), Novosibirskiy 88 (k-14031, Novosibirsk region), Metis (k-13915, Tomsk region), Talisman (k-14785, Tyumen region), Local (k-4509, Gorgia), Sprint 3 (k-14659, Sverdlovsk region). The maximum decrease of characters was observed in the following samples: Local (k-2680, Bashkortostan), Krasnoobsky (k-13953, Novosibirsk region), Avena desnuda (k-14704, Peru), Astor (k-11379, the Netherlands), Perona (k-13478, the Netherlands).

In the group of salt resistant samples of barley the display of characters under study on the background of salinization relatively to control was positive, negative and neutral. In three samples: Jubilant (k-29889, Czech Republic), Cheliabinsky 95 (k-30450, Cheliabinsk region), Cheliabinsky 1 (k-30819, Cheliabinsk region) in the conditions of stress the increase of roots amount was observed. Botanic form (k-24853) from Germany and the sample Sokol (k-30827) from Rostov region responded to the effect of stress factor by significant increase of shoot length. According to the length of roots salt resistant samples in most of the cases were at the level of control. These data may indirectly indicate the relative convenience of the samples from this group for the cultivation at salt soils.

The display of all morphological characters under salinization in samples Acha (standard, Novosibirsk region), Staly (k-30212, Belarus), Gorinsky (k-3081, Belgorod region), as well as in the sample from Ethiopia (L. AHOR2547/63) was the smallest, which allows deriving the conclusion about their low salt resistance. Seeds of the sample Brenda (k-30464, Germany) in the experimental variant did not form germs, whereas in the control they gave good results.

It is possible to judge about the character of growth and development of germs according to the characteristics of germ biomass structure. The ratio of shoot and root mass is determined genetically and reflects the relation between processes of growth and development of organs. Under plant adaptation to stress conditions the coordination of shoots and roots growth aimed to the optimal use of resources occurs.

In the structure of barley fresh mass on the provocative background the domination of roots with decrease of the shoot portion relatively to control was observed.

The analysis of the average ratio of absolutely dry mass of roots and shoots in barley collection of samples under study showed that in control variant shoots loosed the mass more than roots, which led to the increase of the portion of roots in dry biomass. In the variant with NaCl volume fraction of roots in dry biomass decreased relatively to fresh mass and the fraction of shoots consequently increases, which is probably related with higher accumulation of moisture by roots under salinization and its active loss when drying (Figure 17.1).

In the collection samples of oat some peculiarities were detected in the structure of fresh biomass. In 15 samples in the control biomass structure was characterized by the domination of aboveground biomass over root mass. Sample Rovesnik (k-14365, Kemerovo region) was characterized by the highest portion of shoots – 67%, whereas in two samples from Omsk region (Orion, k-14422 and Tarsky 2, k-14778), this value was considerably

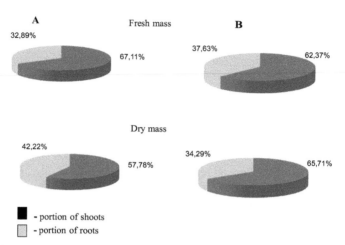

FIGURE 17.1 Ratio of roots and shoots mass in barley samples under study in standard conditions (A) and under salinization (Б).

lower – 41%. In such samples, as Local (k-2680, Bashkortostan), Valdin 765 (k-14574, Krasnodar region) the ratio of roots and shoots was equal to 1. In the biomass structure of 19 samples root mass dominated over shoot mass.

In the solution of NaCl the change of the root to shoot mass ratio relatively to control was observed and the character of this change among samples under study was ambiguous. Consequently the samples were classified into three groups; representatives of each group are shown at Figure 17.2.

About 7 samples with ratio of roots and shoots in control and at the salt background close to 1, which indicates optimal use of resources in early ontogenesis, were attributed to the first group. Such growth type was observed in the following samples: Local (k-2680, Bashkotostan), Orion (k-14422, Omsk region), Skakun (k-13780, Ulyanovsk region), Local (k-4509, Georgia), Anchar (k-14270, Irkutsk region), Zlotniak (k-13523, Romania), CAV 3045 (k-14826, Ethiopia).

Second group is represented by samples with increase of shoot portion with evident inhibition of root system at the provocative background. There were four such samples: Tyumensky golozerny (k-14784, Tyumen region), Tadzhiksky 50 (k-13553, Tadzhikistan), Astor (k-11379, the Netherlands), Borowiak (k-14793, Poland).

According to the response to salinization considerable part of collection (25 samples) was characterized by the acceleration of root system formation

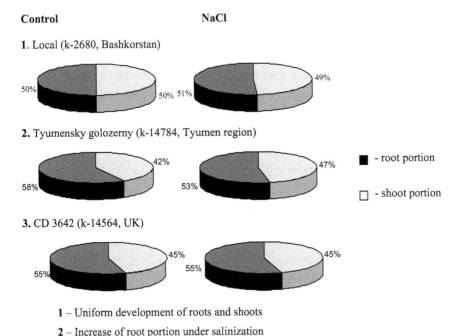

Control   **NaCl**

1. Local (k-2680, Bashkorstan)

50% 50% 51% 49%

2. Tyumensky golozerny (k-14784, Tyumen region)

58% 42% 53% 47%

■ - root portion

□ - shoot portion

3. CD 3642 (k-14564, UK)

55% 45% 55% 45%

1 – Uniform development of roots and shoots

2 – Increase of root portion under salinization

3 – Increase of shoot portion under salinization

FIGURE 17.2   Oat response to salt stress in early ontogenesis according to biomass characteristics.

and slowing of shoot growth. The following samples were attributed to the third group: Rovesnik (k-14365, Kemerovo region), Local (k-8108, Leningrad region), Vagay 2 (k-14786, Altay region), Perona (k-13478, the Netherlands), Local (k-4491, Georgia); the portion of roots in the biomass of germs increased in these samples by 12%, 22%, 14%, 12%, and 16%, respectively.

Our study shows higher inhibition of aboveground system at NaCl background. In the case of roots portion increase in the biomass of oat germ of samples under study relatively to shoot biomass, it is possible to speak about specific adaptive reaction to salt stress expressed in the increase of root system biomass. By this way plant organism more completely uses environmental resources under insufficient water supply of cells, which is observed under salt stress.

On the base of the evaluation of the complex of characters characterizing seeds ability to germinate and morphometric parameters of germs barley and oat collection samples were classified into three groups: salt resistant, sensitive to salinization and salt non-resistant. In barley 16 samples (30,2%) were characterized by high resistance to salinization, strong inhibiting effect of NaCl was observed in 12 samples (22.6%) and 25 samples (47.2%) showed medium resistance to the effect of stress factor.

In the group of salt-resistant 11 samples belong to the subspecies of double-row barley and five to the subspecies of multi-row barley; variety *nutans* dominates (Table 17.2).

According to the results of complex evaluation 36 collection samples of oat were classified into groups relatively to salt stress. 6 samples (16.7%) were highly resistant to salinization: Local (k-14677, Spain), Sprint 3 (k-14659, Sverdlovsk region), Valdin 765 (k-14574, Krasnodar region), Klock 1 (k-13573, Sweden), Local (k-4509, Georgia), Talisman (k-14785, Tyumen region) (Table 17.3). These samples can be recommended to be included in selection and genetic programs as sources of resistance to salinization.

TABLE 17.2 Characteristics of Salt-Resistant Barley Samples

Sample (number in VIR-catalog, origin)	Subspecies	Variety
Cheliabinsky 1(k-30819, Cheliabinsk region)	*Hordeum distichon* L.	*nutans*
Nutans 2419 (k-30536, Samara region)	*Hordeum distichon* L.	*nutans*
Sokol (k-30827, Rostov region)	*Hordeum distichon* L.	*nutans*
Novichok (k-30806, Kirov region)	*Hordeum distichon* L.	*nutans*
Zoriany (k-30496, Ukraine, Vinnitsa region)	*Hordeum distichon* L.	*nutans*
Botanic form (k-24853, Germany)	*Hordeum distichon* L.	*triceros*
Mutant 4033 (k-20225, Germany)	*Hordeum distichon* L.	*medicum*
Annabel (k-30821, Germany)	*Hordeum distichon* L.	*nutans*
Ca 111430 (k-3044, Denmark)	*Hordeum distichon* L.	*nutans*
WW-7024 (k-30445, Sweden)	*Hordeum distichon* L.	*Nutans*
Anadolu 86 (k-30319, Turkey)	*Hordeum distichon* L.	*nutans*
k-19709 (Denmark)	*Hordeum vulgare* L.	*ibericum*
13662/8 (k-30429, Ukraine, Vinnitsa region)	*Hordeum vulgare* L.	*ricotense*
Belgorodsky 95 (k-30449, Leningrad region)	*Hordeum vulgare* L.	*pallidum*
L. AHOR 2553/66 (k-20045, Ethiopia)	*Hordeum vulgare* L.	*grseinigrum*
Hause 563 (k-24811, Germany)	*Hordeum vulgare* L.	*horsfordianum*

TABLE 17.3   Characteristics of Oat Samples Resistant to Salinization

Sample (number in VIR-catalog, origin)	Subspecies	Variety
Local (k-14677, Spain)	*Avena strigosa* Schreb.	*typica*
Sprint 3 (k-14659, Sverdlovsk region)	*Avena sativa* L.	*aurea*
Valdin 765 (k-14574, Krasnodar region)	*Avena sativa* L.	*aurea*
Klock 1 (k-13573, Sweden)	*Avena sativa* L.	*montana*
Local (k-4509, Georgia)	*Avena sativa* L.	*aristata*
Talisman (k-14785, Tyumen region)	*Avena sativa* L.	*mutica*

In the case of strong inhibition of germs growth and development in salt substrate it is possible to speak about low resistance to salt stress. About 7 samples (19.4%) were attributed to the group of salt resistance, 4 of them were representatives of foreign countries and 3 represented regions of Russia: Local (k-2680, Bashkortostan), Astor (k-11379, the Netherlands), Krasnoobsky (k-13953, Novosibirsk region), *Avena desnuda* (k-14704, Peru), Irtysh 13 (k-13924, Omsk region), Vagay 2 (k-14786, Altay region), Local (k-4491, Georgia).

Thus, in order to get informative characteristics of resistance to stress, including salinization, it is useful to apply laboratory evaluation methods, which allow in relatively brief terms and on small laboratory grounds detect forms resistant to unfavorable effects b applying provocative backgrounds. The main advantage of such methods is the possibility to forecast the selection of valuable plants in early ontogenesis [9–11]. In the field conditions salinization also can considerably affect germs, especially under high salt content in upper soil layers.

Salt-resistant samples can be recommended for adaptive selection in the conditions of Northern Trans-Ural region. The obtained results interest researchers when creating the model of variety for local soil and climatic conditions.

17.4   CONCLUSIONS

1. When cultivating barley and oat seeds on the provocative background (0.98% solution of NaCl) inhibition of growth processes in early ontogenesis was observed, which was displayed in the

decrease of quantitative characters (length and mass of roots and shoot). Salt stress evokes considerable slowing of growth until its full stop. According to phenotypic display the inhibiting effect of NaCl is the most expressed in oat germs.

2. The indoubt response of oat samples on the effect of salinization was detected. On the provocative background one part of samples (n=7) was characterized by uniform development of roots and shoots (ratio of their biomasses was close to 1). In four samples the shoot portion increases under salinization on the background of root system inhibition. Considerable part of collection (25 samples) responded to stress factor by active formation of root system under slowed shoot growth.

3. Salt-resistant samples of barley were more often at the control level according to the display of the characters; in some cases stimulation of growth processes was observed. In the germs of samples non-resistant to salinization considerable lag was observed in all parameters (laboratory germination capacity of seeds, morphometric parameters and biomass of germs). In the stress conditions primary root system dominated within the structure of germ fresh biomass, at the same time in the absolutely dry mass the portion of shoots increased.

4. According to the complex study taking into account the display of characters all samples were classified into three groups relatively to salt resistance: salt-resistant, sensitive to salinization and salt non-resistant. The first group is characterized by the most favorable combination of characters (seed germination capacity and germs development).

KEYWORDS

- **germs**
- **morphometric parameters**
- **salinization**

REFERENCES

1. Guzeeva, S. A. Condition of alkaline soils of the south of the Tyumen region and aspects of their mastering. Autoabstract for Candidate of Biological Sci. Tyumen, 2007, 14 p (In Russian).

2. Zonal system of farming the Tyumen Region: Recommendations. Novosibirsk: SB Academy of Agricultural Sciences, Agricultural Research Institute of Northern Zauralye, 1989, 444 p (In Russian).

3. Baranova, E. N. Problems and prospects of genetic-engineering approach to solving the problem of stability of plants to salinity. E. N. Baranova, A. A. Gulevich. Agricultural Biology. 2006, №1, S. 39–52. (in Russian) (In Russian).

4. Batasheva, B. A. Barley plant resistance to salt stress. .B. A. Batasheva, A. A. Al'derov. Agricultural Biology. 2005, №5, 56–60 (In Russian).

5. Polevoy, V. V. Workshop on growth and resistance of plants. V. V. Polevoy, T. V. Chirkov, L. A. Lutova. St. Petersburg: Publishing St. Petersburg University, 2001, 212 p. (in Russian)

6. Lakin, G. F. Biometrics. Moscow, Vysshaja shkola, 1988, 294 p (In Russian).

7. Udovenko, G. V., Semushina, L. A., Sinelnikova, V. N. Features a variety of methods to assess salt tolerance. Methods for evaluating plant resistance to adverse environmental conditions. Leningrad, Kolos, 1976, 228–238 (In Russian).

8. Bome, A.Ya. Variability of quantitative traits of spring wheat under salt stress. The successes of modern science. The Academy of Natural Sciences, 2003, # 3. 60–61 (In Russian).

9. Kolokolova, N. N. Creation of ecological and genetic model of quantitative characters in plant responses to biotic and abiotic stresses Eds. N. N. Kolokolova, N. A. Boma, E. B. Zhelnina, L. M. Sabitova, Bome, A. Ya. Environment and Management of Natural Resources: Abstracts of the International Conference. Tyumen, October 11–13, 2010, The Tyumen, Tyumen State University Publishing House, 2010, 54–56 (In Russian).

10. Bome, A. Y. Investigation of the gene pool *Triticum aestivum, L.* on plant response to low temperatures. Eds. A. Y. Bome, N. A. Bome. Natural and engineering sciences. 2012, №1 (57). 117–121 (In Russian).

11. Bome, N. A. Adaptive and productive properties of *Triticum aestivum, L.* N. A. Bome, T. F. Ushakova, A. Ja. Bome, E. V. Zuev. Envonmemt and natural resource management. V. International Conference. Abstracts. Tyumen, 1–3 October, 2014, Tyumen, Tyumen State University Publishing House. 2014, 43–45 (In Russian).

CHAPTER 18

RESISTANCE TO IMPACT OF ENVIRONMENT FACTORS OF HYBRID FORMS SOFT SPRING WHEAT (*TRITICUM AESTIVUM* L.)

ELENA I. RIPBERGER and NINA A. BOME

CONTENTS

ABSTRACT

The aim of the given research was to study field seed germination and biological resistance of the parent variety plants and hybrids F_1–F_4 of the soft spring wheat in the natural and climatic conditions of the Western Siberian Lowland. Sharply changing climatic conditions and the variety of soil types are the peculiarities of the given territory. This stipulates the necessity of the selection of cultivars with a wider adaptive potential.

The hybrid forms were created and studied within four years (2010–2013). From them samples were singled out which have a wider adaptive capability according to the index of the field seed germination and biological resistance of the plants in the vegetative periods distinguished by the hydrothermal conditions.

18.1 INTRODUCTION

In accordance with the Federal law of the Russian Federation "About the environment protection" [1] the quality of favorable environment provides resistant functioning of natural ecological systems in different natural and natural-antropogenic conditions. The Tyumen Region is characterized by great soil resources potential and belongs to the zone of risky agriculture. It is determined by strict natural and climatic conditions and low bioclimatic potential [2, 3].

The agriculture in the Tyumen Region is directed at the extension of the spring wheat sown areas [4]. In the forming severe natural conditions the special value is acquired by the cultivars of the agricultural plants with broad adaptive potential. For the creation of the selective materials in permanently changing environmental conditions the adaptive selection has a dominant meaning [5].

The aim of the given research was to study seed germination and field biological resistance of the parent and hybrid plant forms of soft spring wheat in the conditions of the Tyumen Region South.

18.2 MATERIALS AND METHODOLOGY

About 10 hybrid combinations of soft spring wheat served as objects of the research: Cara x ♂ Lyutestsens 70, ♀Cara x ♂ Skent 1, ♀Cara x ♂ Skent 3, ♀Hybrid x ♂Cara, ♀Hybrid x ♂ Lyutestsens 70, ♀Hybrid x ♂ Skent 1, ♀Hybrid x ♂ Skent 3, ♀ Lyutestsens 70 x ♂Skent 1, ♀ Lyutestsens 70 x ♂ Skent 3, ♀ Skent 3 x ♂ Skent 1, and their parent forms: Cara, Hybrid, Lyutestsens 70, Skent 1, Skent 3.

The hybrid forms were received by us in 2009 on the experimental district of the biological station "Lake Kuchak" of Tyumen State University.

Incomplete diallel crossings were carried out. The hybridization method with the use of methodical references of V.F. Doropheev was used [6]. The cultivars for hybridization were selected according to the complex study of the collective fund of the soft spring wheat. It was conducted on the experimental plot of Tyumen supporting point of N.I. Vavilov Research Institute of Plant Industry.

The initial samples extracted from the collective fond of N.I. Vavilov Research Institute of Plant Industry varied in ecological and geographical descend Kazakhstan (Lyutestsens 70, Skent 1 and Skent 3) and Mexico (Cara and Hybrid) as well as the variants: erithrospermum Korn. – cultivars Cara, ferrugineum (Alef.) Manf., Hybrid, and lutescens (Alef.) Manf. – cultivars Skent 1, Skent 3, Lyutestsens 70.

The research took place during four vegetation periods (2010–2013) within the precincts of Tyumen State University biostation "Lake Kuchak" which is situated in the Nizhnetavdinski district of the Tyumen Region. The experimental plot is situated on the border between the two agro-climatic zones: subtaiga and north wooded steppe. The soil of the district is cultivated, sod-podzol, sandy. The soil analysis was carried out on the basis of the laboratory ecotoxicology of the Tobolsk complex scientific station of the Ural RAS Department. The gross content of the elements in the samples of the soil was defined. Atomic emission method with the application of spectrometer OPTIMA-7000DV was used.

The acidity of the soil on the experimental spot in the salty extract accounted for 6.6 (alkalescent). The content of humus in soil is 3.67%. The solid residue (the indicator of the soil salinization) is 0.47%, in the norm 0.30%. The amount of the anions accounted for (mg-eq): Cl^- – 0.43±0.00; SO_4^{2-} 0.2±0.00; HCO_3 – 0.23±0.01. Cations (mg-eq): Mg^{2+} – 1.66±0.04; Ca^{2+} – 6.86±0.06. The content of the biogenic substances (mg/kg): NH_4^+ – 19.5±0.12; NO_2 – 9.15±0.73; NO_3 – 18.8±0.32; H_2PO_4. and HPO_4 – 433.3±34.51. The gross content of macro- and microelements (mg/kg): As – 2.09; Ca – 3362.33; Cd – 25.02; Co – 17.52; Cr – 92.27; Cu – 55.41; Fe – 3553.51; Mg – 1125.37; Mn – 382.64; Mo – 68.61; Ni – 61.84; Pb – 38.99; Sr – 29.69; Zn – 402.52.

The content of the chemical elements in the soil did not exceed maximum permissible concentration in accordance with the hygienic standard 2.1.7.2014–06 of the Russian Federation [7].

According to the average long-standing data analyzed by A. S. Ivanenko [8], there are following climatic peculiarities of the Lower-Vartovsk district of the Tyumen Region: firstly, the indicator of hydrothermal coefficient (HTC), reflecting the natural provision of the territory with moisture varying within the limits from 1.2 (humid) up to 1.5 (semi-arid); secondly the amount of the precipitation 220–240 mm during the vegetation period; thirdly average provision with productive moisture 35 mm in the spring crops sowing period on the depth of 0–20 sm; fourthly the general humidity deficit is 100 mm for the summer period (May-August) in the average year (V.S. Mezentsev and I.V. Karnatsevich, cited according to [8]); fifthly the sum of active temperatures more than 10°C accounted for 1875.0°C during120 days.

In the summer field experiment 2010 a technique of each plant individual assessment was used. The sowing of the material under investigation was carried out in blocs with the inclusion of parental and hybrid forms: P_1; F_1; P_2 (the experimental division location example). The square of nutrition for each plant accounted for 10×20 sm.

During the vegetation periods 2011, 2012 and 2013 hybrids were assessed according to families (a family is a posterity of one plant). The depth of the soft spring wheat sowing accounted for 5–6 sm. The experiments were laid according to the techniques [9, 10]. The terms of the sowing 2010–6.05; 2011–10.05; 2012–2013–9.05.

For the biological resistance (survivability of the plants) correlation between the amounts of the kept for the gathering plants to the number of sown seeds expressed as a percentage.

According to the HTK indicator vegetative periods in the years of the investigation varied from weakly arid (2010 and 2013), arid (2009 and 2011) to extremely arid (2012). The sum of active temperatures more than 10°C in the vegetative period in 2009 accounted for 1674.0°C, in 2010–1943.7°C, 2011–1755.8°C, 2012–1950.7°C and 2013–1847.9°C.

The characteristic feature of the vegetative period of spring wheat is the uneven distribution of the precipitations. The deficit of the moisture on the background of the enhanced air temperatures was noted during the exposure to separate phonological phases of plant development. It was especially pronounced in 2012. According to the quantitative characteristics (maximum temperature, periods without precipitation) record meanings were noticed compared to average long-standing data.

The average long-standing meanings were taken as a norm of the temperature (°C) and the amount of precipitation according to the information from the Tyumen weather station.

18.3   RESULTS AND DISCUSSION

Inconstancy of factors and complex influence on the plant organism during its growth and development are characteristic of the environment [11–13]. Cultivated plants must have a wide adaptive potential in response to the forming conditions of the environment to provide big, qualitative and stable harvest.

The field germination and survivability of the plants during the vegetative period can be considered as two very important interrelated features defining the ability of the seeds to germinate in the developed conditions and complex resistance of the plants to the environment factors.

The basic requirement to the seeds is the possibility to germinate, give lavish sprouts. They are not only to survive in the presence of unfavorable environment factors, but also properly grow and develop as well. In opinion of P.I. Zhukovski [14], A. I. Nosatovski [15], A. A. Kovalenko [16], D. F. Askhadulin [17], M. E. Mukhordova and E. G. Mukhordov [18] the field seed germination depends on a great number of factors: the quality of the sown seed material, laboratory germination, coarseness of the seeds, the seed material maintenance condition, traumatizing of the seeds during the harvesting, lesion of the seeds with bacteria and phytopathogenic fungi, treatment of the seed material before the sowing, the depth of the seed embedding, grain-size soil contain, the extent of the soil humidity, mineral and organic fertilizer application, biochemical content of the grains, species and variety peculiarities.

In the Tyumen Region low field germination of the seeds is one of the basic problems in the agriculture [8, 19]. On this basis, the indicator of seed germination was considered by us as one of the key in the evaluation of the hybrid and initial forms.

The influence of the weather factors on the germination of the seeds was noted in the experiment during the analysis of average meanings of the field germination of the seed cultivars and species of soft spring wheat in the researches. At initial cultivars in the field conditions the germination of the seeds varied from 70.5 to 91.0%, at hybrid forms from 68.0 to 84.9%.

At the conditional distribution of the initial forms and hybrid 68.0 to 84.9% combinations on the field germination of the seeds into 4 groups (with low, middle, high and very high field germination of the seeds) it was set that the majority of the studied spring wheat samples in the years of the researches (2010–2013) were characterized by high and very high indicators. In the conditions of 2013–60.0% of the forms were attributed to the group with the middle indicator of the field seed germination. In the period of the sowing and sprouts 60.0% the drop in temperature by 1.6°C was noted (the lowest temperature −1.9°C was recorded for 13.05.2013) in the combination with a lot of (140.9% to the norm) precipitation (HTK = 1.02 – the conditions were characterized as slightly arid). The complex manifestation of these factors can be considered as the cause for the seed field germination indicator reduction (Figure 18.1).

The comparative analysis of the initial material and hybrids of four generations allowed to disclose some differences in the number of full value sprouts.

So the germination of the seeds and formation of the sprouts in May 2010 (from 06.05. to 30.05.) took place at the average daily temperature of 13.0°C (average long standing meaning 10.6°C, which was called

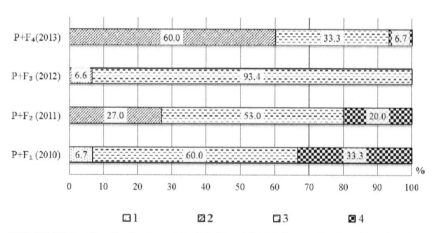

FIGURE 18.1   The distribution of the hybrid and initial forms of soft spring wheat into groups on the field seed germination, % [Note: 1 – low (<50%); 2 – middle (51–70%); 3 – high (71–90%); 4 – very high (>90%); P – parental forms; F_1–F_4 – hybrid forms].

by us "norm"). Despite the fact that the amount of the precipitation in the 1 and 2 decades of the month was minimal (1.0 and 0.4 mm, accordingly), HTK in the period of the sowing and germinating accounted for only 1.24 (slightly arid), the appearance of the sprouts was harmonious.

Taking into account that higher temperature is necessary for the formation of the sprouts than for the seed germination we can say the temperature regime was favorable. It was enough moisture for the swelling of the sprouts as the seeds used it from their spring supplies.

During the comparison of the parental and hybrid forms it was discovered that in F_1 a group of hybrids appeared (20.0%) with low field germination of the seeds, which can be connected with less supply of nutrients in small seeds (Figure 18.2). The inhibition of the growth can be caused by underdevelopment and traumatization of the hybrid caryopsis endosperm, which is the supplier of the nutrients to the seed bud [16].

The sum of the precipitation for May 2011 accounted for only 71.6% with respect to average long-standing meaning with its uneven distribution according to decades of the month. In the first decade 3.0 mm fell, in the second – 0.7 mm and in the third – 6.9 mm on the background of high temperature (by 1.3°C higher). The seeds received necessary amount of moisture for the germination only in the first decade of June (37.9 mm). Swelling and seed germination of parental and hybrid (F_2) forms took place when there was not enough moisture in the soil. The definition of the germinated seed amount took place in two terms: the sowing 10.05, the calculation 7.06 and 12.06. Such influence of abiotic factors could become a cause for the uneven appearance of the sprouts (look Figure 18.2).

The distribution of the initial cultivars and hybrids into the groups with different indicators of the field germination discovered that the part of the samples with high and very high meanings of the given character decreased in comparison with the year 2010. The majority of the samples (53.0%) were included in the group with very high field germination of the seeds (look Figure 18.2). At the initial cultivars the given groups accounted for 60.0% in total.

Enhanced temperature was characteristic of May 2012 (by 2.4°C higher than norm) and deficit of the precipitation (33.9% to the norm) (HTK = 0.25, dry period from 9.05 to 24.05). The total amount of the precipitation varied from 0.3 to 4.0 mm, the number of the days with

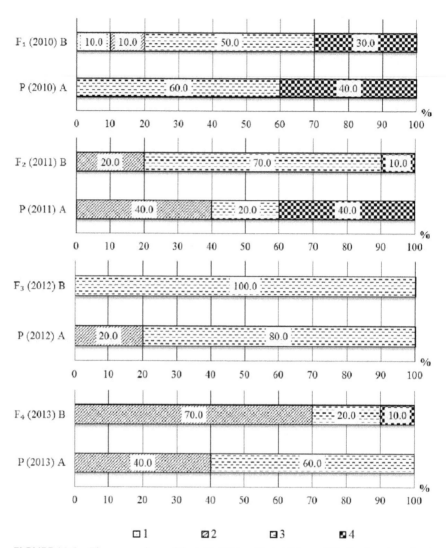

FIGURE 18.2   The comparison of the initial (A) and hybrid forms (B) of soft spring wheat on the field seed germination, 2010–2013 [Note: 1 – low (<50%); 2 – middle (51–70%); 3 – high (71–90%); 4 – very high (>90%); P – parental forms; F_1–F_4 – hybrid forms].

precipitation accounted for 8. The seed germination, the formation of the sprouts took place in warm weather with slight amount of precipitation.

The field germination of hybrids (F_3) varied from 72.8 to 85.0%. On the basis of these meanings they were attributed to one group with high meanings of the character. At the initial cultivars the variation of the given character fitted in the limits from 62.0% to 82.0%. The exposure of the two sample groups with middle (20.0%) and high (80.0%) germination indicates the differences in the passing of metabolic processes in the germinating seed bud and less expressed adaptive features of the sprout at the transformation from mesotrophic to autotrophic nutrition (look Figure 18.2).

As it was said above, the hydrothermal conditions of the sowing and sprouting period (9.05. – 2.06) 2013 were characterized as slightly arid (HTK = 1.02) with low temperature meanings (by 1.6°C) with respect to the average long standing meaning. Even despite difficultly developed conditions for the formation of the sprouts in the conditions at hybrid F_4 forms a group with field germination higher than 90.0%. The rest 90.0% of the hybrid forms were attributed to the groups with middle and high meanings of the given indicator (look at Figure 18.2).

The conducted investigations manifest extensive variation of the seed field germination indicator in different vegetative periods with drastically differing hydrothermal regime. Exciting interaction of the system genotype-environment was noted in 2012 (HTK in the period of sowing-germination (from 9.05. to 24.05) accounted for 0.25 – arid). It took place during the influence of high temperature (by 2,4°C higher in respect of the average long standing meaning) and low amount of the precipitation (only 34.0% with respect to the average long standing meaning). In the developed conditions 93.4% of the samples are attributed to the group with high indicator of the field seed germination and low variation of the character (2 groups of the samples: with the average germination – 6.6%, high – 93.4%). During the impact of the lowered temperatures (by 1.6°C less than the norm) in the combination with a lot of precipitation (140.9% to the average long standing) in 2013 variation of the given field seed germination indicator was noted: with the average germination 60.0%, high – 33.3% and very high – 33.3% (look at the Figures 18.1 and 18.2).

In the years of the investigation the variation of the average biological resistance indicators accounted for 45.4 up to 63.2% at initial cultivars and 48.0 up to 54.9% at hybrids (F_1–F_4).

The excess average daily temperatures are characteristic of the vegetative period 2010 in May (1.7°C), June (1.2°C) and August (2.1°C) in combination with the shortage (May – 6.7%, July – 40.9%, August – 25.2% to the norm) and excess of the precipitation (June – 130.5% and September – 110.5% with respect to the norm). On the whole the conditions of the vegetative period were characteristic as slightly arid (HTK=1.05) with the excess of the maximal meaning of the total amount of active temperatures more than 10°C for soft spring wheat by 243.7°C (the norm according to K. A. Flyagsberger [20] 1500–1750°C). The reduced to –5.3°C temperature noted on 21.05 in the period of the sprout formation could be a limitative factor for young juvenile plants. The minimal temperature of spring wheat plant growth in the period of the sprouts and formation of the vegetative organs accounts for 4–5°C according to N. N. Tretyakov and A. S. Loseva [21]. The lack of moisture was observed phytopathogenic disease and damage caused by pests. It reflected in the biological resistance indicators. At 73.0% of the studied material the survivability of plants was low and middle (Figure 18.3).

F_1 hybrids are more resistant compared to roditelsikim forms. The appearance of the forth group of plant including survivability 10.0% of the form serves as a proof (Figure 18.4).

Vegetative period of 2011 (with the exception of July) was characterized by enhanced temperature. The shortage of moisture was noted in May (71.6% to the norm), July (32.9%), August (52.2%) and September (15.3%). Hydrothermal index in the developed conditions of 2011

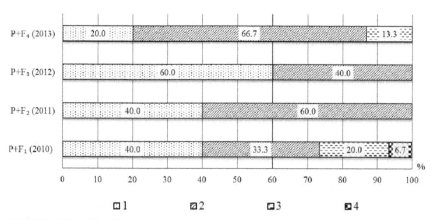

FIGURE 18.3 The distribution of the hybrid and initial forms of soft spring wheat according to the biological resistance, % [Note: 1 – low (<50%); 2 – middle (51–70%); 3 – high (71–90%); 4 – very high (>90%); P – parental forms; F_1–F_4 – hybrid forms].

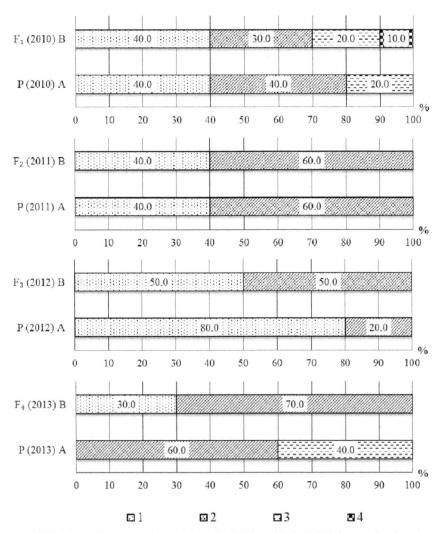

FIGURE 18.4 The comparison of the initial [A] and hybrid [B] forms of soft spring wheat on biological resistance of the plants during the vegetation, 2010–2013 [Note: 1 – low (<50%); 2 – middle (51–70%); 3 – high (71–90%); 4 – very high (>90%); P – parental forms; F_1–F_4 – hybrid forms].

accounted for 0.96 [22]. Excess amount of precipitation (52.0 mm, which is 49.0% higher than the norm) in the second decade of June with the temperature excess by 1.4°C contributed to the creation of favorable conditions for the development of phytopathogenic fungi.

During the distribution of the soft spring wheat samples into groups of plant survivability a common consequence was revealed. It shows itself in the equal correlation of the groups in the analysis of the whole set of samples as well as of the hybrids F_2 separately and parental cultivars. In 40.0% of the cases low ability to withstand the impact of the unfavorable factors of the environment (look Figures 18.3 and 18.4).

The conditions for the passage of all the ontogenesis stages in 2012 can be characterized as extremely arid (HTK = 0.50) with the sum of active temperatures more than 10°C accounted for 1950.7°C. In June the average daily temperature exceeded the average long standing meaning by 4.1°C. The sum of precipitation accounted for 63.7% of the norm. The distribution into the decades was uneven. The essential temperature fluctuation from 7.6°C (08.06.) to 32.3°C (21.06.) could be attributed to stressful influence on the plants in that period. The average daily temperature for July was equal to 21.4°C with the average long standing meaning 18.6°C, the amount of the precipitation – 24.3 mm with the norm 84 mm, the amount of days with precipitation – 8. The conditions for the ripening and aging of the corns in August developed unsuccessfully. Despite the fact that in the second decade the major amount of precipitation fell (166.5% to the norm), they could not grade the consequences from the draft. The average daily temperature for the month was high and accounted for 17.7°C (norm 14.5°C), the sum of the precipitation – 29.3 mm (50.5% from the norm). In the extremely arid conditions of 2012 the investigated samples were divided into two groups with low (60.0%) and middle (40.0%) biological resistance of the plants. 50.0% of the created hybrid forms had middle survivability of the plants (look at Figures 18.3 and 18.4).

The excess of the average daily temperature in June (0.2°C), July, (1.0°C), August (1.4°C) and varying distribution of the precipitation from overwetting in May (140.9%) and July (142.4%) to shortage – in June (62.0%) and August (62.7%) with the fluctuations of the temperature in June 1.9°C (2.06.) and 32.0°C (20.06) were the peculiarities of the vegetative period in 2013 in the tillering period – stem elongation. This period is very important for the formation of generative organs. The indicator of the hydrothermal index existed within the limits of the norm for the given agro-climatic zone (HTK = 1.19).

On the whole the group with low survivability of soft spring wheat predominated according to the studied material (look at the Figure 18.3). Enhanced resistance to stress factors was characteristic of the hybrids F_3. Equal correlation of the groups with low and middle survivability was observed while 80.0% of parental cultivars experienced considerable depression of the growth processes (look at the Figure 18.4).

On the basis of the received data we can note that the indicator of the biological resistance is in the strong dependence on genetic peculiarities of the studied material as well as from the biotic and abiotic factors of the environment. It can also refer to the indicator of the field seed germination. I.B. Phakhrudenova and G.A. Loskutova [23] inform about the dependence on the survivability indicator, which depends on the peculiarities of the variety and definite combination of the outdoor environment factors.

According to the standpoint of A.A. Zhuchenko [5], L.G. Kosulina with co-authors [11], E.I. Koshkina [24], E.A. Tumentseva with co-authors [25] the samples characterized by stability of the selectively valuable characters under the changing environment conditions. On the basis of the given by us long standing researches in this relation the hybrid forms have been singled out: ♀Lyutestsens 70 x ♂Skent 1, ♀Hybrid x ♂Cara and ♀Skent 3 x ♂Skent 1.

18.4 CONCLUSIONS

The dependence of the field seed germination indicators and plant biological resistance of the parental and hybrid forms on the hydrothermal conditions of the vegetative period and genotypic peculiarities of the studied material was revealed on the basis of four-year researches.

According to our data in difficult conditions of 2010 on the complex characters manifestation hybrid combination ♀Lyutestsens 70 x ♂Skent 1 was singled out. 100% of the sprouts were extracted from it and no plant death was recorded. In 2013 the given hybrid combination was characterized by very high indicators of the field seed germination (90.4%). Biological resistance was middle. It accounted for 63.5%.

According to the data of 2011 the best results were received at these two hybrid forms: ♀Lyutestsens 70 x ♂Skent 3 (field germination – 83.5%, survivability – 65.3%) and ♀Hybrid x ♂Skent 1 (field germination – 79.5%, survivability – 68.1%). In 2012 the best results were received at the hybrid ♀Skent 3 x ♂Skent 1 (field germination – 84.0%, survivability – 55.0%).

According to averaged data three hybrid combinations were singled out within four years of the study in the conditions of the Tyumen region: ♀Lyutestsens 70 x ♂Skent 1, ♀Hybrid x ♂Cara and ♀Skent 3 x ♂Skent 1.

The use of diallel crossings in the selection programs allows to create, assess and suggest genetically new material for the adaptive selection. It is especially important in the always changing environment conditions.

KEYWORDS

- **cultivars**
- **hybridization**
- **sod-podzol**
- **soil salinization**
- **variety**

REFERENCES

1. The Federal law from 10.01.2002 №7-FL "On Protection of Environment". http://www.rg.ru/2002/01/12/oxranasredy-dok.html (In Russian).
2. Sapega, V. A. Bioclimate potential of the soil-climatic zones of the Northern Trans-Urals/ V. A. Sapega, G. S. Tursumbekov, N. N. Zhuravleva, S. V. Sapega. Bulletin of the Tyumen state University. 2011, №4, C. 41–44 (In Russian).
3. Russian-German agrarian and political dialog [Official site]. URL: http://www.agrardialog.ru/prints/details/id/74 (In Russian).
4. The information about the state of the agro-industrial complex of the Tyumen Region in 2011–2013. November 14, 2013, The official site of the Russian Federation Ministry of Agriculture. URL: http://www.mcx.ru/documents/document/v7_show/25618.363.htm (accessed: 07.03.2014) (In Russian).

5. Zhuchenko, A. A. The adaptive system of the plant selection (ecological-genetic basis): Monography. In two volumes./A. A. Zhuchenko. Moscow. The editorial board of the Russian University of Peoples' Friendship 2001, T.1. 780 p.
6. Dorofeev, V. F. Plant bloom, pollination and hybridization./V. F. Dorofeev, U. P. Lapteva, N. M. Chekalin. Moscow. Agro-industrial publishing house. 1990, 144 p (In Russian).
7. Hygienic standards 2.1.7.2041–06 "Maximum allowable concentration of the chemical substances in the soil". http://www.tehlit.ru/1lib_norma_doc/46/46714/ (In Russian).
8. Ivanenko, A. S. Agro-climatic conditions of the Tyumen Region: study guide/ A. S. Ivanenko, O. A. Kulyasova. Tyumen. The editorial board of the Tyumen State Agricultural Academy. 2008, 206 p (In Russian).
9. Gradchaninova, O. D. Methodical suggestions on the world wheat collection/ O. D. Gradchaninova, A. A. Philatenko, M. I. Rudenko. Leningrad. Publishing house of the, N. I. Vavilov Research Institute of Plant Industry. 1987, 28 p (In Russian).
10. Dospechov, B. A. The methodology of the field experiment (with the basis of the statistic research procession)/B. A. Dospechov. Moscow. Agro-industrial publishing house. 1985, 351 p (In Russian).
11. Kosulina, L. G. Physiology of the plant resistance towards unfavorable environment conditions. Eds. L. G. Kosulina, E. K. Lutzenko, V. A. Aksenova. Rostov-on-Don. 2011, The publishing house of Rostov-on-Don State University. 2011, 236 c (In Russian).
12. Zverev, A. T. The basic ecology laws/A. T. Zverev. Moscow. The publishing house Paganel. 171 p (In Russian).
13. Baumhauer, R., Kneisel Ch., Möller, S., Schütt, B., Tressel, E., Physische Geographie 2: Klima-, Hydro-, Boden-, Vegetationsgeographie. Darmstadt. Wissenschaftliche Buchgeselschft. 2011, 152 S. (In German)
14. Zhukovski, P. M. Wheat in the USSR. Botanic characterization of wheat– 171 p. Zhukovski. Moscow-Leningrad. 1957, 532 p (In Russian).
15. Nosatovski, A. I. Wheat; Moscow. The State publishing of the agricultural literature. 1950, 407 p (In Russian).
16. Kovalenko, A. A. The field seed germination is the pledge of the future harvest [Electronic resource]/URL: http://rosselhoscenter.com/stati-6/2599-vskhozhest-semyan 07.10.2013 (In Russian).
17. Askhadulin, D. F. The impact of the field germination on the spring rape seeds harvest. Eds. D. F. Askhadulin, L. N. Shayakhmetova, E. A. Prishchepenko. Proceedings of 5-th International Conference of the Yong Scientists and Experts "Perspective Trends of Researches in Breeding and Crop Management of Oil Crops." 2009, 19–21 (In Russian).
18. Mukhordova, M. E. The system of the genetic determinants of the field seed germination of soft spring wheat. Eds. M. E. Mukhordova, E. G. Mukhordov. The bulletin of Altai State agricultural University. 2013, №9, 5–8 (In Russian).
19. Bome, N. A. The resistance of the cultural plants towards unfavorable factors of the environment. Eds. N. A. Bome, A. Y. Bome, A. A. Belozerova. Tyumen. The publishing house of Tyumen State University. 2007, 192 p (In Russian).

20. Phlyaxberger, K. A. Wheat/Moscow-Leningrad: The State Publishing House of the Agricultural Literature. 1938, 296 p (In Russian).
21. Tretyakov, N. N. The Physiology and Biochemistry of the Cultural Plants. N. N. Tretyakov, E. I. Koshkin, N. M. Makrushin et al. Moscow. Kolos. 2000, 640 p (In Russian).
22. Semyonova, S. M. Methods of the climate change consequences assessment for the physical and biological systems. Moscow. Russian Hydrometeorology. 2012, 511 p (In Russian).
23. Phakhrudenova, I. B. The influence of the weather conditions on the field germination and survivability of the plants of hard and soft wheat in different soil-climatic conditions of the Northern Kazachstan. Eds. I. B. Phakhrudenova, G. A. Loskutova. The bulletin of the Altai State agricultural University. 2011, №12, 39–41 (In Russian).
24. Koshkin, E. I. The physiology of the agricultural crop resistance: textbook. Moscow: Drofa, 2010, 638 p (In Russian).
25. Tyumentseva, E. A. The changeability of the main stem length of the Triticum aedtivum, L. forms in the conditions of the Tyumen Region [Electronic resource]. N.A. Bome. The modern problems of the Science and education. 2013, №1, URL: http://www.science-education.ru/107-r8498 (In Russian).

CHAPTER 19

COMPARATIVE TRIALS OF VARIETY SAMPLES OF EASTERN GALEGA (*GALEGA ORIENTALIS* LAM.)

VERA I. BUSHUYEVA, MARINA N. AVRAMENKO, and
CATHERINE S. ANDRONOVICH

CONTENTS

ABSTRACT

The chapter presents the characteristics of variety samples of *Galega orientalis* on morphological and economically valuable traits in the competitive test.

The description of the variety samples on the phenotype, a comparative assessment on the length of the growing season, plant height, yield of green mass, seeds, foliage, dry matter content, the biochemical composition of forage and content of radionuclides in it were carried out.

It was found out that the studied variety samples of *Galega orientalis* differed significantly from each other in color of flowers, leaves, seeds and other morphological characteristics.

According to the results of a comprehensive evaluation of the economically useful traits variety samples SEG-7, SEG-10, SEG-12 were characterized by higher rates.

19.1 INTRODUCTION

Eastern galega (*Galega orientalis* Lam.) as perennial legume is of practical importance in feed production. Different kinds of cheap and highly nourishing animal feeding stuffs are produced from it making animal products of a higher quality at a low cost. *Galega orientalis* differs from other forage crops by its longevity of economic use, high productivity and excellent fodder value. Using rationally agro-climatic conditions during the growing season it forms the earliest green fodder for animals in spring and the last in late autumn when there is particularly acute shortage of juicy nutrient feed in fodder production. *Galega orientalis* has a significant impact on soil fertility and is an excellent precursor for other crops [1].

To improve efficiency in the use of the crop in the Republic of Belarus selective breeding is carried out, varieties Nesterka, Polesskaya, Nadezhda and Sadruzhnosts were created that are included in the State Register and approved for cultivation for commercial production [2]. However, all currently cultivated in production varieties of *Galega orientalis* are characterized by the same morphological features, have a blue coloration of flowers and dark green leaves, making it difficult to identify the differences among varieties during the patent examination in order to provide them with legal protection [3].

19.2 ANALYSIS OF THE SOURCES

In accordance with the Act of Accession of the Republic of Belarus to the International Convention and its entry in the UPOV – International Union of states-parties for the Protection of New Varieties [4], *Galega orientalis* is included in the list of plant species whose varieties are protected at national and international levels. To provide legal protection to new varieties of *Galega orientalis* state testing is carried out to identify varieties and patentability [5]. Identification test is to estimate variety according to the criteria of distinguishability, uniformity and stability on the basis of determining the characteristic and distinctive morphological characters, the degree of manifestation of which can be precisely described [6].

According to article №1 of the International Convention for the Protection of New Varieties of Plants [4] the concept of "variety" is determined by the manifestation of the phenotype. For variety identification it is necessary to identify phenotypic differences. To understand the differences among varieties a visual evaluation by direct plant features as well as assessment of genetic molecular markers directly concatenated with these symptoms are used [7].

Therefore, one of the most pressing problems in selection of *Galega orientalis* is creation of varieties, which are characterized not only by genetic, but also by phenotypic differences. A new variety must be clearly differentiated from any other at least by one trait, be uniform and characterized by a relatively stable degree of variation of quantitative or qualitative characteristics, show their stability in years [8].

Breeding work on the creation of such varieties has been carried out for many years at the Department of Breeding and Genetics of the Belorusian State Agricultural Academy (BSAA). New source material for breeding created by hybridization, polyploidy and mutagenesis, and characterized by a wide range of variability of morphological and economically useful traits is being evaluated in the nurseries of selection process [9, 10].

The purpose of this research was to make a comparative morphological and economic biological characteristic of the obtained variety samples of *Galega orientalis* in the competitive variety testing.

19.3 MATERIALS AND METHODOLOGY

The research was conducted at the experimental field of the Department of Genetics and Selection of the BSAA in 2011–2013. The objects of the study were 11 new variety samples of *Galega orientalis*: SEG-1, SEG-2, SEG-3, SEG-4, SEG-6, SEG-7, SEG-8, SEG-9, SEG-10, SEG-11 and SEG-12, which have been evaluated in a competitive test.

The laying of the nursery was carried out in 2010. The area of the plot was 16 m², fourfold replication, randomized location of the plots. The shape of the plots was rectangular. Sowing was carried out manually with row spacing of 30 cm. Phenological observations of the variety samples were done, morphological characteristics, plant height, yield of green mass and seeds, foliage, dry matter content were evaluated. Biochemical composition of forage and its content of radionuclides were determined.

Description of variety samples by morphological features was carried out in accordance with the procedure of testing *Galega orientalis* varieties for distinguishability, uniformity and stability.

Green mass was cut manually and weighed accurate within 1 kg. Seed production was determined by the elements of the structure of seed yield by analyzing the test sheaf of 25 productive stems. Count of yield of green mass and seeds was carried out by a summed method.

Analysis of plant samples for dry matter content and the foliage was performed in the budding phase – beginning of flowering. The dry matter content was determined by drying the green mass to completely dry state in a drying oven at the temperature of 100–105°C. With the shrinkage factor (the quotient of the fresh grass mass in the middle trial to its dry mass) the mass of absolutely dry matter from the count area was determined.

Foliage was calculated by the proportion of leaves in the total mass of sprout. In 1kg of green grass the mass of leaves with petioles was determined and expressed as a percentage.

To analyze the biochemical composition of fodder in the budding-flowering phase in each variety sample plant samples were selected, dried to completely dry matter and analyzed in the chemical environmental laboratory of the BSAA.

The content of radionuclides strontium-90 and cesium-137 in the plants and soil were determined by the method of scientific production enterprise "Atomteh" on the gamma-beta spectrometer MKS-AT 1315.

19.4   RESULTS AND DISCUSSION

As a result of phenological observations differences among variety samples in the duration of the phases of development and the growing season were revealed. Spring growth was due to the characteristics of the meteorological conditions in the early spring period and in general in the nursery in 2011 began on April 15, in 2012 – April 6, and in 2013 – April 21. Duration of the periods from the onset of spring regrowth till the budding phase varied in the variety samples from 34 to 41 days, and till the flowering phase – from 35 to 43 days. The length of the growing season depending on the variety sample was in 2011–107–114 days, in 2012–79–86 days, in 2013–82–91 days. Variety samples SEG-4 and SEG-12 were characterized by the shortest period of vegetation in all the years of research, and the longest were SEG-1 and SEG-8.

The differences among variety samples in plant height, the shape of the bush, bushiness, the number of internodes on the main stem, its thickness, color and pubescence, color of leaves, their length and number, in the form of leaflets, their length, width, presence of spinelet on them, and also color of flowers and seeds were determined (Table 19.1).

It was found that the plant height in the second year of herbage life varied in the variety samples from 79 to 106 cm. Variety sample SEG-6 was the most dwarfish with an average height of 79 cm, and SEG-4 was the tallest (106 cm).

Variety samples with upright and half-upright forms were isolated by the type of the bush. Thus, variety samples SEG-1, SEG-2, SEG-8 and SEG-9 had an upright shrub, and in all of the rest it was half-upright. The number of the stems per bush depending on the variety sample ranged from 40 to 64 pieces. Variety sample SEG-12 was characterized by the highest bushiness.

The average number of internodes on the main stem was depending on the variety sample 8–13 pieces. The highest rate (13 pcs.) was observed in the variety sample SEG-1.

The thickness of the stem varied from 5.3 mm in thin stem sample SEG-1 to 8.4 mm in SEG-9 with a thick stem.

In addition, variety samples differed in the degree of pubescence of the stem ranging from mild to moderate, and in the intensity of its coloration with anthocyan (from its absence to intensive manifestation).

Differences among variety samples in the color of leaves, their length and the number of leaflets were identified. Thus, the color of leaves in the variety sample SEG-1 is light green, in SEG-4 and SEG6-dark green with

TABLE 19.1 Characteristics of the Morphological and Economically Useful Traits in Variety Samples (SEG-1 – SEG-12) in the *Galega orientalis* in Competitive Variety Testing

Trait	SEG-1	SEG-2	SEG-3	SEG-4	SEG-6	SEG-7	SEG-8	SEG-9	SEG-10	SEG-11	SEG-12
Plant height, cm	96	87	86	106	79	87	98	96	92	92	85
Type of bush	Upright	Upright	Half-upright	Half-upright	Half-upright	Half-upright	Upright	Upright	Half-upright	Half-upright	Half-upright
Number of stems in the bush, pcs.	58	42	46	40	60	48	56	46	49	53	64
The number of internodes, pcs.	13	12	9	12	9	9	10	9	11	9	8
Diameter, mm	5.3	6.0	7.8	5.5	6.5	7.5	6.0	8.4	7.0	7.5	7.8
Pubescence	Average	Weak	Weak	Weak	Average	Weak	Weak	Average	Average	Weak	Average
Anthocyan coloring	—	Medium	Weak	Weak	Weak	Weak	Medium	Medium	Weak	Medium	Medium
Leaf coloring	Light green	Green	Green	Dark green	Dark green	Green	Dark green	Green	Green	Green	Green
Length of the leaf, cm	19.6	20.3	20.0	22.3	19.1	19.9	23.8	18.8	21.0	25.3	24.4
Number of leaflets, pcs.	10	12	14	13	16	15	13	13	11	13	13
Form of leaflet	Lanceolate	Ovate	Ovate	Ovate	Ovate	Ovate	Large ovate	Ovate	Ovate	Lanceolate	Lanceolate
Length, cm	6.5	6.1	6.0	6.7	5.9	6.4	6.0	5.8	7.2	6.2	6.9
Width, cm	3.1	2.7	3.2	3.5	3.4	3.3	3.5	3.1	3.7	3.0	4.1

TABLE 19.1 Continued

Trait	SEG-1	SEG-2	SEG-3	SEG-4	SEG-6	SEG-7	SEG-8	SEG-9	SEG-10	SEG-11	SEG-12
The presence of spinelet	+	+	—	+	—	—	—	—	—	—	—
Color of the corolla	White	Lilac	Light blue	Blue	Light blue	Dark blue	Dark purple	Light pink	Light blue	Pink	Cream
Availability of anthocyan on the calyx	—	+	—	—	+	+	+	—	+	+	+
Seed color	Light Yellow	Olive	Olive	Olive	Olive	Olive	Yellow	Olive	Yellow	Olive	Olive

anthocyan, in all the rest it was green. The leaf length varied depending on the variety samples from short in SEG-9 (18.8 cm) to long in SEG-11 (25.3 cm). Variety sample SEG-1 had the least number of leaflets (10 pcs.), and SEG-6 – the highest (16 pcs.).

Variety samples differed from each other in the form of leaflets from lanceolate to broad ovate, the length ranging from 5.8 to 7.2 cm, the width – from 2.7 to 4.1 cm, and the absence or presence of spinelet (Figure 19.1).

In the variety samples SEG-1, for example, the leaflets were with a spinelet, lanceolate, length 6.5 cm, width 3.1 cm. In SEG-9 leaflets were ovate, length – 5.8 cm, width – 3.1 cm, without a spinelet.

Variety samples differed in flower color the most contrasting. Thus, variety sample SEG-1 was characterized by a white color, SEG-2 – lilac, SEG-3 – light blue, SEG 4 – blue, SEG-6 – blue with anthocyan, SEG-7 – dark blue, SEG-8 – dark purple, SEG-9 – light pink, SEG-10 – light blue, SEG-11 – pink, SEG-12 – cream.

It should be noted that regular link was revealed between the intensity of color of flowers and vegetative organs of the studied variety samples. Thus, albiflorous variety sample SEG-1 has light green stems and leaves without pigmentation, while in the dark violet flowering SEG-8 they are dark green.

Differences among variety samples were found in the silhouette of leaves varying from open, intermediate to the closed one, degree of mani-festation of the flower color – from dark purple to light blue, the color of leaves – from dark green to light green.

In the variety sample SEG-1 the leaf color is light green, in the rest it varied from green to dark green. Almost all the variety samples were

FIGURE 19.1   Variation of forms and tints of the color of leaflets in various variety samples of *Galega orientalis.*

characterized by the open silhouette of the leaf, an exception was SEG-10 whose silhouette was closed.

Seed color in different variety samples varied from light yellow, yellow to olive green. In albiflorous SEG-1 it was light yellow, in dark violet flowering SEG-8 and light blue flowering SEG-10 – yellow, in all the rest – olive green.

According to the dynamics of linear growth conducted in the period from the beginning of spring growth till flowering it was found that on the average during three years the daily increase in the variety samples was: in the phase of branching −1.5–2.0 cm, budding – 2.4–7.0 cm, flowering – 0.6–2.9 cm (Figure 19.2).

Intensive daily gain was recorded in the budding phase and it was the highest in the variety sample SEG-12 (7.0 cm).

The most significant economic beneficial features of variety samples are the height of the plants in the phase of cutting maturity and yield of green mass. As a result of the evaluation it was found that in 2012 the plant height depending on the variety sample was 90–120 cm, in 2013–95–125 cm (Table 19.2).

The tallest both in 2012 and in 2013 with the plant height of 120 and 125 cm, respectively, were variety samples SEG-4, SEG-10, SEG-12, and undersized (90 cm) was variety sample SEG-11.

All the studied variety samples are characterized by a high yield potential of green mass ranging between 65.0 to 105.0 t/ ha. In 2012, the third year of herbage life, it was 65.0–87.0 t/ ha in the variety samples, and in the fourth year – 72.0–105.0 t/ ha. Variety samples SEG-7 (87.0 t/ ha)

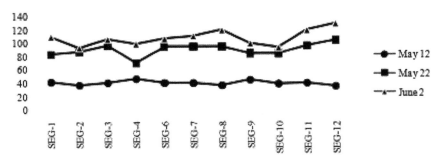

FIGURE 19.2 Dynamics of linear growth of the variety samples of different species of *Galega orientalis.*

TABLE 19.2   Characteristics of the Variety Samples of *Galega orientalis* of Different Species in Plant Height (cm) and Yield of Green Mass, t/ha. 2012–2013

Variety samples	2012		2013		Average yield, t/ha
	cm	t/ ha	cm	t/ ha	
SEG-1	115	73.0	120	88.0	81.0
SEG-2	110	73.0	108	86.0	80.0
SEG-3	110	75.0	108	85.0	80.0
SEG-4	120	70.2	125	96.0	83.0
SEG-6	110	76.0	110	90.0	83.0
SEG-7	118	87.0	120	102.0	95.0
SEG-8	100	79.0	105	83.0	81.0
SEG-9	100	65.0	100	72.0	69.0
SEG-10	120	80.0	125	105.0	93.0
SEG-11	90	66.0	95	84.0	75.0
SEG-12	120	85.0	125	104.0	95.0
LSD_{05}		5.2		6.5	

and SEG-12 (85.0 t/ ha) were more high yielding in 2012, and SEG-7 (102.0 t/ ha), SEG-10 (105.0 t/ ha) and SEG-12 (104.0 t/ha) in 2013. The highest yields of green mass on average during two years were obtained from variety samples SEG-7 (95.0 t/ ha), SEG-10 (93.0 t/ ha), SEG-12 (95.0 t/ha).

Evaluation of the variety samples by foliage showed that the trait varied in the range from 41.6 to 56.6% depending on the variety sample. On average, during two years the best by this indicator were variety samples SEG-7 (54.5%) and SEG-6 (56.6%).

Variety samples also differed in dry matter content. In 2012 this figure amounted to 16.0–23.3% among variety samples, and in 2013–18.7–27.8%. The highest content of dry matter in 2012 (23.3%) was observed in albiflorous variety sample SEG-1, and the lowest (16.0%) in pink flowering SEG-11. In 2013 variety sample SEG-12 was characterized by a higher rate (27.8%), and SEG-1 – the lowest (18.7%).

The biochemical composition of green mass of the variety samples of *Galega orientalis* was studied in the budding -beginning of flowering phase. Variety samples were evaluated by the content of crude protein, fat, ash, fiber, nitrogen-free extractives, phosphorus, calcium, sugar, carotene in the dry matter of green mass.

It was found that the crude protein content depending on the variety sample ranged from 7.5 to 19.7%. It was the highest in the variety sample SEG-12 (17.4%), SEG-6 (17.6%) and SEG-1 (19.7%) (Table 19.3).

On other parameters variation in the variety samples were in the following ranges: fat- 1.1–3.43%; ash – 3.98–6.26%; fiber – 22.5–32.6%; NFE– 32.93–49.91%; Ca – 0.564–0.09%; P_2O_5 0.51–0.0%; sugar – 2.58–4.29%; carotene – 25–57 mg/kg.

An important characteristic of variety samples is their resistance to the accumulation of radionuclides. This problem has become more urgent in connection with the accident at the Chernobyl nuclear power plant. At present, the main dose-forming radionuclides are cesium-137 and strontium-90, which accumulate in forage and being consumed by animals come in food –milk and meat. It should be noted that the content of radionuclides in plant feed produced after the accident at the Chernobyl nuclear power plant is strictly controlled, and at the same time varieties resistant to the accumulation of radionuclides have acquired a special significance for fodder production.

Plant and soil samples in each plot were selected for radiometric analysis by standard methods. Collection of plant samples was carried out in the budding –beginning of flowering phase.

Analysis of the soil samples showed that the density of the soil contamination of the trial plot with cesium-137 was 15 kBq/m². Strontium-90 was not detected.

Radiometric analysis of the plant samples showed that the variety samples differed considerably in cesium-137 content in the feed mass and the ratio of its transfer from the soil to plants. Levels of cesium-137 in the feed mass on average varied in the range from 0.8 to 9.1 Bq/kg in the two years depending on the variety sample (Table 19.4).

Variety samples of *Galega orientalis* SEG-1 (0.8 Bq/kg), SEG-12 (1.5 Bq/kg), SEG-8 (3.2 Bq/kg) and SEG-7 (3.7 Bq/kg) were characterized by a lower content of cesium-137 in the green mass. The highest rate was observed in the variety samples of SEG-3 (9.1 Bq/kg), SEG-11 (8.8 Bq/kg) and SEG-10 (8.5 Bq/kg). In the blue flowering standard variety sample SEG-4 this figure was intermediate and was

TABLE 19.3 Chemical Composition of Dry Matter in the Variety Samples of *Galega orientalis* (2013)

Variety sample	Protein, %	Fat, %	Ash, %	Fiber, %	NFE, %	Ca, %	P_2O_5, %	Sugar, %	Carotene, mg/kg
SEG-1	19.7	2.24	3.98	26.97	32.93	0.599	0.90	3.94	42
SEG-2	14.1	2.67	5.58	24.53	43.02	0.757	0.78	3.19	29
SEG-3	7.5	1.21	5.59	26.79	49.91	0.809	0.58	2.58	25
SEG-4	16.6	1.51	4.31	27.31	41.27	0.594	0.68	3.97	47
SEG-6	17.6	1.90	4.37	24.29	42.84	0.564	0.72	3.19	25
SEG-7	9.9	1.01	5.61	30.05	44.43	0.748	0.51	3.66	32
SEG-8	16.1	1.40	6.26	22.05	45.19	0.804	0.69	3.19	52
SEG-9	15.1	3.43	5.03	25.63	41.81	0.656	0.65	3.97	42
SEG-10	14.4	1.51	5.81	28.79	40.49	0.753	0.73	3.51	50
SEG-11	11.2	2.01	4.97	24.34	48.48	0.654	0.61	4.29	25
SEG-12	17.4	1.14	4.44	32.16	35.86	0.582	0.59	3.82	57

TABLE 19.4 The Accumulation of Cesium-137 by Variety Samples of *Galega orientalis* (Average for 2012–2013)

Variety sample	Flower color	Specific activity		Transition coefficient, Bq/kg: kBq/m²
		K-40, Bq/kg	Cs-137, Bq/kg	
SEG-4	Blue	766	5.9	0.39
SEG-1	White	760	0.8	0.05
SEG-2	Purple	520	6.2	0.41
SEG-3	Blue	474	9.1	0.61
SEG-7	Dark blue	647	3.7	0.25
SEG-8	Dark purple	689	3.2	0.21
SEG-10	Light blue	776	8.5	0.57
SEG-11	Pink	686	8.8	0.59
SEG-12	Cream	800	1.5	0.10
LSD_{05}			0.45	0.08

5.9 Bq/kg. SEG-2 with a lilac color of flowers was at the standard level of cesium-137 content in the feed mass (6.2 Bq/kg).

A similar pattern can also be seen at a transition rate of cesium-137 from soil to plants, whose variation among variety samples ranged from 0.05 to 0.61 Bq/kg: kBq/m².

The highest rate of transition was observed in the variety samples with blue flowers SEG-3 (0.61 Bq/ kg: kBq/m²), pink flowers – SEG-11 (0.59 Bq/kg: kBq/m²) and light blue flowers SEG-10 (0.57 Bq/kg: kBq/m²), which significantly exceeded by this indicator the standard variety sample SEG-4 (0.39 Bq/kg: kBq/m²). In the variety sample SEG-2 with lilac flowers transition coefficient was 0.41 Bq/kg: kBq/m² and was up to standard.

Variety samples SEG-1 with white (0.05 Bq/kg: kBq/m²) and SEG-12 with cream flowers (0.1 Bq/kg: kBq/m²) were characterized by the lowest transition coefficient. Variety samples with dark purple SEG-8 (3.2 Bq/kg: kBq/m²) and dark blue color of flowers SEG-7 (3.7 Bq/kg: kBq/m²) were characterized by the lower transition rate than the standard.

The obtained results showed that the accumulation of cesium-137 in the studied variety samples of *Galega orientalis* depends not only on the genotype, but also on their phenotype. Albiflorous SEG-1 and cream flowering SEG-12 contained least of radioactive cesium-137. Both variety samples are a valuable source material for the creation of resistant to radioactive cesium cultivars of *Galega orientalis.*

Comparative evaluation of variety samples for seeds was carried out on the analysis of the structural elements of seed yield in 2012 and 2013, and at the same time height of the plants was taken into account. According to the results of the average of the two years it was found that each of the elements of the seed yield structure is characterized by its seed yield variability and the degree of variation. Thus, the average number of stems per 1 m² depending on the variety sample ranged from 55 to 98 pieces with a coefficient of variation ($V = 17.5\%$), which indicates an average characteristic variation (Table 19.5).

Plant height was characterized by mild variation ($V = 7.1\%$) and depending on the variety sample ranged from 112 to 139 cm. Variety sample SEG-12 was the tallest (139 cm), and SEG-2 – the most undersized (112 cm).

Depending on the variety sample one stalk formed from 5 to 13 trusses, from 95 to 177 beans, from 173 to 409 pieces, or from 1.2 to 3.2 g seeds.

Variety samples SEG-8 (2.8 g) and SEG-10 (3.2 g) formed the greatest number of seeds on the stem. Weight of 1000 seeds depending on the variety sample was 6.5–7,0 g. Variety samples SEG-4 and SEG-9 had larger seeds, and the mass of 1000 pieces of 7.5–7.7 g respectively. Insemination of beans varied among variety samples from 1.0 to 3.2 pieces. Variety sample SEG-10 was characterized by a higher insemination of bean (3.2 pcs.). Strong variation ($V = 55.0\%$) was observed of the given trait. Strong variation was also noted in the number of trusses ($V = 26.3\%$) and seeds ($V = 30.9\%$) on one stalk.

All variety samples formed high seed yield, which was 0.63–1.77 t/ha. SEG-6 (1.41 t/ha), SEG-8 (1.77 t/ha), SEG-10 (1.57 t/ha) and SEG-12 (1.56 t/ ha) were the highest yielding in seeds variety samples.

19.5 CONCLUSIONS

Studied in a competitive test variety samples of *Galega orientalis* significantly differed in color of flowers, leaves, seeds and other distinctive morphological features that can be used in patent examination to identify the

TABLE 19.5 Elements of Structure and Seed Productivity of the Variety Samples of *Galega orientalis* (Average in 2012, 2013)

Variety sample	Number of stems, pcs	Height, cm	On one stalk				Seeds in a pod, pcs	Weight of 1000 seeds, g	Seed yield t/ha
			Trusses, pieces	Beans, pieces	Seeds				
					pieces	g			
SEG-1	84	137	13	156	173	1.2	1.0	7.0	0.77
SEG-2	56	112	8	95	200	1.4	2.2	6.5	1.05
SEG-3	64	120	7	123	239	1.7	1.9	6.9	1.14
SEG-4	57	126	7	138	185	1.4	1.4	7.5	0.7
SEG-6	98	114	6	125	229	1.6	1.7	7.0	1.41
SEG-7	64	125	9	177	388	1.9	2.2	6.8	1.13
SEG-8	75	126	8	153	409	3.2	2.8	7.3	1.77
SEG-9	59	118	10	174	262	2.0	1.8	7.7	0.63
SEG-10	57	115	9	135	389	2.8	3.2	6.9	1.57
SEG-11	55	127	5	128	262	1.9	2.0	7.2	1.18
SEG-12	80	139	11	148	268	1.9	1.9	7.1	1.56
Xmin	55	112	5	95	173	1.2	1.6	6.5	0.63
Xmax	98	139	13	177	409	3.2	3.2	7.7	1.77
$\bar{X} \pm S\bar{x}$	68±3.6	124±3	8±0.6	141±7	273±26	1.9±0.2	2.0±0.3	7.1±0.1	1.15±0.1
V, %	17.5	7.1	26.3	16.8	30.9	31.5	55.0	4.2	29.5

varieties according to the criteria of distinguishability, uniformity and sta-bility. The evaluation of economically useful traits highlighted best variety samples characterized by high performance.

Variety samples SEG-4, SEG-7, SEG-10, SEG-12 were characterized by higher yield of green mass on average over two years (93.0–95.0 t/ha). These samples differed by more rapid average daily gain in the budding-beginning of flowering phase (7.0 cm/day) and plant height (120–125 cm). Variety samples SEG-7 (54.5%) and SEG-6 (56.6%) were best in foliage. The highest content of dry matter was observed in variety samples SEG-1 (23.3%) and SEG-12 (27.8%). Protein content was highest in variety samples SEG-12 (17.4%), SEG-6 (17.6%) and SEG-1 (19.7%). Variety samples of *Galega orientalis* SEG-1 (0.8 Bq/kg), SEG-12 (1.5 Bq/kg), SEG-8 (3.2 Bq/kg) and SEG-7 (3.7 Bq/kg) were characterized by a lower content of cesium-137 in the green mass. Variety samples SEG-6 (1.41 t/ha), SEG-8 (1.77 t/ha), SEG-10 (1.57 t/ha) and SEG-12 (1.56 t/ha) had the highest seed yield.

KEYWORDS

- anthocyanin
- foliage
- protein
- radionuclides

REFERENCES

1. P. T. Pikun, M. F. Pikun, E. I. Chekel et al. Feed Production: Nontraditional Crops and the ways of their Solution: monograph/Vitebsk: Vitebsk State Academy of Veterinary Medicine, 2005, 119 p (In Russian).
2. State Register of Varieties and Arboreal and Shrubby Species. Ministry of Agriculture and Food of the Republic of Belarus, State Inspection for Testing and Protection of Plant Varieties [Ed. V. A. Beynya]. Minsk, 2012, 204 p (In Russian).
3. Bushuyeva, V. I. *Eastern Galega*: monograph. 2nd ed., Ext. V. I. Bushuyeva, G. I., Taranukho/Minsk. Ekoperspektiva, 2009, 204 p (In Russian).

4. On the Accession of the Republic of Belarus to the International Convention on the Protection of Plant Varieties: the Law of the Republic of Belarus from 29.06.2002 №115–3. Consultant Plus: Belarus. Technology 3000/Open Company "YurSpektr," Nat. Center for Legal Inf. Rep. Belarus. Minsk, 2012 (in Russian).

5. Semashko, T. V.Patenting of Plant Varieties in the Republic of Belarus: an analytical review; State Inspection of the Republic of Belarus for Testing and Protection of Plant Varieties. Minsk, 2011, 36 p (In Russian).

6. On Patents for Plant Varieties Law of the Republic of Belarus of 13 April 1995 №3725-XII; rev. and ext.. Plant Varieties [electronic resource]. Minsk, 2011 (in Russian).

7. International Convention on the Protection of New Varieties of Plants. Geneva: International Union for the Protection of New Varieties of Plants, 2004, 28 p (In Russian).

8. Methods for Variety Testing for distinguishability, uniformity and stability. State Inspection for Testing and Protection of Plant Varieties, Ministry of Agriculture and Food of the Republic of Belarus Minsk. Information Centre of the Ministry of Finance of the Republic of Belarus, 2004, 274 p (In Russian).

9. Bushuyeva, V. I. Genotypic Variation in Eastern Galega and its Use in Breeding Patentable Varieties. Science and Innovation. 2007, №1 (47), 37–41 (In Russian).

10. Bushuyeva, V. I. Using the Gene Pool of *Eastern Galega* to Identify Varieties. News of the National Academy of Sciences of Belarus. Series of Agrarian Science. 2008, №1, 61–67 (In Russian).

PART V

ANTHROPOGENIC PRESSURE ON ENVIRONMENTAL AND PLANT DIVERSITY

CHAPTER 20

PLANT RESPONSE TO OIL CONTAMINATION IN SIMULATED CONDITIONS

NINA A. BOME and REVAL A. NAZYROV

CONTENTS

ABSTRACT

The chapter deals with response of perennial gramineous (awnless brome, red fescue) and leguminous (red clover) grasses to influence of hydrocarbons at different stages of ontogenesis in laboratory and field conditions.

The study has revealed high sensitivity of awnless brome and red fescue to oil pollution by their laboratory seed germination rates. Variability of quantitative characters of germs in laboratory trial and dynamics of plants growth in vegetation vessels depended on their species and oil concentration. The observations show that treatment of seeds with hydrocarbons and oil soil pollution can result both in growth inhibition and growth stimulation.

20.1 INTRODUCTION

The global experience shows that human exploitation of oil fields is associated with contamination of soil and surface water and finally results in transformation of plant and animal life. About 2% of total oil production is released into the environment [1].

Spillage of oil and oil products are caused by various factors: uncontrolled flowing of exploratory wells, leakage of columns in productive wells, slacking of flanged joints of isolation valves, processing facilities amortization, disruption of sludge pit lining, discharge of field waste water. The greatest spillages are caused by oil pipes ruptures due to poor welding, hidden defects of metal, corrosion, tripping-over of truck-laying machines and other procedural violations.

Living organisms, being constantly influenced by oil and its products, have to adapt to such conditions, but this problem is still understudied [2].

Meliorative crops must be resistant to soil pollutants (oil residue, salts), fast growing, dependable in vegetative or seed reproduction in the given climatic, edaphic and hydrological conditions [3].

Regardless of particular goals of biological recultivation, in other words – whether there will be agricultural or forest areas, soil and vegetation cover on the contaminated lands need to be restored, and their productivity must be comparable to that of natural zonal communities. This problem can be solved only on the base of adequate list of plants, development of technique of their cultivation, and accelerated establishment of productive soil [4].

The purpose of this research is to study the response of perennial gramineous and leguminous grasses in ontogenesis to oil contamination.

To achieve the purpose, the following tasks were set: measurement of laboratory challenging and standard seed germination rates; studying of effects of oil pollution on variability of morphometric parameters of seedlings; observation of dynamics of plants growth in vegetation vessels;

analysis of morphometric parameters and phytomass of perennial grasses grown in field conditions.

20.2 MATERIALS AND METHODOLOGY

The field and laboratory experiments involved Shaim oil from Krasnoleninsky oil field opened in 1960. The deposit is located in the Kondinsky district of the Khanty-Mansiysk autonomous area, in the western part of the Kondinsky oil and gas field of the Priuralsk oil and gas-bearing area. The field structure is quite complex, the oil reserves are small [5]. The tectonic structure of Shaim is brachianticlinal fold of uncertain configuration, complicated by two domes – Mulymyinsky and Trehozerny.

Shaim oil shows relatively low specific gravity – 0.819–0.843, low sulfur – 0.29–0.47, considerable ratio of asphalts and resins (silica gel resins – 10.2%, asphaltenes – 0.8%). The content of paraffin is 13.43–17.9% at fusion temperature of 55°C. The light cut yield after oil refining is 50% (Table 20.1).

TABLE 20.1 Characteristic and Components of Shaim Oil From Krasnoleninsky Oil Field Used in the Experiment

Indicator	Value
Density, g/m³	0.8269
Pour point, °C	–2
Fusion point, °C	+55
Sulphurous resins, %	14.0
Silica gel resins, %	10.2
Asphaltenes	0.82
Sulphur, %	0.46
Hydrocarbons, %, of them:	29.32
Paraffins	15.96
Naphthenes	10.26
Aromatic	2.28
Ethylbenzene	0.03

For laboratory and field experiments, we selected seeds of three species of perennial grasses: red fescue – *Festuca rubra* L., awnless brome – *Bromopsis inermis* Leys., and red clover – *Trifolium pratenze* L. (Table 20.2). The seeds were obtained from the Research Institute of Agriculture of the Northern Trans-Urals from the harvest of 2012. The first phase of the study dealt with the biological properties of seeds of perennial grasses (thousand-seed weight and laboratory germination).

Original solution of water-soluble fractions of oil (WSFO) was prepared according to the methods developed by the Siberian Research and Design Institute of Fishery (Tyumen). One part of oil was added to nine parts of distilled water and shaked for 20 minutes. The shake was repeated 3 times, at intervals of 10 minutes for setting-out. Then solution was left for setting-out for 24 hours. The top layer of oil emulsion was pumped out, and the rest of the mixture was filtrated to remove the emulsion.

The experiment to reveal the impact of oil on seed germination and morphometric parameters of seedlings was carried out in the laboratory of biotechnology and microbiology of the Institute of biology of the Tyumen State University.

Each variant had four replicates; the sample size was 50 seeds in each replicate. The seeds were planted in Petri dishes on filter paper moistened with the following oil concentrations (%) – 0.3; 0.6; 0.9. The seeds processed with distilled water served as controls. Precut filter paper was sterilized in the oven at 130°C for an hour. The seeds were evenly laid, covered with lids and placed in the thermostat (TSO-1/80SPU). The germination was carried out at 21°C. The germination rate reflected the amount of normal germinated seeds by their morphometric parameters (length and weight of roots and shoots) [6]. 1216 seedlings were analyzed.

The field studies were conducted at the experimental site of the biological research station "Lake Kuchak" located in subtaiga zone of the

TABLE 20.2 Characteristics of Seeds of Perennial Grasses Cultivars by Size and Laboratory Germination

Cultivar	Cultivar	Thousand-seed weight, g	Germination, %
Awnless brome	Langepas	4.14	90.0
Red fescue	Sverdlovskaya	0.63	90.0
Red clover	Rodnik Sibiri	2.14	93.0

Tyumen region (the border of the northern part of the Tyumen region and the southern part of the Nizhnetavdinsk district). The area is moderately moistened. The annual rainfall is 350–417 mm. The accumulated positive air temperature (temperatures above 10°C) is 1700–1900°C; duration of the period is 114–123 days. There are fairly frequent droughts of low and moderate intensity [7].

The field experiment was established on May 26, 2013 in vegetation vessels of inert material. The vegetation vessels of 2.5 l were filled with soil taken as a substrate. The soil was cultivated, sod-podzol sabulous. The acidity of the soil salt extract amounted to 6.6 – weakly alkaline; the content of humus – 3.67%; dry residue – 0.47%. Soil analysis was carried out on the basis of the laboratory of ecotoxicology of the Tobolsk complex scientific station of the Ural branch of the Russian Academy of Sciences.

The soil was treated with oil of various concentrations (3; 6; 9%). The soil was thoroughly stirred with oil, and water was added up to 60% of maximum water-holding capacity. 50 seeds were laid in each vegetation vessel at a depth of 2–3 cm in a four-fold replicate. The soil moistened up to 60% of maximum water-holding capacity was taken as a control. The vegetative vessels were partly placed in the ground to increase stability and protection from overheating in hot weather.

Monitoring and measurements were conducted throughout the growing season to September 6, 2013. Measurements of the top parts were carried out on June 15, July 25 and September 6. After the last measurement, the plants were cut and transported to the laboratory to estimate their dry and wet phytomass. The plants in vegetative vessels were left for overwintering. The condition of the plants after overwintering was assessed in the spring of 2014.

The statistical data processing was carried out according to the standard methods [8, 9].

20.3   RESULTS AND DISCUSSION

Seed quality is assessed by a complex of various parameters – germination, viability, seedling development – closely related to genotype and conditions of seed formation [10].

The laboratory experiment has revealed the overall pattern in the studied species of perennial grasses, manifested in decrease in seed germination rate with the increase in concentration of water-soluble fractions of oil (WSFO). It should be noted that inhibition of growth processes was more expressed in the red fescue. The laboratory germination of seeds with a high concentration of WSFO (0.9%) was lower than the control by 38%, while the germination of the awnless brome amounted to 18%. The character most vulnerable to inhibitive WSFO effect is shoot length of the red fescue (Table 20.3).

The study has revealed the differences in degree of variability of seedling morphological characteristics of perennial cereal grasses. The awnless brome showed very high variation of shoot length in control (CV = 31.08%) and in the variant with WSFO concentration of 0.6% (CV = 35.16%). The shoot weight showed more stability (CV = 5.99–10.66%). The maximum value of the variation in the germination rates is observed at concentration of 0.6% and amounts to 18.79%; the control value of variation coefficient is 1.81%.

The seed germination rate variation of the red fescue was within the range from low (control, CV = 3.14) to very high (concentration of 0.9%, CV = 57.91%). Such type of variation of the character can be explained by increase of inhibitive influence of WSFO with increase in concentration,

TABLE 20.3 Laboratory Seed Germination and Indicators Morphometric Parameters of Seedlings Perennial Grasses Depending on WSFO Concentration

Concentration, %	Germination, %	Shoot length, mm	Root length, mm
Red fescue			
Control	90.0±0.71	51.3±2.52	20.4±1.16
0.3	80.0±1.87*	42.8±1.64*	20.5±1.14
0.6	66.0±4.45*	39.3±2.18*	22.7±1.71
0.9	52.0±7.53*	46.7±2.07	21.9±1.35
Awnless brome			
Control	90.0±0.41	63.9±3.41	65.0±3.58
0.3	83.0±0.65*	67.9±2.67	74.9±3.26
0.6	75.0±3.53*	71.5±2.13	84.3±3.18
0.9	72.0±2.08*	65.2±2.29	70.3±3.29

*differences with control are statistically valid.

manifesting in the fact that the most of the seeds germinate in favorable conditions, and only the strongest seeds germinate under stress. The coefficient of variation of shoot length was 18.82–33.81%, root length – 35.31–47.62%, no dependence on the concentration of WSFO has been revealed.

In the biomass structure of awnless brome is dominated shoots; their share in all variants of the experiment was over 80%. No significant differences in the proportion of roots to shoots have been noted in the challenging variant with contamination.

The red fescue shows acute inhibition of the root system in response to contamination. With an increase in hydrocarbons concentration, the proportion of roots in the germ reached minimum values (012–2.44%).

For more complete study of oil contamination effects, the field observations of growth of red clover and awnless brome were carried out in the vegetation season of 2013 and after overwintering in spring of 2014.

Significant differences were revealed between the variants of the experiment in terms of the shoot length of red clover at different stages of ontogenesis. The first measuring of plants (2013.06.15) did not show any valid differences between the control and experimental variants. The variation of character during this period was significant, as evidenced by the coefficient of variation (28.69–36.24%) (Table 20.4).

The second measuring revealed the differences between the variants by the given character. The clover plants in the conditions of oil contamination had lower values of shoot length than the controls. The degree of variability of the character in the contaminated variants was higher.

The measurement of September 6, 2013 showed no valid differences in comparison with the control by the shoot length. The plants in vegetative vessels reached 8.5–9.9 cm by this period. The oil-affected plants became more aligned compared to the control and the preceding measurement.

In 2014, the spring after-growing of the clover was registered on May 6–8. Measuring of the shoot length showed that the after-growing was delayed in the variants with oil pollution, which may be indicative of adverse effects of toxicant in the second year of the clover growth. Variability of the character increased in the variants with oil pollution. However, absence of loss of plants in autumn, winter and spring seasons suggests that this species may be used for remediation of oil-contaminated lands.

TABLE 20.4   Influence of Oil Contamination of Soil on the Shoot Length of Red Clover

Measuring date	Variant of experiment, concentration, %	Shoot length, cm	CV, %
June15, 2013	Control	4.3±0.41	30.32
	3	4.9±0.45	28.69
	6	4.8±0.45	31.94
	9	4.0±0.48	36.24
July 25, 2013	Control	10.2±0.46	16.70
	3	7.6±0.41*	17.75
	6	5.3±0.63*	35.64
	9	3.1±0.26*	21.50
September 6, 2013	Control	9.3±1.06	37.36
	3	8.5±0.57	18.95
	6	9.9±0.66	20.04
	9	9.9±0.84	24.61
May 22, 2014		9.8±0.48	26.78
		6.4±0.29*	28.76
	Control	6.8±0.40*	37.72
	3	5.9±0.40*	42.84

* differences with control are statistically valid.

Analysis of control plants growth character showed that more active growth was observed at the beginning and in the end of the growing season. The maximum daily growth rate in the challenging variant with oil contamination at a concentration of 3% was recorded in the first period, at a concentration of 6 and 9% – in the later period (25.07–06.09). The influence of oil pollution could also be observed in phytomass parameters (Figure 20.1). The decrease in plant phytomass in contaminated soil began at the lowest concentration (3%) and continued at higher concentrations (6 and 9).

Similar observations of the growth of awnless brome were performed. The dynamics of growth of this species was of a different nature compared with the red clover. The first measuring of the plants shoot length in the vegetation vessels showed no valid difference (Table 20.5). It should be noted that the variants with oil contamination had much higher coefficient

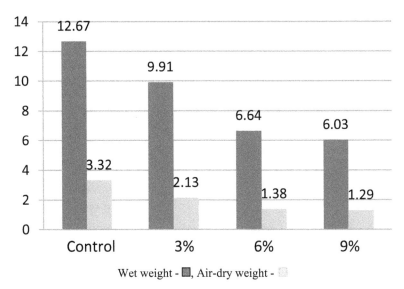

Wet weight - ■, Air-dry weight - ▨

FIGURE 20.1 Phytomass of red clover grown in vegetation vessels with contaminated soil, 1 g per vessel.

of variation of this character, which indicated the differences between the plants in the analyzed sample.

No stimulation of awnless brome growth processes was observed after overwintering. At relatively low concentrations, there were no difference between control and experimental variants, and the character was significantly lower at the maximum concentration.

The measuring of plants on 2013.07.25 showed that the maximum concentration of oil contributed to the valid decrease in shoot length compared with controls. This variant was also marked by high coefficient of variation, while in other cases the variability was weak or moderate.

At the end of vegetation, the inhibition of growth processes was observed at two concentrations (3 and 6%). At the same time, the high level of the soil contamination (9%) contributed to the stimulation effect. In this variant, a sharp increase in the shoot length was observed. In the laboratory experiment, we also noted an increase in the character in the variant with maximum contamination, and this phenomenon was confirmed in pot study. Thus we should draw attention to the fact that the variation of the character was medium in all the variants.

TABLE 20.5   Influence of Oil Contamination on the Awnless Brome Shoot Length

Measuring date	Variant of experiment, concentration, %	Shoot length, cm	CV, %
June 15, 2013	Control	10.8±0.62	17.37
	3	10.5±0.61	19.03
	6	14.2±1.78	33.40
	9	15.5±2.72	43.06
July 25, 2013	Control	17.5±0.73	13.64
	3	17.7±0.46	8.10
	6	16.6±0.68	12.66
	9	8.5±0.55*	23.88
September 6, 2013	Control	22.6±1.03	11.68
	3	17.7±0.71*	12.90
	6	19.3±0.95*	15.09
	9	53.6±1.08*	15.97
May 22, 2014	Control	13.3±0.41	19.79
	3	12.2±0.39	20.13
	6	12.4±0.53	26.83
	9	9.8±0.49*	28.79

*differences with control are statistically valid.

The stimulating effect of oil on red fescue and awnless brome plants was described in the experiments of I. I. Shilova [11]. The author suggests that the following factors may be involved: effects of stimulant contained in oil, improvement of the plants nutrition through the decomposition of organic matter, less intense competition between plants due to thinning caused by oil contamination.

The growth character of awnless brome in vegetation period differed from that of red clover. The relative growth rate of the controls was high in all three measuring periods, reaching maximum values in the first and third periods. The plants treated with 3% solution showed active growth in the first and third periods, the plants treated with 6% solution grew uniformly throughout the growing season, and the plants treated with 9% solution showed the maximum daily growth in the second and third periods.

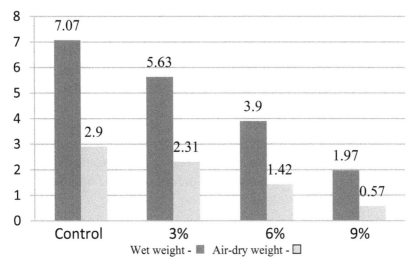

FIGURE 20.2 Phytomass of awnless brome when grown in vegetation vessels with contaminated soil, 1 g per vessel.

The measurement of wet and air-dry phytomass of awnless brome revealed that these parameters decrease on contaminated soils (Figure 20.2), and the inhibitive effect of oil pollution was directly dependent on the concentration, so the plants with a maximum concentration were characterized by the lowest productivity. The same pattern was observed in red clover.

The method of toxicity assessment of oil-contaminated soil by seed germination is used quite frequently [12, 13]. However, such assessment does not provide a clear picture of impact of oil on the vegetative body due to the lack of information about the further growth and development of seedlings. Along with the germination of seeds, in our experiments we also studied the morphological characteristics of seedlings in laboratory and plants in vegetation vessels.

The death of young plants was observed in vegetation vessels after emergence of seedlings with no external symptoms of abnormal development, because plants in the early stages of development are the most sensitive to the impact of oil [14]. Similar results were obtained by other authors [11, 15].

Full and objective assessment of oil contamination requires integrated use of laboratory and field methods. Our research of 1996–2001 performed in the experimental plot at the 707th km of Ust-Balyk-Altemyevsk pipe-line (57° Northern latitude, Yalutorovsk district of the Tyumen region) revealed the indicators that are most sensitive to oil pollution after acci-dental spillage. The indicators were: plant species composition, projective cover, mosaicism, volume of phytomass, which can be used to assess the negative impact on plant communities [16].

Great prospects are related to a complex of analytical methods (FTIR spectroscopy, gas-liquid chromatography, chromato-mass spectrometry) for studying of oil-contaminated cryogenic soils degradation affected by biopreparation based on aboriginal hydrocarbon-oxidizing microflora in Arctic Yakutia [17].

20.4 CONCLUSIONS

1. Laboratory seed germination rate of perennial grasses (red fescue and awnless brome) decreases in challenging conditions of oil contami-nation. The seedlings of awnless brome with larger seeds (thousand grain weight – 4.14 g) are less sensitive to stressful factor compared with the seedlings of red fescue (thousand grain weight – 0.63 g).
2. Morphological characteristics of seedlings (shoot and root length) show moderate to strong variability in the control and experimental variants.
3. In the laboratory experiment, oil contamination in some cases resulted in inhibition of growth processes (shoot length of red fescue), in other cases – in stimulation (shoot length of awnless brome). The results indicate the specificity of the plants reaction to the stress factor.
4. The reaction of red clover plants on soil pollution in the pot study consisted in decrease in shoot length in vegetation season of 2013 (measuring on 07.25) and after overwintering in the spring of 2014 (measuring on 05.22).
5. Soil contamination with oil of 9% concentration resulted in stimu-lation of shoots growth in the plants of awnless brome at the end of vegetation period 2013 (measuring on 09.06) and in signifi-cant decrease in the shoot length after overwintering in the spring of 2014.

KEYWORDS

- awnless brome
- hydrocarbons
- perennial gramineous

REFERENCES

1. Chyzhov, B. E. Forest and oil of the Khanty-Mansiysk autonomous area. Boris Chizhov. – Tyumen. OOO "Izdatelstvo of Yury Mandrika". 1998.-52 pp (In Russian).
2. Petukhova, G. A. Mechanisms of resistance of organisms to oil pollution: monograph. Tyumen. Publishing House of the Tyumen State University, 2008, 172 p (In Russian).
3. Vaver, V. I. Recultivation of oil contaminated soils. V. I. Vaver. Biological resources and natural resource management. Collection of scientific papers. Nizhnevartovsk, 1997, Issue 1. 114–135 (In Russian).
4. Kurnishkova, T. V. Geography of plants with fundamentals of botany/T. Kurnishkova, V. V. Petrov, A. G. Voronova. Moscow. Prosveschenie. 1987, 207 p (In Russian).
5. Bagautdinov, A. K. Geology and development of largest and unique oil and gas deposits in Russia/A. K. Bagautdinov, S. L. Barkov, G. K. Belevich, etc. Vol. 2. Moscow, JSC "VNIIOÈNG," 1996, 352 p (In Russian).
6. Bome, N. A. Biological properties of seeds and phenogenetic analysis of cultivated plants. N. A. Bome, A. A. Belozerova, A.Ya. Bome. Guidance manual. Tyumen State University Publishers, 2007, 96 p (In Russian).
7. Ivanenko, A. S. Agroclimatic conditions of the Tyumen region. Guidance manual. A. S. Ivanenko, O. A. Kulyasova. Tyumen. Tyumen State Agricultural Academy, 2008, 206 p (In Russian).
8. Dospekhov, B. A. Methods of field experiment (with the basics of statistical processing of results of research). Moscow. Kolos, 1979, 416 p (In Russian).
9. Lakin, G. F. Biometry. Moscow. Publisher "Vysshaya Shkola." 1990, 295 p (In Russian).
10. Reimers, F. E. Physiology of seeds of Siberian cultivated plants. Moscow: Nauka, 1974, 142 p (In Russian).
11. Shilova, I. I. Biological recultivation of oil-contaminated lands in taiga zone. I. I. Shilova/Remediation of oil contaminated soil ecosystems. Moscow: Nauka, 1988, 159–168 (In Russian).
12. Petukhov, V. N. Biotesting of soil and water contaminated with oil and oil products using plants/V. N. Petukhov, F. M. Fomchenko, V. A. Chugunov et al. Applied biochemistry and microbiology. 2000, Vol. 36, №6, 652–655 (In Russian).

13. Nazarov, A. V. Effects of soil oil pollution on plants. A. V. Nazarov. Bulletin of the Perm University. 5 (10). The Perm State University Publishers, 2007, 134–141 (In Russian).

14. Kazantseva, M. N. Effects of oil pollution on taiga phytocenoses of the Middle Ob area. M. N. Kazantseva. Author's abstract for a PhD thesis in Biology – Yekaterinburg, 1994, 26 p (In Russian).

15. Maksimenko, O. E. Revegetation dynamics of anthropogenically disturbed Sphagnum bogs in the oilfield in the Middle Ob area. O. E. Maximenko, N. A. Chervyakov, T. Ii. Karkishko et al.. Ecology, 1997, №4, 243–247 (In Russian).

16. Bome, N. A. Vegetational history of the industrial landscape in the Northern forest-steppe of the Tyumen region. N. A. Bome, V. V. Hoteev. Bulletin of the Tyumen State University. №3, 106–111 (In Russian).

17. Glyaznetsova, Yu. S., Zueva, I. N., Chalaya, O. N., Lifshits, S. H., Erofeevskaya, L. A. Evaluation of the Biological Treatment Effectiveness of Oil Polluted Soils for the Yakutian Arctic Region. Biological Systems, Eds, L. I. Weisfeld, A. I. Opalko, N. A. Bome, S. A. Bekuzarova. Biodiversity, and Stability of Plant Communities. Apple Academic Press, 2014.

CHAPTER 21

INFLUENCE OF ANTHROPOGENIC PRESSURE ON ENVIRONMENTAL CHARACTERISTICS OF MEADOW HABITATS IN THE FOREST AND FOREST-STEPPE ZONES

ANNA A. KUZEMKO

CONTENTS

ABSTRACT

The changes of the environmental characteristics (soil moisture, soil reaction and nutrient content in the soil) in habitats of the nine alliances of the *Molinio-Arrhenatheretea* class along a gradient of anthropogenic transformation were studied using synphytoindication method with

Ellenberg indicator values. It was established that the general trends of edaphic properties changes caused by anthropogenic pressure are decreasing of soil moisture, increasing of soil reaction and rise of nutrient content in the soil.

21.1 INTRODUCTION

Changes of grassland vegetation due to anthropogenic transformation caused by mainly grazing and mowing, many times discussed in the literature [1, 2]. In many publications it pointed out that anthropogenic pressure on plant community changes also environmental properties of habitat: grazing and trampling lead to soil compaction and enrichment with mineral elements, mowing, on the contrary, causes withdrawal of mineral nutrients and drainage of habitat. However, in most cases, these conclusions are hypothetical. However, there are a variety of nowadays methods and techniques that allow more accurately reveal the changes that taking place in habitats due to its economic exploitation. In this regard, it should be mentioned the synphytoindication technique, the one direction of the bioindication, where plant communities used as environment indicators. Application of this technique for study of vegetation dynamics is a perspective area in contemporary geobotany [3].

In this context, the aim of the present study was to reveal a change in edaphic parameters of meadow habitats in Forest and Forest Steppe zones of Ukraine along a gradient of anthropogenic transformation.

21.2 MATERIALS AND METHODOLOGY

Materials for the study were relevés of grassland vegetation of Forest and Forest-Steppe zones of Ukraine carried out by different authors from 1932 to 2010 (in total 3124 relevés of 22 authors, including 998 own relevés). As a result of the performed classification 2122 relevés were classified within the *Molinio-Arrhenatheretea* class with 3 orders, 9 alliances, and 33 associations [4].

Degree of anthropogenic transformation of communities was determined using the phytocenosis (plant community) destruction index.

Biological meaning of this index is in the fact that a strong external influence can disrupt the internal balance of phytocenosis, thereby disturbed the completeness regime [5], which prevents the penetration of unusual species (destructors) into the phytocenosis. This index was proposed by B.A. Bykov [6], earlier used by L.S. Balashov [7] for the evaluation of the state of meadow communities of Ukrainian Polesye and modified by the author of the present paper in order to improve its accuracy [8]. Modified phytocenosis destruction index calculated by the formula:

$$K_d = \frac{\sum_{i=1}^{Pd} n_i}{\sum_{i=1}^{Pf} n_i} 100$$

where n_i – number of plant species of a certain group; P_d – cover of species-destructors (synanthropic species except occasional apophytes); P_f – cover of all species of phytocenosis.

Species-destructors were identified based on the characteristics of synanthropic species provided by J. Kornaś and adapted for the flora of Ukraine by V.V. Protopopova [9].

In accordance with the values of the phytocenosis destruction index the six classes of digression were assigned (Table 21.1), which correspond respectively to the six stages of digression.

Quantitative characteristics of the investigated alliances of the *Molinio-Arrhenatheretea* class are presented in the Table 21.2.

TABLE 21.1   Correspondence Between the K_d Values and the Classes of Digression

Class	Range of the K_d values, %
0	0
I	1–20
II	21–40
III	41–60
IV	61–80
V	81–100

TABLE 21.2 Distribution of Relevés on Syntaxonomic Alliances and Stages of Digression

Alliance	Total number of relevés	Digression class					
		0	**I**	**II**	**III**	**IV**	**V**
Agrostion vinealis	219	0	139	59	19	2	0
Trifolion montani	212	0	53	97	39	15	8
Arrhenatherion elatioris	36	0	21	14	1	0	0
Festucion pratensis	603	1	192	275	113	21	1
Cynosurion cristati	98	0	29	31	18	13	7
Deschampsion caespitosae	410	15	213	145	27	8	2
Molinion caeruleae	88	17	69	2	0	0	0
Alopecurion pratensis	124	2	85	27	7	3	0
Calthion palustris	263	50	190	17	5	1	0

For assessment of the environmental properties of habitats along the digression gradient used Ellenberg Indicator Values (EIV) [10] for main soil factors (soil moisture – Hd, soil reaction – Rc, and nutrients content in the soil – Tr), calculated for each relevé in JUICE program [11].

To identify general trends in changes of habitat properties along the digression gradient were constructed graphics in Excel 2007 using polynomial trend with six degrees of freedom. Statistical data processing was carried out in the Statistica 7.0.

21.3 RESULTS AND DISCUSSION

In the analysis of trends of studied environmental parameters changes for communities of the *Agrostion vinealis* alliances, attracts attention a slight decrease in moisture in the direction of habitat transformation. In this is noteworthy reduction in oscillation amplitude of values of this factor along the gradient (Figure 21.1a). Soil reaction remains practically unchanged, but there is a considerable reduction in the oscillations amplitude of values of this factor (Figure 21.1b). However, the content of nutrients in the soil increased significantly (Figure 21.1c).

Communities of the *Trifolion montani* alliance also characterized by decreased levels of habitat moisture already at the beginning of the gradient, further this parameter are almost unchanged. Also not

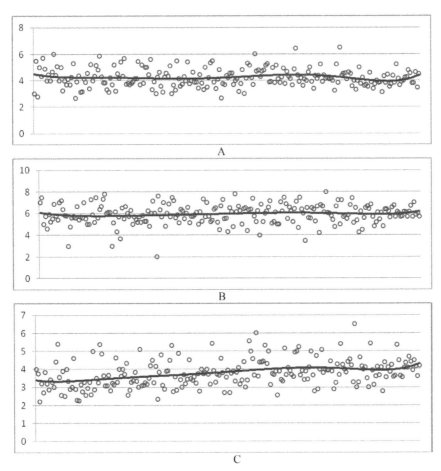

FIGURE 21.1 Changing of environmental properties of habitats for phytocenoses of the *Agrostion vinealis* alliance along the digression gradient [Note. Here and in Figures 21.2– 21.9 relevés arranged on the X axis, and the EIV – on the Y axis; A – Hd, B – Rc, C – Tr].

observed changes in soil reaction, as well as reducing the oscillation amplitude of values for this factor. Nutrients throughout the gradient remain practically unchanged, and at the end of the gradient – increases dramatically.

For the *Arrhenatherion elatioris* alliance, which represented by the least number of relevés in our dataset, mainly at the initial stages of digression, revealed a slight decrease in the moisture level at the end of the gradient, significant fluctuations in the level of soil reaction with a general

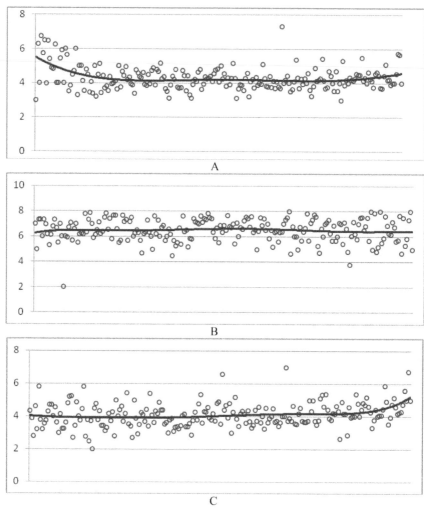

FIGURE 21.2   Changing of environmental properties of habitats for phytocenoses of the *Trifolion montani* alliance along the digression gradient (see the note for Figure 21.1).

tendency to its reduction, as well as fluctuations in the nutrients content with a general trend to increasing (Figure 21.3).

The *Festucion pratensis* alliance contrary represented the largest number of relevés in the dataset and characterized by a slight decrease in moisture along the gradient, constancy of soil reaction with decreasing of oscillation amplitude and a slight increase in the nutrient content of the soil (Figure 21.4).

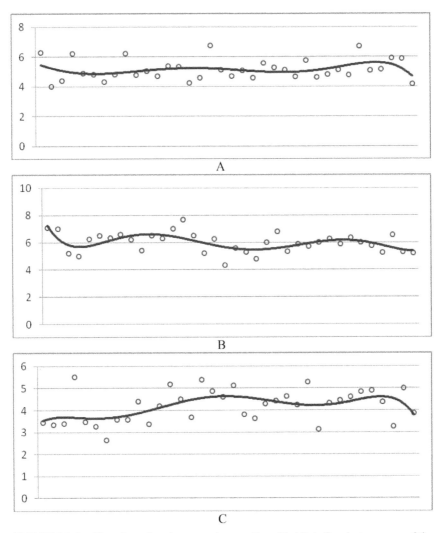

A

B

C

FIGURE 21.3   Changing of environmental properties of habitats for phytocenoses of the *Arrhenatherion elatioris* alliance along the digression gradient (see the note for Figure 21.1).

For the *Cynosurion cristati* alliance revealed slight increase of the moisture level with a significant reduction in the oscillations amplitude of values of this factor along the gradient as well as significant increase of soil reaction and nutrient content (Figure 21.5).

The *Deschampsion caespitosae* alliance characterized by decrease of moisture level along the gradient, especially dramatic in the early

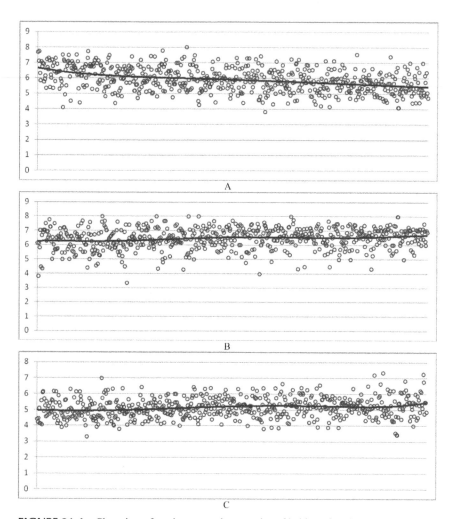

FIGURE 21.4   Changing of environmental properties of habitats for phytocenoses of the *Festucion pratensis* alliance along the digression gradient (see the note for Figure 21.1).

stages of digression, the constancy of soil reaction and a slight increase in nutrient content, more pronounced at the end of the gradient (Figure 21.6).

The *Molinion caeruleae* alliance characterized by a decrease in soil moisture along the gradient and an increase of oscillation of amplitude of this factor values. The soil reaction along the gradient varies considerably, but at the beginning it has been a dramatic decline, which is

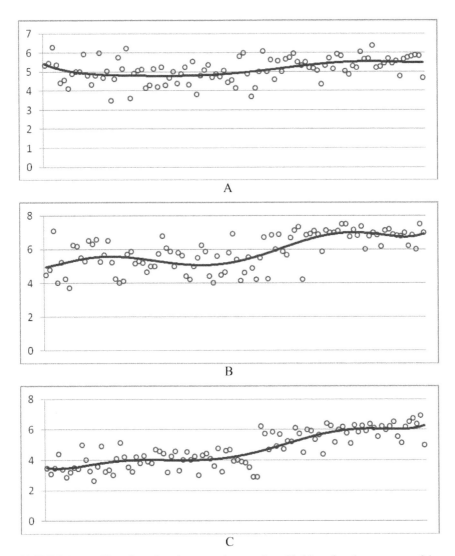

FIGURE 21.5 Changing of environmental properties of habitats for phytocenoses of the *Cynosurion cristati* alliance along the digression gradient (see the note for Figure 21.1).

probably explained by the formation of tree-shrub communities with no anthropogenic pressure and subsequent activation of podzolic process, which causes an increase of soil acidity. The content of nutrients in the soil is gradually increases (Figure 21.7).

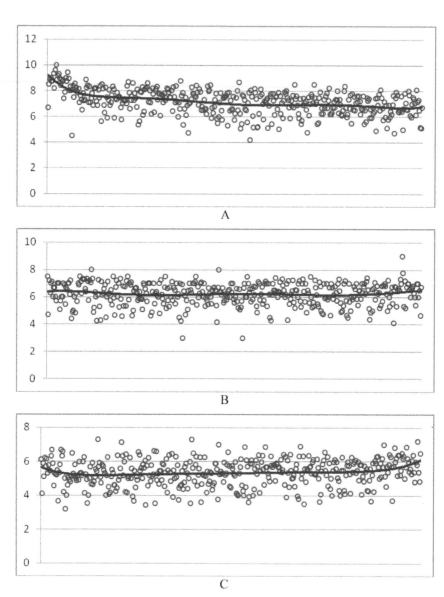

FIGURE 21.6 Changing of environmental properties of habitats for phytocenoses of the *Deschampsion caespitosae* alliance along the digression gradient (see the note for Figure 21.1).

For the *Alopecurion pratensis* alliance revealed moisture reduction, especially noticeable at the beginning and the end of the gradient, a slight

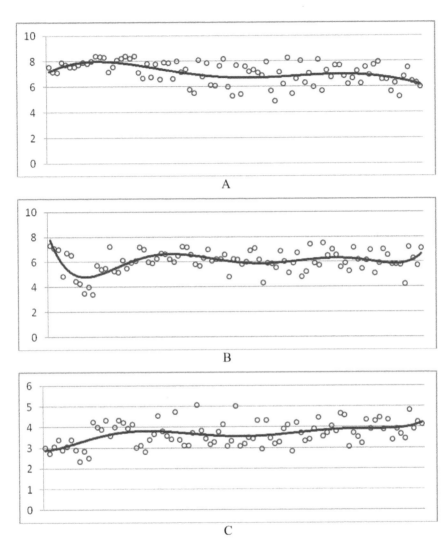

FIGURE 21.7   Changing of environmental properties of habitats for phytocenoses of the *Molinion caeruleae* alliance along the digression gradient (see the note for Figure 21.1).

increase of the soil reaction, as well as a gradual but nevertheless quite significant increase of nutrient content of the soil (Figure 21.8).

The same peculiarities of edafotop parameters dynamics along the digression gradient are characteristic for the *Calthion palustris* alliance (Figure 21.9).

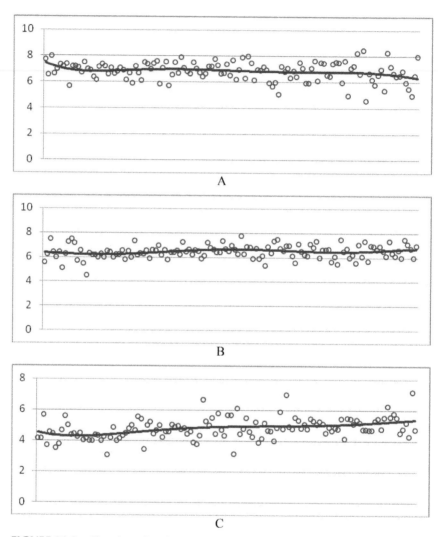

FIGURE 21.8 Changing of environmental properties of habitats for phytocenoses of the *Alopecurion pratensis* alliance along the digression gradient (see the note for Figure 21.1).

For statistical testing of the identified trends were calculated arithmetic mean values for each factor within the different stages of digression (Tables 21.3–21.5).

As can be seen from the Table 21.3 for the majority of the analyzed alliances could note gradual decrease of moisture level from the initial to

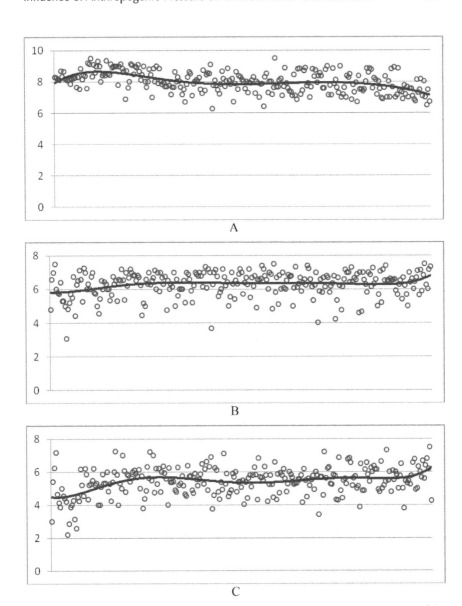

FIGURE 21.9 Changing of environmental properties of habitats for phytocenoses of the *Calthion palustris* alliance along the digression gradient (see the note for Figure 21.1).

the final stages of digression. In some cases there may be a slight deviation from this rule. So for the *Deschampsion caespitosae* alliance revealed a slight increase of the mean for soil moisture at the IV stage of digression

TABLE 21.3 Statistical Parameters of the Synphytoindication Evaluation of Soil Moisture (Mean ± Standard Deviation) for Different Stages of Anthropogenic Digression of Syntaxonomic Alliances of the *Molinio-Arrhenatheretea* Class

Alliance	Digression classes					
	0	I	II	III	IV	V
Agrostion vinealis	—	4.24±0.61	4.20±0.46	4.16±0.36	3.96±0.51	—
Trifolion montani	—	4.57±0.80	4.16±0.40	4.23±0.41	4.26±0.30	4.52±0.51
Arrhenatherion elatioris	—	5.05±0.56	5.31±0.47	4.14±0.00	—	—
Festucion pratensis	7.67±0.00	6.23±0.63	5.85±0.61	5.56±0.53	5.52±0.64	4.73±0.00
Cynosurion cristati	—	4.95±0.53	4.85±0.47	5.40±0.32	5.53±0.30	5.49±0.44
Deschampsion caespitosae	8.83±0.49	7.35±0.74	6.88±0.69	6.79±0.68	6.87±0.69	5.95±0.81
Molinion caeruleae	7.76±0.33	6.96±0.77	6.16±0.16	—	—	—
Alopecurion pratensis	7.16±0.60	6.88±0.55	6.82±0.80	6.33±0.39	6.50±1.00	—
Calthion palustris	8.50±0.43	7.94±0.52	7.46±0.40	7.20±0.42	6.69±0.00	—

TABLE 21.4 Statistical Parameters of the Synphytoindication Evaluation of Soil Reaction (Mean ± Standard Deviation) for Different Stages of Anthropogenic Digression of Syntaxonomic Alliances of the *Molinio-Arrhenaheretea* Class

Alliance	Digression classes					
	0	I	II	III	IV	V
Agrostion vinealis	—	5.87±0.72	5.9±0.74	6.13±0.33	5.85±0.15	—
Trifolion montani	—	6.52±0.72	6.50±0.64	6.53±0.77	6.30±0.86	6.42±1.21
Arrhenatherion elatioris	—	6.05±0.72	5.93±0.35	5.22±0.00	—	—
Festucion pratensis	6.11±0.00	6.30±0.64	6.49±0.58	6.56±0.52	6.65±0.43	6.91±0.00
Cynosurion cristati	—	5.28±0.70	5.36±0.73	6.70±0.58	6.85±0.26	6.77±0.38
Deschampsion caespitosae	6.46±0.67	6.20±0.68	6.18±0.67	6.50±0.69	6.38±0.45	5.71±1.04
Molinion caeruleae	5.50±1.08	6.16±0.64	6.41±0.68	—	—	—
Alopecurion pratensis	5.93±0.33	6.46±0.43	6.51±0.48	6.62±0.46	6.58±0.39	—
Calthion palustris	5.98±0.67	6.35±0.55	6.27±0.49	6.75±0.50	7.33±0.00	—

TABLE 21.5 Statistical Parameters of the Synphytoindication Evaluation of Nutrient Content (Mean ± Standard Deviation) for Different Stages of Anthropogenic Digression of Syntaxonomic Alliances of the *Molinio-Arrhenatheretea* Class

Alliance	Digression classes					
	0	I	II	III	IV	V
Agrostion vinealis	—	3.64±0.60	3.99±0.55	4.14±0.35	3.91±0.29	—
Trifolion montani	—	3.93±0.65	4.00±0.51	4.30±0.52	4.34±0.50	4.97±0.61
Arrhenatherion elatioris	—	4.04±0.70	4.41±0.40	3.85±0.00	—	—
Festucion pratensis	4.40±0.00	4.97±0.50	5.21±0.48	5.25±0.50	5.44±0.57	4.87±0.00
Cynosurion cristati	—	3.74±0.53	4.20±0.62	5.57±0.49	6.00±0.31	6.12±0.60
Deschampsion caespitosae	5.66±0.52	5.23±0.64	5.41±0.59	5.83±0.50	5.88±0.53	6.68±0.82
Molinion caeruleae	3.30±0.52	3.75±0.45	4.19±0.06	—	—	—
Alopecurion pratensis	4.16±0.02	4.69±0.49	5.04±0.30	5.24±0.40	5.41±1.19	—
Calthion palustris	4.90±0.89	5.52±0.58	5.66±0.60	6.49±0.52	4.23±0.00	—

with subsequent sharp decrease in V stage. The same is true for the *Alopecurion pratensis* alliance. Exceptions are the *Trifolion montani* and *Arrhenatherion elatioris* alliances, for which there is no moisture reduction from the initial to the final stage of digression. For the *Cynosurion cristati* alliance founded an inverse trend of the moisture level raising from the initial to the final stage of digression, which obviously can be explained by excessive soil compaction due to overgrazing, which in turn causes some waterlogging.

With regard to the dynamics of soil reaction within the stages of digression for the *Arrhenatherion elatioris* alliance observed steadily diminishing, and for the *Festucion pratensis* and *Molinion caeruleae* alliances revealed a gradual rise. Increasing of this factor parameters from the initial to the final stages of digression observed for the *Agrostion vinealis, Cynosurion cristati, Alopecurion pratensis, Calthion palustris* alliances, although for individual stages of digression this pattern can be broken. For the *Trifolion montani* and *Deschampsion caespitosae* alliances no regularity of soil reaction dynamics along the digression gradient not revealed.

Analysis of the dynamics of the soil nutrient content by the stages of digression showed a gradual rise of the factor parameters along the gradient for the *Trifolion montani, Cynosurion cristati, Molinion caeruleae, Alopecurion pratensis* alliances. General tendency to increase of nutrient content within the digression stages is characteristic also for other alliances, but for the individual stages this trend can be broken.

For revealing the relationship between the values of the phytocenosis destruction index and the indicator values of the studied environmental factors the correlation analysis was carried out. The results of this analysis are presented in Table 21.6.

As shown in Table 21.6, there is a clear statistically significant negative correlation of K_d with soil moisture for the *Festucion pratensis, Cynosurion cristati, Deschampsion caespitosae, Molinion caeruleae, Calthion palustris* alliances. Positive correlations of K_d and soil reaction are characteristic for the *Festucion pratensis, Cynosurion cristati, Alopecurion pratensis,* and *Calthion palustris* alliances. Also positive correlation of K_d with nutrient content of the soil is characteristic for all analyzed alliances except *Arrheantherion elatioris.* The greatest absolute value of the correlation coefficient of K_d with edaphic factors obtained for the *Cynosurion cristati*

TABLE 21.6   Values of the Correlation Coefficients Between K_d and Indicator Values of the Studied Environmental Factors for Alliances of the *Molinio-Arrhenatheretea* Class

Alliance	Hd	Rc	Tr
Agrostion vinealis	−0.06	0.06	**0.29**
Trifolion montani	−0.11	−0.02	**0.30**
Arrheantherion elatioris	0.08	−0.28	0.29
Festucion pratensis	**−0.35**	**0.15**	**0.19**
Cynosurion cristati	**0.37**	**0.60**	**0.80**
Deschampsion caespitosae	−0.36	0.01	**0.17**
Molinion caeruleae	**−0.37**	0.08	**0.32**
Alopecurion pratensis	−0.18	**0.18**	**0.40**
Calthion palustris	**−0.40**	0.12	**0.21**

*Marked correlations are significant – bold at $p < 0.05$.

alliance, which includes communities of mesophytic pastures where significant anthropogenic pressure is a major limiting factor providing long-term and stable existence. For the *Arrheantherion elatioris* alliance found no significant correlation of K_d values with indicator values of studied edaphic factors that can be explained by low anthropogenic pressure, mainly through mowing and perhaps by insufficient number of relevés for the alliance in the dataset.

Thus, for the alliances of meadow vegetation revealed the following patterns of changes in environmental properties of habitats along a digression gradient: decreasing of soil moisture, increasing of soil reaction and increasing of nutrient content of the soil. These regularities can be explained first of all by soil compaction caused by trampling due to overgrazing and recreation, which changes the water permeability of soil and grass stand xerophytization respectively. Changes in the physical properties of soil promotes changes in its chemical composition, in particular occurs slowdown of podzolic process, and as a consequence the loss of humic acids content. This may explain the revealed trend of increasing soil pH due to anthropogenic transformation of habitats. Increasing of nutrients content in the soil along the gradient of digression can be explained by the accumulation of waste products of animals as a result of grazing transformation or by contamination during recreational transformation. In favor of this idea is supported by the fact that communities of predominantly with pasture use (*Cynosurion cristati, Deschampsion caespitosae* alliances) characterized by a more sharp increase of this parameter than communities with primarily haying use (*Arrhenatherion, Molinion, Calthion* alliances).

Decline of the oscillation amplitude of the values of analyzed edafotop properties can be explained by the leveling of ecological peculiarity of habitats under the anthropogenic pressure.

21.4 CONCLUSIONS

The results obtained in the study can be used in organization of environmental management and monitoring of natural grasslands, for the development of optimal regimes of grazing, mowing and recreation.

KEYWORDS

- grazing
- soil nutrients
- synphytoindication
- syntaxonomic alliances

REFERENCES

1. Ramenskiy, L. G. Problems and Methods of the Plant Cover Studying. Selected Works. Leningrad. Nauka (*Science*). 1971, 334 p (In Russian).
2. Grasslands in Europe of high nature value. P. Veen, R. Jefferson, J. de Smidt, and, J. van der Straaten [Eds.] Zeist. Netherlands, KNNV Publishing, 2009, 320 p. (In English)
3. Didukh Ya.P. Fundamentals of bioindication. Kyiv. Naukova dumka. 2013, 344 p. (In Ukrainian)
4. Kuzemko, A. A. Classification of grassland vegetation of the Polesie and Forest-Steppe of Ukraine using statistical methods. Collection of articles and lectures IV All-Russian School-Conference "Actual problems of Geobotany" (1–7 October 2012). Ufa. Publishing Center "Media Print". 2012, 227–233 (In Russian).
5. Kurkin, K. A. Ecological and coenotic regime of the meadow biogeocoenosis completeness. Problems of biogeocoenology. Moscow. Nauka (*Science*). 1973, 137–148 (In Russian).

6. Bykov, B. A. Pastures and hayfields of Kazakhstan (Classification). Alma-Ata. Nauka (*Science*). 1969, 71 p (In Russian).
7. Balashov, L. S. Anthropogenic changes of the meadows of Ukrainian Polesie. Ecology. 1991, №1, 3–9 (In Russian).
8. Kuzemko, A. A. Anthropogenic transformation degree of true meadow plant communities of Forest and Forest-Steppe zones of Ukraine. Indigenous and introduced plants. 2006, №2, 29–34. (In Ukrainian)
9. Protopopova, V. V. Synanthropic flora of Ukraine and ways of its development. Kyiv. Naukova Dumka. 1991, 200 p (In Russian).
10. Ellenberg, H. Zeigerwerte der Gefässpflanzen Mitteleuropas. Scripta geobotanica. Gottingen. 1974, №9, 197 p. (in German)
11. Tichý, L. Juice, software for vegetation classification. J. Vegetation Sci. 2002, №13, 451–453.

CHAPTER 22

DYNAMICS OF THE FLORISTIC DIVERSITY OF MEADOWS AS A STABILITY FACTOR OF HERBACEOUS ECOSYSTEMS

RAFAIL A. AFANAS'EV

CONTENTS

ABSTRACT

In the evolutionary development herbaceous ecosystems elaborated the ways of protection from excess weight of wild large herd phytophages. Absence of this protection would lead not only to the disappearance of grassland vegetation but also to loss of soil due erosion processes. The author of this study believes that general protection measures include decrease of natural pasture productivity as a result of the change of vegetation cover, the manifestation of forage herbs toxicity under unfavorable growth conditions and at destruction of the sod due to ungulates – the appearance of weeds which have poisonous and repellent properties and so not consumed by the animals; then the normal herbage restores.

22.1 INTRODUCTION

Using a system approach the author gives a new interpretation of the known facts about grassland vegetation reaction to overload of pastures by animals. The reduction in pasture bioproductivity due to the change of vegetation cover, toxicogenic defense reactions against eating of different plant species under adverse weather conditions and abundant appearance of inedible weeds in localities where with destroyed sod are evolutionary developed return reactions of herbaceous ecosystems to the demolition. From these positions we ought to consider mass appearance of weeds on cultivated land or while drastic meadow improving. Ecosystems perceive the violation of natural turf as impact of graminivorous animals and try restore status quo using their funds developed by the evolution. First of all these are meadow or field weeds.

22.2 MATERIALS AND METHODOLOGY

Materials of investigation were published messages of geobotany, phytosociology, ecology, meadland farming and other sciences associated with the study of vital activity of plant communities and also the results of own monitoring the state and dynamics of ley phytocenoses.

The methodological principles of generalization of the materials were the system approach [1, 2] and the actualism that is widely used in geology [3]. The system approach is to consider of a plant community and his biotope as a functional unit having the properties of self-regulation, protection from external actions through the development of appropriate responses. The method of the actualism is to recreate the history of development of the ecosystems based on the study of modern processes and conditions of their functioning as functional properties of plant communities have been developed in the process of long evolution. This fact is the basis for the retrospective assessment of the conditions and nature of these properties formation.

22.3   RESULTS AND DISCUSSION

According to modern data system is a dynamic organization of living and non-living elements in which there are external and internal circulations of substances determining its functional stability [4]. The most important condition for the stability of any system of the material world is its flexibility, the ability to direct the efforts of its parts (subsystems) and the whole system back to where danger, to avert this danger and save themselves. To some extent this property of the systems describes Le Chatelier's principle: "If a system is to be any impact then as the result of current processes the equilibrium will shift in such a direction that the impact will decrease " [5, p.185].

Herbaceous ecosystems (biogeocenoses) are forms of existence of living matter, specifically plant cover of the Earth. They meet all the requirements of a system as multi-component, self-regulatory "purposefully functioning structures capable of resolving problems in certain external conditions" [1, p.75]. So far herbaceous ecosystems have been studied mainly on specific aspects of their functioning without sufficient generalization of accumulated data. Meanwhile enormous factual material allows approaching its broad-scale understanding at the system level and gets "new system measurements, new genetic parameters of reality" [2, p.14].

The aim of our work was to evaluate the diverse properties of biogeocenoses from a system approach perspective and to identify cause-effect relationships in the dynamics of their floristic composition under the influence of external and internal factors.

Going directly to the statement of research materials, note that from the time of Leonardo da Vinci the main role in the formation of the soil cover along with other factors was given to higher plants including herbs [6]. Pedologists in particular L.O. Karpachevsky [7] believe that the formation of modern soil cover dates back to the Cretaceous period, when angiosperms are widely spread that is about 100 millions years ago, forming deciduous forests and meadows. Perhaps modern meadows differ from those that existed 100 millions years ago but their existence they are since the time and during this period meadow ecosystems developed adaptive response to external influences that threaten their existence. Thus it is necessary to distinguish adaptation of separate types of plants and adaptation, characteristic for the entire community of these species, for example, phytocenoses as a whole. The system analysis allows you to select at least three categories of these adaptations that can be considered as factors of stability of herbaceous ecosystems. Though it may seem paradoxical, but all of these factors are aimed combating against the destruction of the soil cover, the preservation of its fertility, reducing the loss of plant nutrients into the environment. And more specifically, that they are directed against the negative impact on grassland cenoses primarily by large herbivores as well as natural phenomena causing the destruction of the sod.

The role of perennial grasses in the protection of soil from water and wind erosion is widely known [8]. However, until now, remained essentially out of sight scientists such function of herbaceous ecosystems as protection of soil from pasture erosion, for example, from destruction under the influence of large herbivorous animals. In natural conditions the wild herbivorous consume on average about 10% of the biomass of natural pastures [9], which corresponds to the general biological law of the energy pyramid [10]. However, when feeding or migration herd animals especially ungulates (buffaloes, bisons, tarpans, aurochs, antelopes, saiga, horses and other animals) could repeatedly be situations of heavy grazing, destruction of sod, which in combinations with weather anomalies – torrential rains, hurricanes – was supposed to lead to the destruction of the soil cover, death of ecosystems. In the process of biological evolution to survive could only such communities that due to the principle of natural selection have developed adequate defense remedies primarily of soil that was the basis of existence of the plant communities and also grassland landscape in general. Soil as bioinert body formed under the influence of vegetation, is unable to resist the active external actions, in particular the anthropogenesis [11]. Therefore, in the historical past, the soil could not develop appropriate protective mechanisms against the effects of major phytophages and the role of defenders of the soil from destruction in biogeocenoses was given to living beings – plants. Consider these mechanisms more with attraction of well-known facts.

The first is the reduction in the productivity of plant community with increased heavy grazing of above-ground biomass, which is designated as pasture digression. It is well known from the theory and practice of modern grassland agriculture [12]. Thus according to I.V. Larin [13] and scientists to which he refers, with increasing systematic stocking in forest-meadow area first of all disappear high and semi high perennial grasses, forming the largest biomass: common timothy, tall oat grass, awnless brome, meadow fescue, red clover, meadow foxtail. These plants towering over the other attract animals in the first place. With frequent grazing such plants rapidly waste away and fall out of the grass cover giving way to low-growing and less productive grass – Kentucky bluegrass, fine bent grass, white clover and others which without encountering competition for light and nutrition from high grass begin to dominate in the grass cover.

At the further increase of stocking the stand composition changes more rapidly due to different eat ability of plants. In these cases the plants, which are eaten most, also waste away and fall out of the grass cover. According to long-term observations, plants, grazed down 6–7 times during the summer, died or severely become sparse. On pastures remain inedible plants or plants, which are eaten not much: prostrate knotweed, silverweed, ladies'-mantle, dandelions, plantains and the like. The nutritional value of pastures for large herbivorous is falling, thus, to almost zero. The speed of pasture degradation depends on the stocking per unit of forage areas: the higher it is, the shorter the life cycle belonging to eatable species. Such dynamic change of grassland plants during high stocking from herbivores in all soil-climatic zones goes through procedure. The difference lies only in species composition of plant communities, replacing each other, and in the rate of substitution of one type by others. T.A. Rabotnov [14] pointed out that in England as a result of long unregulated grazing of sheep on pastures grew Nardus, bracken, heather, for example, they lost feed value.

The universality of the above-mentioned patterns, logically explainable by biological features of the different types of plants, from system approach should be seen as a defensive reaction of biogeocenoses from destruction of their grazing animals by decrease of stocking and in more general terms by reducing the population of herbivores in the region due to the lack of pasture forage. When the stocking of cattle on pastures decreases degraded pastures are recovered in the process of the so-called demutation [10]. However, an excessive stocking that may arise due to any reasons may not only lead to the degradation of pasture but also to pasture erosion, for example, to complete destruction of ecocenosis and destruction of the soil cover. So in modern conditions when excessive unregulated stocking of cattle in the mountains (the Caucasus, Altai, Buryatia) sod failure (formation of paths) may exceed 60% of pasture area and soil washout from deprived of vegetation places – more than half of its power. According some data [15], the annual reduction of soil profile in the Eastern regions of the Caucasus for this reason averaged 0.8 mm, for example, it decreased by 1 cm every 13 years. More striking manifestation of pasture erosion is observed on the Black Lands (Caspian lowland), where on winter pastures many years was converted cattle from different regions of the Northern Caucasus and Transcaucasia. As the result of excessive stocking here was almost completely destroyed not only the vegetation and soil, occurred desertification areas, up to the formation of drift sand.

But the decline, depression of pasture productivity is "the first line of defense" of grassy ecosystems. The second factor, or the mechanism of their sustainability, also directed against large herbivorous animals, triggering by the breach of the normal functioning of perennial grasses due to unfavorable weather conditions: long cold, drought, or vice versa waterlogging of the soil. In these conditions, ungulate excessive stocking on weakened plant communities may also lead to their death, sod destruction with subsequent erosion of soil cover. Protective function of herbaceous ecosystems in such situations is expressed in development in the organism of normal forage plants various substances, toxic for animals. This phenomenon is well known from literature [16, 17]. First of all, this accumulation in the herbs of nitrates by a violation of the synthesis of proteins and also retardation of the growth of plants due to adverse weather conditions: hailstorm and similar anomalies. The excessive consumption of such plants causes nitrate-nitrite toxicosis of animal. It is established that feed containing more than 0.07% $N-NO_3$, dangerous for animals and a doubling of the concentration can be fatal. In these conditions in plants of different families including Gramineae, Fabaceae, Cyperaceae hydrocyanic acid is formed by splitting of cyanogenic glycosides by relevant enzyme into glucose and hydrocyanic acid, which is a potent poison. Are marked many other toxicosis, caused by different glycosides, alkaloids,

saponins, essential oils and other substances, resulting in elevated concentrations in plants at violation of the normal processes of growth and
development. For example glycoside cumarin which is harmless in normal conditions and found in melilots slow drying of wet habitats makes
to dicumarin – highly toxic compound that causes the death of animals
within 2–3 days. See the numerous cases of poisoning of herbivores due
to eating plants affected by fungal diseases that infect weakened fodder
plants. Characteristically, the most sensitive to the action of toxic compounds are the animals that came from other habitats (migrants), pregnant
animal, young animals and the animals weakened due to starvation. Often
there is the death of suckling calves because of the switch to the milk toxicants contained in the forage of the cows but not rendered visible harm to
the health of the latter. The toxicity of many plants depends of the habitat,
the phase of development and other factors that have been studied not
enough. However, it is obviously that this property was developed by fodder grasses not accidentally and, at the system level, is the protection of
biocenoses from herbivores with the deterioration of habitat conditions.

And, finally, on the last, the third obvious way to protect herbaceous
ecosystems from destruction of the soil cover at the expense of weeds. It
is in effect when the first two methods appear insufficient and animals, for
whatever reasons, violate the integrity of the sod. It is the mass appearance
of plants, usually called weeds. Among them are poisonous and indel-
ible plants with distinct properties deter herbivores: toxicity, thorniness,
hairiness, the presence of coarse stems, sharp smell and the like. To this
group belong common thistle, plumeless thistle, sow thistle, black henbane, water hemlock, larkspur, aconite, white hellebore, datura – almost
all the weeds of our gardens and arable lands – a total of more than 700
species, or about 15% of the floristic diversity of the natural pastures. The
largest number of poisonous and noxious species of plants there are in the
next families: Euphorbiaceae – 98% (29 species), Solanaceae – 97% (29
species), Equisetaceae – 81% (9 species), Ranunculaceae – 54% (117 species). Quite a lot of them are in Cruciferae (Brassicaceae) – 37% (60 species), Polygonaceae – 37% (39 species), Liliaceae – 26% (34 species). In
the families Gramineae (Poaceae) and Fabaceae (Papilionaceae) there are
5% (25 and 28 species respectively), Cyperaceae – 1% (1 species). Thus,
from more than 1000 species of Gramineae, Fabaceae and Cyperaceae

only 54 species are dangerous for animals, whereas in the of miscellaneous herbs, which also include the weeds, there are about 700 species. Species diversity makes adequate herbage reactions to external influences depending on environmental conditions of their existence, including soil, intra- and interspecific, weather, phytosanitary, by a number of phytophages, and, as we can see, large herbivores.

In violation of the integrity of sod by ungulates animals weeds quickly fill the gaps in the grass, preventing further appearance of animals on damaged areas. Striking the adequate responses of ecosystems to the negative impact of the animals: the greater the harm caused to the grass, the sharper repellent properties of weeds, which have experienced animals. In our time clearly this reaction can be traced about sheep enclosures, temporary stock stands of animals, where the degree of damage to turf decreases from the center of damage to the habitat periphery. The author of this study had to observe the emergence of a dense bed of a black henbane on the place of multiple milking in the valley of the small river in the Yaroslavl region, where the turf in the previous year was completely damaged on an area of about 100 m². On the periphery of the bed increased common thistles and musk thistles, and in the process of removal from the center of the bed their number and habitus respectively decreased. It is characteristic that in a year on the place of former bed was not one of the weed; ring there was only green carpet of grass, although when seeding of black henbane millions of seeds of this plant were in the soil. And this is no accident. V.R. Williams [18] described in detail the change of tall weeds consisting of various weeds, to grasses with abandonment of arable land in fallow in the steppe zone. The fallow period with prevalence of tall weeds, which usually lasted one year replaced with the couch grass fallow (5–7 years), which in turn transformed into a solid fallow, consisting mainly of loose bunchgrasses (15–20 years), with a gradual transformation of the stipa steppe with the advantage of firm bunchgrasses. In conditions of formerly widespread in the South of Russia fallow land-use system the restore of natural phytocenoses, appropriate to soil and climatic conditions of the steppe, lasted, so 20–30 years. On a similar scheme, but with a different floristic composition and duration of cenogenesis procenoses (intermediate cenoses) changed on the meadows of the humid zone in case of damage or destruction of the meadow sod by animals or technical means (with anthropogenesis). For example, on a small potato field (Tver region), previously fertilized by manure, procenoses of tall weeds a year after the end of treatment consisted of wormwood – Artemisia vulgaris L. (Figure 22.1), the following year, mainly from willow herb – Chamerion angustifolium L. (Figure 22.2).

It weeds inhabit first of all arable land, because they are perceived by wildlife as areas with broken sod, requiring protection and restoration of natural vegetation – grass stand or forest depending on environmental conditions. If in modern conditions people define the stocking, in the nature it regulate themselves herbaceous communities, to be exact – herbaceous ecosystems, or ecosystems, including soil and local biota. These property ecosystems have developed in the process of long evolution (cenogenesis) and natural selection of systems: ecosystems failed to produce adequate protection, – have disappeared from the face of the earth as disappeared grass and soil on the Black Lands. But this is the guilt of man; nature was powerless to fight against it. In natural condition, figuratively speaking, poison and antidote were developed simultaneously, and ecosystem natural selection happened, obviously, on the same principle as natural selection of individual species of plants or animals. Due to this in nature has been preserved dynamic equilibrium; the example of this equilibrium is the ecological balance in the system of predator-prey. The same can be

FIGURE 22.1   Wormwood (Artemisia vulgaris).

FIGURE 22.2   Willow herb (Chamerion angustifolium).

said about the dynamic equilibrium that exists between herbaceous communities, on the one hand, and grazing animals, on the other. In this case as a "predator" are herbivores (consumers), and in the role of "prey" – fodder grasses (producers). Upon termination of grazing occurs recovery (demutation) of productive phytocenoses that used in practice by providing pasture rest. Compilation of materials on the biological productivity of natural forage lands shows that it is inversely proportional to the intensity of grazing and is a regulator to stocking.

Interestingly, in phytocenological system of protection of soils from pasture erosion fit some insects. According O.S. Owen [19], in dry years in America locust destroyed up to 67% stock of phytomass on natural pastures, and the cattle had to be distilled in other places that he did not die of hunger. There was observed a direct link between stocking and number of locusts, the cause of which is the ecologist could not explain. In normal hydration or with a moderate stocking during the dry years the impact of the locusts was practically invisible or less significant. Similar phenomena have been observed previously in Russia, in particular in the Baraba steppe of Western Siberia [20]. From these facts it follows that the protection of herbaceous ecosystems. From these facts it follows that the protection of herbaceous ecosystems from stocking involved not only the vegetation, but also Orthoptera of this biotope, creating serious competition for large herbivores, reducing their numbers in case of increased threat to ecotope by the latter.

Appearing on the places of damaged sod, weeds treat juvenile undergrowth grass almost paternal care. Otherwise you will not say. First, it is the protection of seedlings of slow-growing perennial grasses from trampling by animals, second, from their grazing in young, immature age, thirdly, the accumulation of nutrients, especially nitrates, formed by mineralization of the destroyed sod, prevent them from losses due to leaching and denitrification, and, finally, the programed destruction of weeds order to give the living place to the next vegetation formations, for example, cereal grass, passing it "inherited" nutrients, accumulated in the plant residues.

Although in the nature after community of tall weeds would grow usually wheatgrass as a dramatic example of neutrophils, under the canopy of weeds with no less success you can grow types of forage grasses. This is evidenced by how scientific expertise and a wide practice of meadow grass cultivation. In particular, in the field experiment conducted in the state farm "Voronovo" of the Moscow region, on the fertile land of loam soil with coverless sowing in pure form or in mixtures of more than 10 types of cultivars of cereals and legumes, including oligotrophic, characteristic for the poor habitats (slough grass), in the year of planting white pig weed was growing abundantly. Weeding it manually on half of the area and leaving the other half before the end of vegetation did not reveal any significant difference in the condition of the herbage and yield of perennial grasses in any of subsequent years of research. These facts point to the specificity (commensalism) of relationships one-biennale weeds with perennial grasses developed under their canopy regardless of the floristic composition of each group: oligo-, meso- or eutrophes formed on the relevant fertility soils. However, fallow of tall weeds are able, apparently, to some extent, to control the floristic composition of the procenosis, which goes for a change. Research conducted in Timiryazev Moscow Agricultural Academy (V.A. Zvereva, unpublished data) on the effects of the extract from the seeds of sosnovsky cow parsnip (*Herackleum sosnowskyi*) on the germination of common valerian (*Valeriana officinalis*), St. John's wort (*Hypericum perforatum*), snow-on-the-mountain (*Euphorbia marginata*), green amaranth (*Amaranthus retroflexus*) and hare'tail grass

(*Lagurus ovatus*) showed that for two species of plants extract had inhibiting effect, for two other had stimulating effect and for one – additive. It should also be that the consistent successions (changes) of procenoses (intermediate cenoses) on the ruins of the sod from tall weeds to stable (climax) phytocenosis aimed at achievement of a definite purpose – to hold and accumulate in biogeocenoses formerly accumulated elements of mineral nutrition. This can be seen in the changing attitude of plants to have in the soil mobile nutrients, in particular nitrogen. The most demanding of them weeds such as mugwort (*Artemisia vulgaris*) white pig weed, common thistles, plumeless thistles, sow thistles, black henbane etc. High consumption of nitrogen weeds-eutrophes indicates at least such fact as the content of crude protein, not inferior legume grasses, from 18 to 28% (calculated on the dry weight).

From the system point of view, this change of plant groupings in place of the destroyed phytocenoses can be explained any otherwise than evolutionary developed way of herbaceous ecosystems to restore the "status quo" with the least loss of mobile plant nutrients from the destroyed sod formed by mineralization of its organic residues. One-biennale weeds, possessing powerful starting-growth and developing the greater weight, catch nitrogen and ash elements and when death passed them to subsequent herbaceous procenoses. Already loose bunchgrasses, which change rhizomatous grasses begin to inhibit the processes of mineralization of organic substances in soil that leads to a gradual accumulation of humus, and firm bunchgrasses with the prevalence of mycotrophic nutrition type complete the immobilization of nutrients transferring its main part in the organic form.

Significant role in the retention of nutrients in the soil at destruction of natural meadow turf by ungulates also plays a soil microflora. Now it is known [21], that in untilled soil at decomposition of plant residues most immobilization mineral soil nitrogen or nitrogen fertilizers [$N-NO_3$ and $N-NH_4$] by soil microorganisms occurs in the first 10–12 days, thus preventing its infiltration losses in the underlying soil and denitrification in the form of gaseous nitrogen forms. This process is accompanied by mineralization of organic substances in soil and plant residues with the release of mineral nitrogen, which is consumed by another group of soil microflora. With the advent of vegetation on the site of the destroyed natural turf role of soil microflora in the retention of nutrients from exogenous losses are gradually decreasing. According to S.P. Smelov [22], the number of microorganisms, mineralizing nitrogen of soil organic matter on the roots of loose bunchgrass – meadow fescue during four years of observations decreased from 9.2 to 1.4 milliards per 1 g of dry roots. Similar results were obtained with timothy. For the seventh year of life separate species of herbs accumulates from 34 to 47 tons of humus on 1 hectare transforming into immobile state 1.5–2 tons of nitrogen, from 0.5 to 1 ton of ash matter. From the results of our studies [23], it is also obvious that consort communications in biogeocenoses between herbs microflora and soil aim to the retention of nutrients in the soil including inhibition of nitrification, and more in sabulous in comparison with loamy. With equal doses of nitrogen fertilizers and almost the same removal of nitrogen with grass yield on irrigated pasture consisted mainly of cockfoot Dactylis glomerata L., and approximately equal to the content of mineral nitrogen in soil ratio nitrate form to ammonium one in the upper layer of sabulous on the average for vegetation period amounted one to five, whereas in loamy – no more than one to two. Otherwise, the environment-forming role of biocenoses was reduced to a maximum retention of mineral nitrogen in its sphere adapting to the habitat nature (ecotope). Thus, even in the artificially created agro-coenosis of cultural pastures under irrigation of sabulous where nitrates bigger risk of loss with infiltration waters, compared with loamy and where the best aeration, it would seem, must strengthen the processes of nitrification, the main fund of mineral nitrogen was presented ammonium form. Something of the kind also mentioned B.N. Korotkov [24] in his lysimeter studies with ^{15}N.

Characteristically, biological processes in the soil to some extent adapted by physical-chemical (bioinert) properties of these soils. According to our research, in the initial samples of poor sabulous and loamy soils the content of ammonia nitrogen was significantly higher in the loam at approximately equal to the content in both soil nitrate nitrogen (Figure 22.3). However, with sterilization samples and when washing them with a solution of ammonium nitrate followed by rinsing with distilled water it was found that ammonium stronger recorded sabulous soil than loamy (Figure 22.4). In other words, microbiological and physico-chemical processes in soils occur unidirectional – to hold nutrient elements in this context, nitrogenous

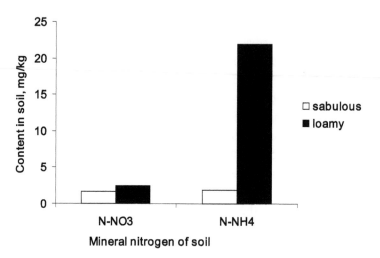

FIGURE 22.3   The content of mineral nitrogen in sabulous and loamy soils before application of nitrogen fertilizers.

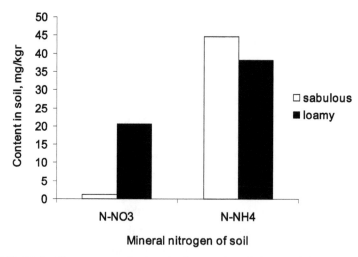

FIGURE 22.4   The content of mineral nitrogen in sabulous and loamy soils after application of nitrogen fertilizers.

substances, in strategic biogeocenoses and, above all, in their subsystems – soils, which are characterized as bioinert systems.

About that ecosystems, including soils, are self-regulatory systems, evidence and other facts. V.V. Kidin [21] in the experiments with ^{15}N established that with increasing doses of mineral fertilizers higher consumption of cultural plants increases the degree of immobilization of nitrogen by the soil microflora forming humus, on the one hand, and gaseous nitrogen losses of soil and fertilizer on physico-chemical and biological denitrification, on the other hand. It is known also, that the lack of soil nitrogen in natural plant communities appear accumulators of nitrogen – leguminous plants – clover, alfalfa, and other species, reducing then its abundance in phytocenoses in favor of grass species and diversity of herbs. It follows that the soils as if support within certain limits, the contents of the mobile nitrogen, the most sensitive than phosphate and potash, for the climax (relatively stable) plant communities.

On the whole, analysis of the results of research on retention in the system soil-plant of mineral forms of soil nitrogen shows that nitrate nitrogen is held mainly by soil microflora and plants: the soil microflora by placing it in an organic form (new growth of humus), plants due to the increase of biomass of weeds and the following plant associations, and also due to accumulation of nitrates (NO^-_3) in plant biomass. At that plants can accumulate in their biomass nitrates, except for the reproductive organs, without any functional limitations. In the soil environment, as shown above, nitrates can be lost due to denitrification and infiltration to groundwater. Ammonium nitrogen (NH^+_4) is held mainly by the soil, as in plants it cannot accumulate due manifestations of properties toxic to plants. It follows that between agents (subsystems) of the system soil-plant role of the main depot for nitrate nitrogen is given to the plant, for ammonium nitrogen – to soil.

In all these phenomena clearly one could trace antientropic trend of the herbaceous ecosystems in the soil, on the other hand, functioning, for example, accumulation and structuring of matter and energy (overground and underground mass of plants, biota and humus) in the number and proportion ensuring minimum loss into the environment. Soil cover of our planet was formed and stored due to this property of herbaceous (and forest) ecosystems. And not the last role in these processes played the ability of plant formation resist destructive effects on the soil by large herbivorous, their floristic biodiversity with severe soil-protecting functions.

22.4 CONCLUSIONS

In general biological sense, the sustainability of ecological systems is interpreted as the ability to resist the action or to return to the initial state after exposure. In this paper does not discuss factors of stability of herbaceous ecosystems associated with adaptive responses of plant communities on the change of soil, hydrological, meteorological and other conditions, which is the subject of synecology and is described by many geobotanists [14, 20 and 25]. The subject of discussion in the work are the reactions of herbaceous ecosystems produced in the process of the evolutionary development of such ecosystems and aimed at soil conservation as a primary basis of their existence. Opened us at the system level, this aspect of the sustainability of natural ecosystems bases on known facts, directly or indirectly pointing to the specific, essentially passive, but adequate counteraction of phytocenoses to ungulates in danger of destroying the sod and soil cover. In general relations between herbivorous animals and phytocenoses obey the laws of biological systems "predator-prey" [10], where the role of "predator" belongs to herbivorous, and "prey" – phytocenoses. We found also the reason, functional predetermination of successive change of procenoses in the process of demutation (recovery) of destroyed vegetation which consists in evolutionary developed expediency of preservation and transfer of "inherited" moving nutrients produced when organic matter mineralization of former turf. When this was first shown soil-protecting role of weeds in nature, contrary to the established in the agriculture opinion of them as the plunderers of soil fertility [26]. We disclosed antientropic trend of successions herbaceous communities leading ultimately to the accumulation and structuration of mater and energy in ecosystems, ensuring their dynamic stability and, consequently, creation and preservation of soils in areas with grass vegetation.

From system positions seem unreasonable recommendations for implementation in forage production new fodder crops from plant species one way or another related to procenoses of tall weeds, for example, weeds with severe soil protection properties. An example is recommendation about cultivation for fodder mentioned above. Sosnovsky cowparsnip [27] which did not lead to increased production of fodder but caused the blockage this species many habitats suitable for its growth, in particular roadsides of roads and railways in the European part of Russia, other habitats with increased moisture and focal accumulation of nutrient elements.

KEYWORDS

- animals
- ecosystems
- erosion of soil
- pastures
- perennial grasses
- plants
- weeds

REFERENCES

1. Sagatovsky, V. N. The experience of creation of a categorical apparatus of the system approach. Philosophical Sciences. 1976, №3, 75 (in Russian).
2. Afanas'ev, V. G. System approach and society. Moscow: Publishing House of Political Literature, 1980, 14 (in Russian).
3. General biology with principles of historical geology. Moscow: Higher School. 1980, 4 (in Russian).
4. Chernikov, V. A., A. V. Aleksashin, A. V. Golubev. Agroecology. Moscow: Kolos. 2000, 14 (in Russian).
5. Glinka, N. L. General chemistry, Leningrad: Chemistry. 1976, 185 (in Russian).
6. Krupenikov, I. A. The history of pedology. Moscow: Science. 1981, 327 p (In Russian).
7. Karpachevsky, L. O. Soil and pedosphere in space coordinates. Agrarian Science. 1995, №4, 46–48 (in Russian).
8. Pavlovsky, E. S. Soil-protecting significance of natural forage lands. Natural forage resources of the USSR and their use. Moscow: Science. 1978, 74–78 (in Russian).
9. Rakitnikov, A. N. The use of natural forage resources as a factor of development of agriculture. Moscow: Kolos. 1978, 35–47 (in Russian).
10. Reymers, N. F. Nature management. Moscow: Thought. 1990, 152 (in Russian).
11. Dokuchaev, V. V. Sustainability of soils to natural and anthropogenic influences: Abstracts of All-Russian Conference. Moscow: Soil Science Institute. 2002, 489 p (In Russian).
12. Andreev, N. G. Meadland farming. Moscow: Publishing House of Agricultural and Industrial Literature. 1985, 83–85 (in Russian).
13. Larin, I. V. Grassland science and pasture farming. Moscow; Leningrad: Publishing House of Agricultural Literature. 1956, 63 (in Russian).

14. Rabotnov, T. A. Meadland farming. Moscow: Publishing House of Moscow State University. 1974, 349 (in Russian).
15. Erizhev, K. A. Mountain hay lands and pastures. Moscow: Spring. 1998, 320 p (In Russian).
16. Vilner, A. M. Forage poisonings. Moscow: Kolos. 1974, 408 p (In Russian).
17. Dimitrov, S. Diagnostics of poisoning of animals/S. Dimitrov, A. Dzhurov, S. Antonov. [Ed. V. A. Beskhlebnov. Translation from Bulgarian: K. S. Bogdanov] Moscow: Publishing House of Agricultural and Industrial Literature. 1986, 284 p (In Russian).
18. Williams, V. R. Collected papers. Vol. 3. Moscow: Publishing House of Agricultural Literature. 1949, 132–135 (in Russian).
19. Owen, O. S. The protection of natural resources. Moscow: Kolos. 1977, 179–180 (in Russian).
20. Kurkin, K. A. Studies of meadow dynamics as systems. Moscow: Science. 1976, 287 p (In Russian).
21. Kidin, V. V. Fundamentals of plant nutrition and application of fertilizers. Part 1. Moscow: Publishing House of Timiryazev Moscow Agricultural Academy. 2008, 415 p (In Russian).
22. Smelov, S. P. Theoretical foundations of grassland science. Moscow: Kolos. 1966, 121–126 (in Russian).
23. Afanas'ev, R. A. Fertilizer of intensive irrigated pastures in the Nonchernozemic zone of the RSFSR. Summary of the doctoral thesis. Scriveri: Latvian Institute of Agriculture and Rural Economy. 1987, 44 p (In Russian).
24. Korotkov, B. I. Regulation of water and nutrient regime of soils, organization and use of irrigated pastures: Irrigated pastures and hay lands in the Nonchernozemic zone. Moscow: Publishing House of Russian Agricultural Literature. 1984, 9–12 (In Russian).
25. Sharashova, V. S. The sustainability of grassland ecosystems. Moscow: Publishing House of Agricultural and Industrial Literature. 1989, 239 p (In Russian).
26. Afanas'ev, R. A. Soil protection function of weed grasses in ecosystems. Agricultural Biology. 1983, №3, 11–15 (In Russian).
27. Vavilov, P. P. New forage plants. Field crops of USSR. Moscow: Kolos. 194, 59–63 (in Russian).

CHAPTER 23

BOTANICO-GEOGRAPHICAL ZONING OF THE UPPER DNIEPER BASIN ON THE BASE OF THE J. BRAUN-BLANQUET VEGETATION CLASSIFICATION APPROACH

YURY A. SEMENISHCHENKOV

CONTENTS

ABSTRACT

In this chapter, the characteristic of botanico-geographical subprovinces of Upper Dnieper basin within the Russian Federation on the basis of data on the distribution of forest vegetation syntaxa established by J. Braun-Blanquet is done. The borders of allocated subprovinces largely correspond to the boundaries of communities of the alliance rank with a set of specific associations and subassociations and areas of some tree edificators. The groups of a differential species for the four subprovinces are detected.

23.1   INTRODUCTION

Upper Dnieper basin – the third largest transboundary basin of the Europe – covers an area of about 100.5 km² in six regions of the Russian Federation. This part of the basin extends from the North-West to South-East for over 600 km. Forest vegetation of the region with a total area of about 3.5 million hectares, is the result of long anthropogenic transformation and an indicator of the overall status of nature.

The role of the Dnieper basin as an important phytogeographical abroad have repeatedly noted in the literature. By floristic zoning of

A.L. Takhtadzhyan [1], this region lies within the Eastern European region, Circumboreal province, Holarctic floristic kingdom, which combines Central-Russian part of Sarmatian floristic province. Following the zoning of H. Meusel et al. [2], the region belongs to the Sarmatian province of the Middle-European floristic region. It should be noted that in this part of the Dnieper basin the boundaries of some significant tree edificators lies: *Acer campestre, A. tataricum, Alnus incana, Carpinus betulus, Picea abies*, as well as a number of significant by botanico-geographical positions shrubby and herbaceous species. In the literature reflected the substitution of *Picea abies* and *Quercus robur* and complex relationships of indigenous tree species: *Tilia cordata, Acer platanoides, Quercus robur, Picea abies* on the latitudinal gradient in this region [3].

Essential to identify the main trends in the development of vegetation, its rational use and protection is a botanico-geographical zoning. It aims to identify geographic patterns of vegetation. While on the first place its territorial structure, showing the impact of climatogenic factors, florogenesis and phytosociogenesis of the vegetation cover.

Earliest materials to botanico-geographical zoning of the region based mainly on data of dominant vegetation classification [3–5]. The boundaries of the chorions of the rank of province and subprovince in the published zoning schemes carried out conventionally enough and need to be clarified in relation to the accumulation of a large amount of new information on the distribution of plant communities and species of plants.

In our paper we discuss the possibility of using the classification of forest vegetation on the basis of J. Braun-Blanquet [6] approach to describe and clarify the boundaries of the basic units of botanico-geographical zoning in the basin of the Upper Dnieper. This method is based on the classification of complete floristic composition of plant communities. The syntaxa established based on it, have a well-defined geographical and ecological volume. Therefore, the data on the distribution of syntaxa and their combinations can adequately describe botanico-geographical patterns of territory and ecological diversity of habitats.

23.2   MATERIALS AND METHODOLOGY

From 2002 to 2014 author conducted floristic and geobothanical survey of the basin of the Upper Dnieper within the Belgorod, Bryansk, Kaluga,

Kursk, Orel, Smolensk regions of Russia. Based on more than 2700 rele-vés made by the author, as well as phytocenarium of the Bryansk State University materials and available data in the literature the forest veg-etation syntaxonomy on the basis of J. Braun-Blanquet [6] developed 38 associations and 16 subassociations established within 15 alliances, 11 orders and 6 classes of vegetation.

During the zoning used floristic and phytocoenoic features of zonal and azonal-zonal vegetation syntaxa. In this zoning was carried out on the basis of data of the potential vegetation, which generally corresponds to a set of territorial "epytaxons." Each of the designated chorologic units cor-responds to a certain combination of syntaxon of the rank of alliance and association with a set of lower-level units.

An important basis for our zoning became regional schemes of natural, forest growth and floristic zoning, generalized maps and plans of forest vegetation, forest management regulations and plans on individual forest areas, as well as surveys of the flora of administrative regions in the basin.

Based on authors survey implemented in 2002–2014, analyzing her-barium materials (LE, MW, BRSU, OHHI, etc.), analysis of the available literature on the flora of the study area and adjacent areas identified differ-ential species of plants, the distribution of which must be considered when conducting botanico-geographical boundaries. This is, first and foremost, tree and shrub edificator species of some areas of the study region, as well as to a large extent reflect the climatic and ecological potency typical for installed zoning units habitats. On the other hand, it is some species of herbaceous plants, differentiating selected phytochorions of the province and subprovince rank.

Names of vascular plants are done following S.K. Cherepanov's survey [7].

23.2.1   CHARACTERISTIC OF THE BOTANICO-GEOGRAPHICAL SUBPROVINCES OF THE UPPER DNIEPER BASIN

Features of the flora as well as the patterns of distribution of plant commu-nities of zonal and azonal-zonal types allowed to specify the boundaries of existing provinces and subprovinces (Figure 23.1). Below given their characteristics, performed on the basis of our studies.

FIGURE 23.1   Botanico-geographical zoning of the Upper Dnieper basin (within the Russian Federation; the basin boundary in bold dotted line) [*Legend:* Eurasian taiga region, Northern European taiga province. 1 – Valdai-Onega subprovince. Zonal vegetation types: composite spruce, broad-leaved forests. European deciduous forest region, Eastern-European province. 2 – Polessian subprovince. Zonal vegetation types: northern deciduous forests (with a little participation of spruce), hornbeam and spruce-deciduous forests. 3 – Middle-Russian subprovince Zonal vegetation types: deciduous forests without spruce. Eurasian steppe region, Eastern-European steppe province. 4 – Middle-Dnieper subprovince. Zonal vegetation types: deciduous forests without spruce, meadow steppes].

23.2.2   EURASIAN TAIGA REGION, NORTHERN EUROPEAN TAIGA PROVINCE, VALDAI-ONEGA SUBPROVINCE

This subprovince enters the territory of the Upper Dnieper basin its south-eastern edge and covers the Smolensk region and north-west of the Bryansk region. North-western boundary within the basin runs along the watershed of the Dnieper and the Zapadnaya Dvina and coincides with the south-eastern border of the Valdai glaciation. This corresponds to a suppositive line of geobotanical districts of oak- coniferous forests subzone of neighboring Belarus [8], forest-cultural districts of mixed forests subzone [9] differentiation. Southeastern boundary of the subprovince within the basin extends approximately along the line "Vyshkov – Novozybkov – Unecha – Mglin – Zhukovka – Bryansk – Karachev." From the south and southeast it limited by the north-western edge of the opolyes landscapes of Sudost right bank (Bryansk region).

In the northern part of the subprovince the zonal association is the nemoral spruce forests with the participation of broad-leaved species – ass. *Rhodobryo rosei–Piceetum abietis* Korotkov 1986 (alliance *Querco–Tilion* Solomeshch et Laiviņs ex Bulokhov et Solomeshch 2003) provided a large number of variants [10, 11]. A distinctive feature is the presence here nemoral species: *Acer platanoides, Carex pilosa, Corylus avellana,* not characteristic of such communities in the northern and north-eastern parts of the association area [3, 12]. The most typical are *Stellaria nemorum* var., *Galeobdolon luteum* var., *Oxalis acetosella* var., *Hepatica nobilis* var.

The predominance of spruce forests of this type in this area is reflected in the map of natural vegetation of Europe [13], where they are classified as Baltic-north-west-Sarmatian mixed deciduous-spruce forests (*Picea abies, P. abies* × *P. obovata, Tilia cordata, Acer platanoides, Quercus robur,* in the west part with *Carpinus betulus*) with *Corylus avellana, Euonymus verrucosa, Galeobdolon luteum, Stellaria holostea* (category D19). Widespread mesophytic gray alder changes of this association (ass. *Rh. r.– P. a. Alnus incana* facies), as well as community of stream forests with Alnus incana domination of ass. *Scirpo sylvatici–Alnetum incanae* Semenishchenkov 2014. Southern boundary of distribution of the gray alder forests passes north-west of the Bryansk region [11].

In general, the distribution and abundance of gray alder forests can be considered characteristic of the subprovince. The southern boundary of the watershed alder spread approximately coincides with the northern boundary of broad-leaved (deciduous) zone (Figure 23.2). In neighboring Belarus the southern boundary of *A. incana* distribution considered the northern boundary of the hornbeam-oak-coniferous forests subzone [14, 15].

In this part of the basin from the north come by small fragments the communities of taiga spruce forests (alliance *Piceion* Pawłowski et al. 1928), an extensive area which lies to the north of the European part

FIGURE 23.2   Boundaries of the most significant tree edificators in the Upper Dnieper basin. Border of species distribution marked by dotted lines. The basin boundary in bold dotted line

of Russia. These forests are found absently differ besides dominance of mosses and boreal shrubs also the presence of a number of nemoral species, which is not typical for taiga zone: ass. *Linnaeo borealis–Piceetum abietis* (Caj. 1921) K.-Lund 1962 and ass. *Melico nutantis–Piceetum abietis* K.-Lund 1981. Considerably wider on the well-watered depressed areas represented sphagnum-spruce forests of ass. *Sphagno girgensohnii–Piceetum abietis* K.-Lund 1981. Overall in the territory such forests of the taiga types can be considered extrazonal.

Broad-leaved-spruce forests in the southern part of subprovince match the southern line of broad-leaved-spruce forests characterized by the sod-omination of spruce and broad-leaved species on the watershed areas [3].

The most common zonal association in southern part of the subprov-ince (mostly north-west of the Bryansk region) is *Mercurialo perennis–Quercetum roboris* Bulokhov et Solomeshch 2003 with the subass. *M.–Q. piceetosum abietis* subass. nov. prov. (alliance *Querco–Tilion*). These communities represent the mesophytic nemoral linden-oak forests with a small spruce participation. Communities of the ass. *Rhodobryo–Piceetum* found only in small patches. Here, at the southern border of subtaiga sub-zone, these forests can be considered zonal, although they are very differ-ent from the more typical northern spruce nemoral forests by the role of some nemoral species and, in particular, edificators: *Quercus robur, Tilia cordata, Acer platanoides, Corylus avellana.*

Communities of ass. *Melico–Piceetum, Linnaeo–Piceetum* and *Sphagno–Piceetum* are very rare and are not typical.

Zonal position also affects the composition of azonal-zonal forest veg-etation. Within the subprovince on outwash plains widespread forests of the alliance *Dicrano–Pinion sylvestris* (Libb. 1933) Mat. 1962, which represents acidophyte moss-shrub pine and spruce-pine forests. A char-acteristic feature of pine forests is a significant part in the communities of *Picea abies*. The most prominent association of this alliance here is ass. *Vaccinio vitis-idaeae–Pinetum* Caj. 1921 with subass *V.–P. piceeto-sum abietis* Bulokhov et Solomeshch 2003. In most mesophytic habitats represented ass. *Corylo–Pinetum* Bulokhov et Solomeshch 2003 with participation of spruce and deciduous trees. For habitats with moist soils the typical association of pine forests is ass. *Molinio caeruleo–Pinetum* (E. Schmid. 1936) em. Mat. (1973) 1981, in which usually spruce present.

In general, the main arguments to the attribution of the described area to the Eurasian taiga region following:

1. This area is located within the area of *Picea abies*, serving major edificator in forest communities.

2. The main type of the potential vegetation is composite spruce forests with an admixture of broad-leaved trees, the role of which increases in the direction from north to south.
3. Currently on the area nemoral composite spruce forests dominate.
4. The recovering of the softwood forests (birch, aspen and gray alder) forests, as a rule, run to spruce composite nemoral forests. When this gray alder changes are not typical for the European broad-leaved forest zone within the basin. In plantations in this region is also commonly occur nemoral spruce forests.
5. Position of the subprovince correlated with geobotanical zoning schemes of the European part of the USSR, the system of typological units of the European vegetation map, silvicultural zoning of the Nechernozemye region, as well as the latest offerings to the botanico-geographical zoning of Russia, according to which this territory falls into the subtaiga subzone, Eurasian taiga zone, Circumboreal region.

Differential species of the Valdai-Onega subprovince, not passing within the Upper Dnieper basin for its south-eastern border: *Aconitum septentrionale, Alnus incana, Arctostaphylos uva-ursi, Botrichium matricarifolium, Empetrum nigrum, Geranium phaeum, Oxycoccus microcarpus, Polystichum braunii, Ranunculus lanuginosus, Rubus chamaemorus, Swertia perennis.*

23.2.3   EUROPEAN BROAD-LEAVED FOREST REGION, EASTERN-EUROPEAN PROVINCE, POLESSIAN SUBPROVINCE

Subprovince territory lies to the southeast of the Valdai-Onega subprovince; in southeastern limited western spurs of the Middle-Russian Upland [3, 4, 8]. On this territory combines different types of landscapes: polessies, subpolessies, predopolyes, opolyes.

Deciduous forests with spruce (alliance *Querco–Tilion*), which on the map of natural vegetation of Europe [13].

are depicted as the Baltic-South Sarmatian linden-oak forests (*Quercus robur, Tilia cordata*), partly with *Picea abies* (category F 70) dominated on the area.

Typical association is ass. *Mercurialo–Quercetum*, which is divided into three geographical subassociations: northern with spruce – *M.–Q. piceetosum abietis*, southern – without it – *M.–Q. typicum* and, conditionally, the western – *M.–Q. carpinetosum betuli* Bulokhov et Solomeshch 2003 – oak forests with spruce and hornbeam.

For the subprovince characteristic communities of acidophyte oak forests, often mixed with *Pinus sylvestri*s, – ass. *Vaccinio–Quercetum roboris* Bulokhov et Solomeshch 2003 (alliance *Vaccinio–Quercion* Bulokhov et Solomeshch 2003), which are presented in an upland areas of polessies and subpolessies landscapes in the central part of the basin (Bryansk region). In the Vegetation map of Europe [13], these forests are designated as North Central European-Sarmatian pine-oak forests (*Quercus robur, Pinus sylvestris*) with *Tilia cordata*, partly with *Picea abies*, with *Euonymus verrucosa, Potentilla alba, Chamaecytisus ruthenicus* (category F 13).

The xeromesophyte oak forests of ass. *Lathyro nigri–Quercetum roboris* Bulokhov et Solomeshch 2003 are also characteristic for the subprovince. Central part of their natural habitat lies on the extreme west of the Middle-Russian Upland, in moving to the northwest, these forests lose some characteristic thermophilic predominantly forest-steppe species and become more mesophytic. This pattern is fixed in the several geographical variants [16, 17, 18].

The predominant type of vegetation adjacent to the Russian part of the Dnieper basin from Ukraine are the pine forests. Widespread and even dominance in Polessian subprovince pine and pine-oak forests, according to E.M. Lavrenko, should be regarded as purely edaphic phenomenon [4].

Pine forests of the subprovince represented mainly by subass. *Vaccinio vitis-idaeae–Pinetum*. However, on the latitudinal gradient the composition of pine forests varies, and in promoting to the southeast of subprovince on the place of spruce-pine forests come oak-pine forests sometimes with a small admixture of spruce (subass. *Vaccinio vitis-idaeae–Pinetum quercetosum roboris* Bulokhov et Solomeshch 2003).

For habitats with wetter soils the typical ass. *Molinio-Pinetum* and ass. *Corylo-Pinetum* characteristic. In their communities also usually present spruce. On the upland dry areas occasionally distributed lichen pine forests (ass. *Cladonio–Pinetum* Juraszek 1927).

In the lower valley of the Desna river (Bryansk region) presented steeped oak-pine forests: ass. *Peucedano–Pinetum sylvestris* W. Mat. (1962) 1973 and ass. *Veronico incanae–Pinetum sylvestris* Bulokhov et Solomeshch 2003.

23.2.3.1   Differential Species of the Polessian Subprovince, Not Passing Within the Region For Its South-Eastern Border

Andromeda polyfolia, Armeria vugaris, Botrichium virginianum, Calamagrostis purpurea, Carex brizoides, C. brunescens, C. colchica, C. globularis, C. juncella, C. loliacea, C. remota, C. umbrosa, Carpinus betulus, Cinna latifolia, Corynephorus canescens, Dactylorhiza fuchsii, D. longifolia, D. sambucina, D. traunsteineri, Diphasiastrum tristachium, Dryopteris expansa, Eleocharis ovata, Epipogium aphyllum, Equisetum variegatum, Festuca altissima, Galium triflorum, Glyceria nemoralis, Goodyera repens, Holcus lanatus, Hypericum montanum, Hypochoeris radicata, Hypopitis hypophegea, Jovibarba sobolifera, Ledum paluster, Lerchenfeldia flexuosa, Linnaea borealis, Listera cordata, Lycopodiella inundata, Melampyrum sylvaticum, Phegopteris connectilis, Picea abies, Polygala vulgaris, Rubus nessensis, Sieglingia decumbens, Stellaria alsine, S. nemorum.

23.2.3.2   Differential Species of the Middle-Russian Subprovince, Not Passing for Its North-Western Border

Acer tataricum, Aconitum nemorosum, Adenophora lilifolia, Adonis ver-nalis, Allium flavescens, Alopecurus arundinaceus, Alyssum desertorum, Amoria fragifera, Amygdalus nana, Arenaria micradenia, Aristolochia clematitis, Artemisia armeniaca, Asperula cynanchica, Asplenium ruta-muraria, Aster amellus, Astragalus austriacus, Carduus hamulosus, Carex atherodes, C. michelii, C. supina, Centaurea marchalliana, C. ruthenica, Cirsium canum, Clematis integrifolia, Cotoneaster alaunicus, Delphinium cuneatum, Dianthus andrzejowskianus, Echinops sphaerocephalus, Echium russicum, Erigeron poolicum, Fritillaria ruthenica, Fumaria schleicheri, Galatella lynosiris, Gladiolus tenuis, Gypsophyla altissima,

Hieracium virosum, Hyacinthella leucophaea, Hypericum elegans, Juncus gerardi, Jurinea arachnoides, Linum perenne, Lycopus exaltatus, Melampyrum cristatum, Melica altissima, M. transsilvanica, Omphalodes scorpioides, Orobanche purpurea, Polygala sibirica, Polygonum alpinum, Polystichum aculeatum, Potentilla reptans, Ranunculus illyricus, Rosa rubiginosa, Salvia nutans, Salvia testiqola, Scilla sibirica, Senecio schvwetzovii, Serratula coronata, Sisimbrium strictissium, Sium sisarum, Sonchus palustris, Spiraea crenata, S. litwinovii, Stipa pennata, S. tirsa, Thesium arvense, Th. ebracteatum, Tofieldia calyculata, Trinia multicaulis, Vicia pisiformis, Viola accrescens.

23.2.4 MIDDLE-RUSSIAN SUBPROVINCE

The western border of the subprovince is limited by western boundaries of the landscapes of erosion-denudation elevated (200–250 m) loess plains of the western slopes of the Middle-Russian Upland. The soil cover is represented by gray forest soils.

Aboriginal vegetation was represented by Eastern European deciduous (oak) forests, which are preserved in small areas of agricultural land. "Forest-steppe" appearances of these landscapes have received as a result of deforestation. It is a region of ancient agricultural civilization, whose territory is almost completely plowed. On notions of Meshkov [19], the northern boundary of the subprovince outlines the southern limit of continuous distribution of *Picea abies* and its satellites. In general, forest of the subprovince corresponds to the "southern botanico-geographical strip" of Middle-Russian-Subvolga deciduous forests, which are distinguished by the absence of *Picea abies* [3].

Within the Central-Russian subprovince as we move deeper into the zone of deciduous forests and steppe the forests of the alliance *Querco–Tilion* replaced by the alliance *Aceri campestris–Quercion roboris* Bulokhov et Solomeshch 2003 (ass. *Aceri campestris–Quercetum roboris* Bulokhov et Solomeshch 2003). Such forests are widespread in the south and east of the Bryansk region in the Middle-Russian Upland. They are characterized by participation in communities Acer campestre and its satellites, the northwestern border of the area which, according to A.L. Takhtadzhyan [1], coincides with the north-western boundary of forest-steepe zone. At the western border of the area they are rare [20], in general, on the Russian plain sometimes called "vanishing" type of vegetation [12].

Within the Dnieper basin, these forests are divided into geographical subassociations: typical community – the western spurs of the Central Russian Upland and communities with greater participation of some steppe and forest-steppe species in the Central Southern Forest-steppe [21].

This type of forest vegetation on the Vegetation map of Europe [13] classified as North-South Ukrainian-Sarmatian linden-oak (*Quercus robur, Tilia cordata*) forest with *Fraxinus excelsior, Acer campestre, A. tataricum* (category F 71). The subprovince is characterized by broad participation of *Tilia cordata, Fraxinus excelsior* and lack of *Carpinus betulus* in the coenoflora of the forest vegetation [5]. As rare indicator species is also called: *Acer tataricum, Malus sylvestris* s.l., *Pyrus pyraster, Prunus spinosa* [20]. Individual fragments, including on the beams and river valleys, found here communities of subass. *Mercurialo–Quercetum typicum* (without *Picea abies*). In small-preserved forests of subpolessian landscapes, as well as on the slopes of hills and river valleys presented community of xeromesophytic deciduous forests of the ass. *Lathyro nigri–Quercetum*. In the floodplains of small rivers are occasionally found community of ash-oak forest with maple outfield – ass. *Fraxino excelsioris–Quercetum roboris* Bulokhov et Solomeshch 2003 (alliance *Aceri campestris–Quercion*). For river valleys characteristic floodplain deciduous forests – ass. *Filipendulo ulmariae–Quercetum* Polozov et Solomeshch 1999.

For this subprovince the sporadic spread of steppe vegetation of the class *Festuco–Brometea* Br.-Bl. et R. Tx. in Br.-Br. 1949 characteristic.

23.2.4.1   Differential Species of the Middle-Russian Subprovince, Not Passing Over Its South-Eastern Border

Calluna vulgaris, Carex disticha, C. echinata, C. limosa, C. panicea, C. paniculata, Chamaedaphne calyculata, Chimaphila umbellata, Circaea alpina, Corydalis cava, Cruciata glabra, Daphne mezereum, Dianthus borbasii, Digitalis grandiflora, Eleocharis ovata, Galium intermedium,

*Huperzia selago, Hypopitis monotropa, Juniperus communis, Lycopodium annotinum, Moneses uniflora, Neottianthe cucullata, Oxycoccus paluster, Pyrola chlorantha, Pyrola minor, Rhynchosphora albus, Scirpus radicans, Sempervivum ruthenicum, Senecio arcticus, Thymus serpyllum, Trientalis europaea, Vaccinium myrtillus, V. vitis-idaea, Viola uliginos*a.

23.2.4.2   Eurasian Steppe Region, Eastern-European Steppe Province, Middle-Dnieper Subprovince

Covers an area of south-west of Kursk region and north-west of Belgorod region. Following E.M. Lavrenko [5], the eastern boundary of subprovince extends from Kursk to Belgorod and coincides with the western boundary of the Artemisia armeniaca and A. latifolia areas. A.R. Meshkov [19] on the basis of data on the distribution of a broader set of species moves this border to the west of city of Belgorod, roughly along the upper Uda river or Kharkov river.

Here are the typical zonal chernozems in the watersheds distributed. The northern boundary of the subprovince is marked by a continuous spread in the watersheds of gray forest soils, as well as the July isotherm of 20°C and a line of monthly precipitation totals July 90 mm [19].

The distribution of forest-steppe communities of the alliance *Aceri tatarici–Quercion* Zólyomi 1957 (Kursk and Belgorod regions) is characteristic. Northwestern boundary of the area of this alliance coincides with the boundary of the *Acer tataricum* [22]. This alliance has multiple associations represented on the Middle-Russian Upland in upland habitats on the Chernozem zone, and on the slopes of ravines and river valleys.

Along with the forest vegetation for this subprovince the steppe vegetation is zonal [3]. As shown by A.V. Poluyanov [23], the boundary between the European broad-leaved forest and the Eurasian steppe regions, primarily determined by the spread here syntaxa of herbaceous vegetation of the class Festuco–Brometea, zonal for the southern and eastern regions of the Kursk region. Through the line "Ponyri – Kursk – Glushkovo" approximately coinciding with valleys of Tuskar river and Seim river, lies a zone of the areas of a large group of steppe and forest-steppe species condensation: *Salvia nutans, Stipa capillata, S. stenophylla, Astragalus onobrychis, A. austriacus, Asperula cynanchica, Euphorbia seguierana, Polygala sibirica, Clematis integrifolia, Helictotrichon schellianum, Inula ensifolia* and other. Northwest of the border, they are not available or are known from single extrazonal localities; respectively, known from the north-west region syntaxa of the class *Festuco–Brometea* strong floral depleted.

23.2.4.3   Differential Species of the Middle-Dnieper Subprovince, Not Passing Beyond Its North-Western Border

Ajuga glabra, A. laxmanni, Allium inaequale, A. podolicum, Alyssum calycinum, A. diversicaule, A. gmelini, A. lenense, Anchusa leptophylla, Anthemis cotula, Arabis recta, Arenaria biebersteinii, A. procera, Artemisia sericea, Astragalus onobrychis, Asyneuma canescens, Campanula altaica, Carex buekii, C. tomentosa, Centaurea apiculata, Cephalaria uralensis, Cerinthe minor, Chorispora tenella, Clausia alpina, Crepis pannonica, Dipsacus strigosus, Echinops ruthenicus, Ephedra distachia, Eringium campestre, Erysimum marschallianum, Euclydium syriacum, Euphorbia leptocaula, E. sareptana, E. sequeriana, E. subtilis, Gagea granulosa, G. pusilla, Galatella angustissima, G. biflora, G. biflora, G. divaricata, G. punctata, G. virosa, Galium octonatum, Glaux marithima, Goniolimon tataricum, Lathyrus lacteus, Linaria biebersteinii, L. cretacea, L. odora, Linum nervosum, Lycopsis arvensis, Melampyrum argyrocomium, Melica picta, Onosma simplicissima, Orobanche laevis, Peucedanum ruthenicum, Phlomis pungens, Pimpinella tragium, Poa stepposa, Potentilla pimpi-nelloides, Rosa glabrifolia, R. mollis, R. subpomifera, Salvia ethiops, S. nemorosa, S. stepposa, Scheverekia podolica, Scilla bifolia, Scorzonera stricta, S. taurica, Scutellaria supina, Serratula lycopifolia, S. radiata, Silene amoena, S. chersonensis, S. repens, S. wolgensis, Stipa pulcherrima, Theucrium chamaedrys, T. polium, Thesium procumbens, Vincetoxicum cretaceum, Viola ambiqua, V. suavis.

23.3   CONCLUSIONS

Thus, the establishment of botanico-geographical zoning units in Upper Dnieper basin is supported by data on the distribution of syntaxa of different rank and their combinations can be considered as botanical and geographical markers. Actually allocated subprovinces border largely correspond to the areas of syntaxa on the rank of alliance with a set of specific associations and subassociations and boundaries of geographically important species areas.

ACKNOWLEDGEMENTS

The study was supported by RFBR under the research project № 13–04–97510 "Forest vegetation of the Dnieper basin within the Russian Federation."

KEYWORDS

- **edificators**
- **forest vegetation**
- **subprovinces**
- **syntaxonomy**
- **taiga regions**

REFERENCES

1. Takhtadjan, A. L. Floristic regions of the Earth. Leningrad. Science (Nauka). 1978, 41–43 (In Russian).
2. Meusel, H., Jager, E., Weinert, E. Vergleichende Chorologie der zentraleuropäischen Flor. Bd. 1. Text, Karten. Jena. 1965, 583 S. (In German).
3. Vegetation of the European part of the USSR [Eds. S. A. Gribova, T. I. Isachenko, E. M. Lavrenko] Leningrad. Science (Nauka). 1980, 429 p (In Russian).
4. Geobotanical zoning of the Ukrainian SSR. Kiev. 1977, 306 p. (In Ukrainian)
5. Geobotanical zoning of the USSR/Ed. E. M. Lavrenko. Moscow. Academy of Science of the USSR. 1947, 149 p (In Russian).
6. Braun-Blanquet, J. Pflanzensoziologie. 3. Aufl. Wien; N.Y., 1964, 865 S.
7. Cherepanov, S. K. Vasculart plants of Russia and adjacent states. St. Petersburg. World and Family. 1995, 992 p (In Russian).
8. Map of the vegetation of the BSSR. M 1:1000000. Minsk. Science and Technics. 1969, (In Russian).

9. Kurnaev, S. F. Fractional forest vegetation zoning of the Non-Chernozem center [Ed. Zvorykina, K. V.] Moscow. Science (Nauka). 1982, 118 p (In Russian).

10. Semenishchenkov, Yu. A. Ecological variants of the nemoral spruce forests on the South of the Subtaiga zone (Smolensk region). Scientific statements of the Belgorod State University. 2012, Vol. 19, №9 (128). 22–30 (In Russian).

11. Semenishchenkov, Yu. A., Kuzmenko, A. A. Forest vegetation of the moraine and fluvioglacial plains of the Northwest of the Bryansk region [Ed. A. D. Bulokhov] Bryansk. 2011, 112 p (In Russian).

12. East-Europaeae forests: history in Holocene and Present. Vol. 2. Moscow. Science (Nauka). 18–35 (In Russian).

13. Karte der natürlichen Vegetation Europas. Maßstab 1:2–500000. Compiled and revised by, U. Bohn, G. Gollub, Ch. Hettwer. Bundesamt für Naturschutz Federal Agency for Nature Conservation. Bonn-Bad-Godesberg, 2000, (In German)

14. Alekhin, V. V. Vegetation and geobotanical districts of the Moscow and adjacent regions. Moscow. Moscow Society of the Nature Researchers. 1947, 71 p (In Russian).

15. Yurkevich, I. D., Geltman, V. S. Geography, typology and zoning of the forest vegetation of Belarus. Minsk. Science and Technics (Nauka and Technika). 1965, 288 p (In Russian).

16. Semenishchenkov, Yu. A. Discussion questions of the syntaxonomy of the xeromesophyte broad-leaved forests of the South-Western Nechernozemye of Russia. News of the Samara Scientific Centre. 2012, Vol. 14, №1 (4). 1117–1120 (In Russian).

17. Bulokhov, A. D., Semenishchenkov, Yu. A. Botanico-geographical features of the xeromesophyte broad-leaved forests of the alliance *Quercion petraeae* Zólyomi et Jakucs ex Jakucs 1960 of the Southern Nechernozemye of Russia. Bulletin of the Bryansk department of Russian Botanical Society. 2013, №1 (1). 10–24 (In Russian).

18. Semenishchenkov, Yu. A. Forest vegetation of the Upper Dnieper basin within Russian Federation: diversity, ecology, botanico-geographical regularities. Modern botany in Russia. Proceedings of the XIII Congress of Russian Botanical Society. Vol. 2. Togliatti. Cassandra. 2013, 307–308 (In Russian).

19. Meshkov, A. R. Scheme of geobotanical districts of Chernozem Centre. Questions of Geofraphy. 1953, №32, 157–188 (In Russian).

20. Green Data Book of the Bryansk region (plant communities need in protection). Bulokhov, A. D., Semenishchenkov, Yu. A., Panasenko, N. N., Anishchenko, L. N., Averinova, E. A., Fedotov Yu.P., Kharin, A. V., Kuzmenko, A. A., Shapurko, A. V. Bryansk. Bryansk polygraphical association. 2012, 144 p (In Russian).

21. Semenishchenkov, Yu. A. Communities of the alliance *Aceri campestris–Quercion roboris* Bulokhov et Solomeshch 2003 in the Vorskla river basin (Belgorod region). News of Tula State University. Vol. 3. 221–230 (In Russian).

22. Semenishchenkov, Yu. A. Communities of the alliance *Aceri tatarici–Quercion roboris* Zólyomi et Jakucs ex Jakucs 1960 in the Vorskla river basin (Belgorod region). Gerald of Tver State University. Series of Biology and Ecology. 2012, Vol. 28, №25, 54–62 (In Russian).

23. Poluyanov, A. V. Syntaxonomy of vegetation and composition of flora of the South-West of the Central Chernozem'e as a base of the botanico-geographical zoning. Sc.D. Dissertation. Bryansk. Bryansk State University. 2013, 48 p (In Russian).

PART VI

METHODS OF EVALUATION OF THE QUANTITATIVE AND QUALITATIVE CHARACTERS OF SELECTION SAMPLES

CHAPTER 24

SUSTAINABILITY OF AGROCENOSES IN THE USE OF FERTILIZERS ON THE BASIS OF SEWAGE SLUDGE

GENRIETTA YE. MERZLAYA and MICHAIL O. SMIRNOV

CONTENTS

ABSTRACT

In the results of field experiments we established that the use of fertilizers based on sewage sludge when optimizing their doses ensures sustainable productivity of agrocenoses of perennial grasses and fiber flax, increases the fertility of sod-podzolic soils, their biological activity, preserves biodiversity, does not cause heavy metal accumulation in soil and plant production.

24.1 INTRODUCTION

As cities grow dramatically increases the water consumption – up to 300–400 m^3 and more in calculation on one inhabitant. At the same time volumes of sewage water are growing, and when they are processed at treatment facilities accumulate tremendous amounts of sludge whose disposal is a serious ecological problem. According to the accumulated Russian and foreign experience, the most appropriate method for their use is recognized soil method with the cultivation of crops [1–7]. In Russia when wastewater processing annually produces more than 3.5 million tons of dry matter sludge [2]. However, the use of sludge as fertilizer is often constrained due to adverse physical properties of their natural mass, the presence in them due to immature purification technologies increased concentrations of heavy metals in comparison with the standards, as well as due to the lack of reliable results on the effects of fertilizer on the basis of sludge in the system soil-plant. In this regard it is interesting accomplish studies to determine the effectiveness of sewage sludge and fertilizers of them in modern technologies of crop cultivation.

Taking into account the above studies we have been conducted agroecological assessment of the actions of sewage sludge and composts on the basis of the wastes.

24.2 MATERIALS AND METHODOLOGY

Fields experiments with composts produced on the basis of urban sewage sludge were established in Moscow and Vologda regions. In the Moscow region, we accomplished micro field experiment on sod-podzolic clay loam soil when used as fertilizer composts from sewage sludge in Moscow in different periods of storage and wood waste in the amount of 10% (calculated on the dry matter).

Compost 1 was prepared from fermented sewage sludge, coming directly from the filter-presses Kuryanovskaya aeration station, compost 2 – from the sludge of the same station after 10 years of posting on the sludge beds. For comparison, in the scheme of experiment we put options with bedding manure of cattle. All organic fertilizers were applied in two doses: 10 and 35 t/ha (calculated on the dry matter). It is important to note that the chemical composition of the sewage sludge of the Moscow aeration stations is subject to significant changes in time, which is explained by the improvement of the purification technologies and reduction of their discharge into drains mainly due to declines in many industries while market relations. In terms of experiment compost, which was prepared from sludge of long storage, was more polluted with heavy metals than compost of sludge, coming directly from the filter-presses.

Composts from sludge of various storage terms used in the experiment were characterized by high fertilizing value, contained organic matter – 48–52%, total nitrogen – 2–2.1% and had a neutral pH (Table 24.1).

From manure composts differed with less content of organic matter, nitrogen, potassium, but surpassed its phosphorus. Compost based on sludge from the filter-presses on the content of heavy metals met the standards [7]. At the same time compost from long-term storage on the sludge beds was contaminated zinc and cadmium, the contents of which, respectively, at 31 and 49% exceeded the permissible concentrations. The total amount of heavy metals in the compost was 2 times higher than in the compost based on sludge from the filter presses and 10 times higher than bedding manure.

Soil is sodpodzolic clay loam. In the original soil layer 0–20 cm contained 0.8% of organic carbon, 118 mg/kg of mobile P_2O_5, and 119 mg/kg of K_2O at pH 4.6.

When experimenting in 2000 we have sown cocksfoot – *Dactylis glomerata* L. – variety VIC 61 under the cover of spring barley "Zazersky 85." The experiment was carried out under the scheme: 1 – control without fertilizers; 2 – compost 1, 10 t/ha; 3 – compost 1, 35 t/ha; 4 – compost 2, 10 t/ha; 5 – compost 2, 35 t/ha; 6 – manure, 10 t/ha; 7 – manure, 35 t/ha. All organic fertilizers in both doses were applied into soil in 2000, in the following years we studied their aftereffects.

In the Vologda region the studies were carried out in field experiment on sod-podzol middle loam soil, realized on three fields, which were introduced sequentially in 2010, 2011 and 2012. We studied the effect of compost from Vologda sewage sludge after biological treatment and peat in the ratio 1:1, which was applied for fiber flax. In the scheme of experiment were included variants with increasing doses of compost (2, 4, 6 t/ha dry

TABLE 24.1   Chemical Composition of Organic Fertilizers

Indicator	Bedding manure	Compost 1	Compost 2	The permissible content for sludge of groups [7]	
				1	2
Humidity, %	79.8	71.0	53.7		
Dry matter, %	20.2	29.0	46.3		
pH	7.0	7.4	7.2		
Content in dry substance:					
Organic matter, %	70.0	52.0	48.0		
Ash, %	29.8	48.0	52.0		
N, %	2.7	2.0	2.1		
$N_{ammonium}$, %	0.06	0.01	0.01		
$N_{nitrate,}$ %	0.02	0.02	0.04		
P_2O_5, %	2.4	5.3	5.5		
K_2O, %	2.1	0.2	0.2		
C, %	35.1	26.0	24.0		
C:N	13	13	11		
Cu, mg/kg	36	425	1452	750	1500
Pb, mg/kg	6	50	167	250	500
Cd, mg/kg	2	8	42	15	30
Ni, mg/kg	16	104	353	200	400
Zn, mg/kg	160	1743	4589	1750	3500
Cr, mg/kg	60	147	774	500	1000
As, mg/kg	5	11	31	10	20

Note. Compost 1 was prepared from fermented sewage sludge, coming directly from the filter-presses Kuryanovskaya aeration station; Compost 2 was prepared from the sludge of the same station after 10 years of posting on the sludge beds.

matter), variant with mineral fertilizers – NPK, EQ. – 4 t/ha of compost, variant – compost 2 t/ha + NPK, EQ. –2 t/ha of compost, as well as variants with one sewage sludge and with granulated organic-mineral fertilizer, which was prepared of sewage sludge with the addition of nitrogen and potash fertilizers at the 5% of the active substance to the dry weight. The compost contained (on dry matter) 2% of total nitrogen, 0.9% of phosphorus (P_2O_5), 0.3% of potassium (K_2O), and 67% of organic matter with pH 6.3. The content of heavy metals in compost was low (in mg/kg: Cd – 0.9, Zn – 147; Cr – 11.5; Pb – 11; Cu – 42; Hg – 0,097; Ni – 140. These compost indicators meet the standards of the Russian Federation

Arable layer of sod-podzol middle loam soil before fertilizer application was characterized by a weak acid reaction (pH 5.1–5.3), contained 1.5–1.9% of organic matter, 230–290 mg/kg of mobile phosphorus (P_2O_5), 94–113 mg/kg of potassium (K_2O). The content of heavy metals in soil was low. Research in experiments was performed by standard methods [8]. Heavy metal content was determined in accordance with the methodological guidelines [9]. To determine carbon dioxide emissions from soil we used the method of infrared gasometry. Mathematical processing of experimental data was performed by the method of dispersion analysis [10] with the use of computer program STRAZ.

24.3   RESULTS AND DISCUSSION

According to results of researches, in the conditions of Moscow region the use of composting of sewage sludge in high doses compared with the control unlike variants with low doses significantly increased the yield of perennial grasses (LSD_{05}=48 g of fodder units/ha), as evidenced by the data received in microfield experiment on the average for 10 years (Figure 1).

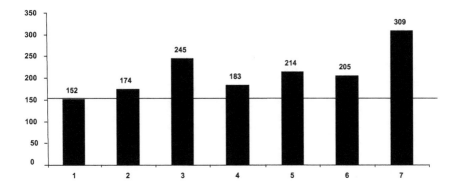

FIGURE 24.1   The yield of perennial grasses on average for 10 years (2000–2009), grams of fodder units with 1 m² [Note. 1 – control; 2 – compost 1, 10 t/ha; 3 – compost 1, 35t ha; 4– compost 2, 10 t/ha; 5– compost 2, 35t/ha; 6 – manure, 10 t/ha; 7 – manure, 35 t/ha].

At the same time bedding manure high doses during this period exceeded grass yield both composts in the same doses. It should be noted significant aftereffect of organic fertilizers in high doses in the next four years (35t/ha – calculated on the dry matter). While the aftereffect of compost from sewage sludge of long-term storage manifested in 2010 and 2013, and the aftereffect of bedding manure – in 2010 and 2012.

The aftereffects of low doses of organic fertilizers (10 t/ha of dry matter) were shorter: the aftereffects of manure continued the first 2 years, the aftereffects of composts – 1 year.

Despite many years of using herbage for hay, in the tenth year of experiment (2009), in variants of high doses of compost and manure up to 32–57% remained cocksfoot 2000–cocksfoot (Table 24.2).

In control, where fertilizers were not applied, the content of cocksfoot in herbage was 51%, and invaded group of motley grass took 48%, for example, agrocenoses with application of high doses of manure and compost from long-term storage on the sludge beds according to the botanical composition approached control. It made little difference from the control on the content of the studied components in the variants on the 13th year of experiment. In this case, most cocksfoot (up 13%) remained in the use of high doses of manure – 35 t/ha of dry matter. Under the influence of manure and composts changed agrochemical soil properties (Table 24.3).

TABLE 24.2　Botanical Composition of Cocksfoot Herbage with Many Years of Hay Use Depending on the Various Types and Doses of Organic Fertilizers, % by Weight

Variant of experiment	10th year of experiment			13th year of experiment		
	Cocks-foot	Wild-growing cereals	Motley grass	Cocks-foot	Wild-growing cereals	Motley grass
Control	51	1	48	2	4	94
Compost 1, 10 t/ha	17	14	69	2	6	92
Compost 1, 35 t/ha	32	6	62	7	4	89
Compost 2, 10 t/ha	14	5	81	2	3	95
Compost 2, 35 t/ha	57	3	40	4	6	90
Manure 10 t/ha	75	1	24	6	1	93
Manure 35 t/ha	56	2	42	13	9	78

TABLE 24.3 Agrochemical Properties of Soil

Variant of experiment	2000, effect	The years of aftereffect				
		2001	2005	2007	2010	2011
pH						
Control	3.8	3.9	3.9	4.4	4.2	4.4
Compost 1, 10 t/ha	3.8	4.1	4.4	4.2	4.3	4.5
Compost 1, 35 t/ha	4.2	4.5	4.8	4.5	4.5	4.6
Compost 2, 10 t/ha	3.9	4.0	4.3	4.5	4.2	4.5
Compost 2, 35 t/ha	4.1	4.5	4.5	4.5	4.4	4.5
Manure 10 t/ha	4.5	4.6	4.7	4.5	4.5	4.5
Manure 35 t/ha	4.5	4.6	4.7	4.5	4.5	4.6
Humus, % C						
Control	0.75	0.72	0.69	0.93	0.99	0.95
Compost 1, 10 t/ha	0.75	0.69	0.64	0.87	0.92	0.92
Compost 1, 35 t/ha	0.79	0.89	0.98	1.01	0.95	1.06
Compost 2, 10 t/ha	0.71	0.69	0.69	0.83	0.86	0.90
Compost 2, 35 t/ha	0.79	0.89	0.81	1.06	0.89	0.93
Manure 10 t/ha	0.77	0.86	0.92	0.92	0.91	0.92
Manure 35 t/ha	0.88	0.95	0.98	0.98	1.02	0.97
Mobile phosphorus (P2O5), mg/kg						
Control	110	110	111	114	105	99
Compost 1, 10 t/ha	180	125	110	205	152	145
Compost 1, 35 t/ha	320	300	310	380	370	322
Compost 2, 10 t/ha	160	140	90	180	148	141
Compost 2, 35 t/ha	220	240	260	415	277	285
Manure 10 t/ha	130	110	90	141	95	123
Manure 35 t/ha	270	220	180	247	168	183
Mobile potassium (K2O), mg/kg						
Control	96	96	55		92	69
Compost 1, 10 t/ha	101	99	39		98	68
Compost 1, 35 t/ha	90	102	39		98	71
Compost 2, 10 t/ha	95	96	35		99	57
Compost 2, 35 t/ha	99	100	42		100	70
Manure 10 t/ha	109	115	50		100	66
Manure 35 t/ha	120	109	98		118	75

In the year of application of all organic fertilizers pH in high doses markedly improved. This pattern remained during the next five years. Then to 7–11th year aftereffect noted the alignment of the pH values for different variants. In general you should specify that the manure and composts from both types of sludge produced a positive effect on the acid properties of the soil due to a neutral or slightly alkaline pH of the fertilizers and high content of calcium hear, which was capable with mineralization of organic matter enter into soil solution. The content of organic carbon in the soil, compared to control was increased from applying high doses of all organic fertilizers used as a year of action, and in the aftereffect years. On the tenth-eleventh aftereffect years in comparison with the year of their introduction in all variants achieved a positive balance of organic substance in the soil. The same regularity was observed in the control variant. In general, we can say that the composting of sewage sludge in agrocenoses of perennial cereal grasses with their long mowing use improved humus status of sod-podzolic soils.

In the analysis of phosphate regime of the soil we marked its optimization under the influence of all kinds of organic fertilizers applied in high (35 t/ha) and low (10 t/ha) doses (except 2005). Content of mobile phosphorus in the soil increased more intensively, as a rule, when applying high doses of compost based on sludge from the filter presses [11]. Application of compost containing potassium lower than manure practically had no effect on change of soil potash regime.

Much attention during the experiment was paid to the study of soil biological activity. Research results in 2008 jointly with the Department of ecology of Russian State Agricultural University – Timiryazev Moscow Agricultural Academy showed (Table 24.4) that composts on the basis of both kinds of sludge did not negatively impact on the microbial destruction of organic matter in the soil (in the layer 0–19 cm).

The total number of microorganisms on BEA and SAA in variants with organic fertilizers was close to the number on the control (11.6 million cells) and varied from 8.3 to 13.3 million cells in 1 g of dry soil. Higher values of this indicator were observed in the variants of long term storage compost and manure used in high doses – 35 t/ha of dry matter.

In determining the activity of the carbon dioxide release from soils in the second year aftereffect of fertilizers we showed statistically significant

TABLE 24.4   The Number of Heterotrophic Aerobic Microorganisms (million/ha dry soil)

Type of fertilizer	BEA	SAA	Σ
Control	4.8	6.8	11.6
Compost 1, 10 t/ha	3.3	6.4	9.7
Compost 1, 35 t/ha	4.6	4.5	9.1
Compost 2, 10 t/ha	4.2	4.1	8.3
Compost 2, 35 t/ha	5.2	8.1	13.3
Manure 10 t/ha	4.4	4.6	9.0
Manure 35 t/ha	5.5	6.6	12.1

Note. BEA – Beef-extract agar; SAA – Starch-and-ammonia agar.

changes of this indicator in relation to the control in versions with application of high doses of manure and both types of compost. And in variants with high doses of composts carbon dioxide emissions was more than when making a similar dose of manure. According the correlation analysis conducted on the basis of the experimental data; we established a connection CO_2 and emissions with humus ($r = 0.87$) and with pH ($r = 0.67$) and grass yield ($r = 0.79$). Studying the activity of the carbon dioxide release from the soil at the end of the experiment (2012) showed (Figure 24.2) that in the variants with the

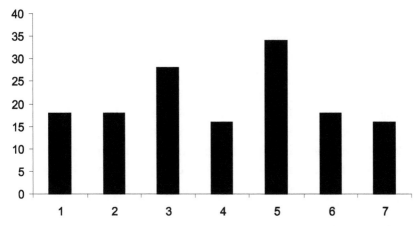

FIGURE 24.2   Carbon dioxide emission from soil (mcg $C\text{-}CO_2$/g·day) depending on the types and doses of manure, 2012 [Note. 1 – control, 2 – compost 1, 10 t/ha; 3 – compost 1, 35t/ha; 4– compost 2, 10 t/ha; 5– compost 2, 35t/ha; 6 – manure, 10 t/ha; 7 – manure, 35 t/ha].

organic fertilizers at the 11th year of their aftereffect the activity has declined sharply compared with the first term of definition (2001).

However, high doses composts from sludge in relation to low doses contributed to the increase of CO_2 emission.

It is important to note that if you apply all studied fertilizers total heavy metal content in the soil will not exceed the standards of the Russian Federation. The heavy metal content in perennial grasses depended on the type and dose of composts, increasing by most indices at application of compost based on long-term storage of sludge with increasing dose, but not beyond acceptable levels.

Thus, on the basis of long-term studies in microfield experiment on sodpodzolic loamy soil we established that the application of organic fertilizers produced by fermentation of sewage sludge with wood waste with the optimization of their doses increased the biological activity and soil fertility, preserved the biodiversity of cenoses for long time, contributed to productive longevity of perennials. The aftereffect of compost based on sewage sludge in doses of 10 tones dry matter /ha was noted in the course of one year, bedding manure – for two years. The composts of sludge and manure when making in high doses – 35 t/ha of dry matter were characterized by a long aftereffect, which could be traced in the course of 10 years or more.

When studying the effect of fertilizers from sewage sludge of Vologda on the yield of flax straw it was found that on average for 3 years experiment (2010–2011) compost application in a low dose of 2 t/ha was not been effective. At the same time increasing the dose of compost in 2 times provided a reliable response in flax straw yield. However, further increasing the dose of compost (3 times) was found to be inappropriate (Table 24.5).

The highest yield of flax straw 27–29.8 c/ha and statistically meaningful yield response in relation to the control without fertilizers at the level of 26–39% were achieved applying granulated organic-mineral fertilizers at a dose of 4 t/ha, sewage sludge of Vologda in pure form in the same dose and combination compost with complete mineral fertilizer in variant 2 t/ha compost + NPK, EQ. t/ha compost.

The use only mineral fertilizers in doses equivalent to the sum of NPK 4 t/ha compost, though gave a true increase in the harvest of flax straw to control, but yielded almost all variants of experiment with organic fertilizers (the exception was only the lowest dose of compost – 2 t/ha).

TABLE 24.5   Influence of Fertilizers on the Straw Yield of Fiber Flax for 3 Years

Type of fertilizer	Yield, c/ha	Response in yield	
		c/ha	%
Control	21.5	–	–
Compost 2 t/ha	23.1	1.6	7
Compost 4 t/ha	25.0	3.5	16
Compost 6 t/ha	26.7	5.2	24
NPK, EQ. 4 t/ha of compost	24.4	2.9	14
Compost 2 t/ha + NPK EQ. 2 t/ha of compost	27.0	5.5	26
Sewage sludge, 4 t/ha	27.0	5.5	26
Granulated organic-mineral fertilizer, 4 t/ha	29.8	8.3	39
LSD_{05}		2.1	

On the yield of flax seeds, with humidity of 12% on average for 3 years (2010–2012) the most significant influence exerted the compost in a doze 6 t/ha and granulated organic-mineral fertilizer that provided by 3.3 and 3.4 c/ha of seed or 43–48% above control, as well as organic-mineral variant (2 t/ha compost + NPK, EQ. 2 t/ha compost) and sewage sludge in pure form in the dose of 4 t/ha where seed yield was 3–3,1 c/ha, that is by 30–35% exceeded the control (Table 24.6).

According to the results of chemical analysis (Figures 24.3 and 24.4), the use of fertilizers based on sewage sludge in all the analyzed variants

TABLE 24.6   Influence of Fertilizers on the Seed Yield of Fiber Flax for 3 Years

Type of fertilizer	Yield, c/ha	Response in yield	
		c/ha	%
Control	2.3	–	–
Compost 2 t/ha	2.5	0.2	9
Compost 4 t/ha	2.9	0.6	26
Compost 6 t/ha	3.4	1.1	48
NPK, EQ. 4 t/ha of compost	2.7	0.4	17
Compost 2 t/ha + NPK, EQ. 2 t/ha of compost	3.1	0.8	35
Sewage sludge, 4 t/ha	3.0	0.7	30
Granulated organic-mineral fertilizer, 4 t/ha	3.3	1.0	43
LSD_{05}		0.4	

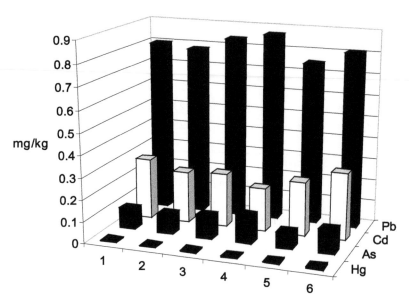

FIGURE 24.3 The content of heavy metals and arsenic in the straw of flax [1 – control, 2 – compost, 4 t/ha, 3 – NPK, 4 – compost + NPK, 5 – organic-mineral fertilizer, 6 – sewage sludge 4 t/ha].

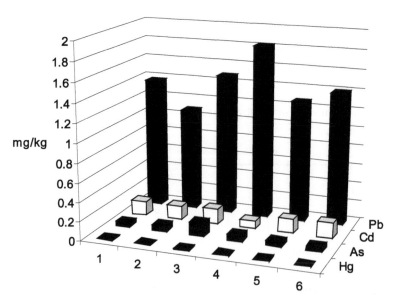

FIGURE 24.4 The content of heavy metals and arsenic in the seeds of flax [Note. 1 – control, 2 compost, 4 t/ha, 3 – NPK, 4 – compost + NPK, 5 – organic-mineral fertilizer, 6 – sewage sludge 4 t/ha].

has not led to the accumulation of heavy metals in plant products. The contents of lead, cadmium, mercury and arsenic in flax straw and flax seeds when fertilizing sewage sludge were in permissible limits.

We have not established appreciable influence of fertilizers produced from sewage sludge, on the accumulation of heavy metals and arsenic in the soil when compared with the control (Table 24.7).

The content of all the investigated heavy metals in the soil of manorial experimental variants with fertilizer application, according to 2012, was below the maximum concentration limits of CC/RAC established in Russia.

24.4 CONCLUSIONS

Thus, the results of studies in field experiments have shown that the use of fertilizers based on sewage sludge in sod-podzolic soils reliably increased the yield of perennial grasses and flax.

TABLE 24.7 Influence of Fertilizers on the Total Contents of Heavy Metals and Arsenic in the Soil, mg/kg

Type of fertilizer	Cu	Zn	Pb	Cd	Ni	Cr	Hg	As
Control	5.5	24.1	5.9	0.35	10.0	8.5	0.026	2.23
Compost 2 t/ha	5.0	22.9	5.1	0.37	9.2	8.6	0.024	2.07
Compost 4 t/ha	5.1	22.9	5.5	0.35	9.3	8	0.023	2.36
Compost 6 t/ha	5.1	22.8	5.1	0.38	8.9	8.2	0.022	1.95
NPK, EQ. 4 t/ha of compost	6.0	25.0	5.7	0.43	9.0	7.9	0.027	2.09
Compost 2 t/ha + NPK, EQ. 2 t/ha of compost	5.8	24.2	5.9	0.39	9.7	8.2	0.027	1.97
Sewage sludge, 4 t/ha	6.7	23.9	5.8	0.40	9.4	8.1	0.024	1.9
Granulated organic-mineral fertilizer, 4 t/ha	5.9	24.1	5.1	0.42	8.6	7.8	0.026	1.94
CC/RAC	33–132	55–220	32–130	0.5–2.0	20–80		2.1	2–10

Note. CC – Critical concentration of every element; RAC – Roughly allowable concentration of every element [7]; EQ. – in a quantity equivalent to 4 (the 6th line) or 2 t/ha (the 7th line) of compost.

When optimizing doses of fertilizers from sewage sludge plant products (hay of perennial grass, straw and seeds of the flax) met Russian standards of content of heavy metals and arsenic.

While the level of heavy metals and arsenic in the soil was low and has no adversely affect on its ecological status.

Regulated application of fertilizers produced by fermentation of sewage sludge, increased biological activity and soil fertility, and contributed to the preservation of biodiversity in agrocenoses, ensuring their productive value.

KEYWORDS

- composts
- fertilizers
- forest vegetation
- organic fertilizers
- soil
- heterotrophic microorganisms
- heavy metals

REFERENCES

1. Pryanishnikov, D. N. The selected works. Vol. 3. General issues of agriculture and chemicalization. Moscow: Kolos, 1965, 639 p (In Russian).
2. Resources of organic fertilizers in agriculture of Russia (Information and analytical reference book) [Ed. Eskov, A. I.]. Vladimir: Russian Research Institute of Organic Fertilizers and Peat (in Russian), 2006, 200 p (In Russian).
3. Sanitary regulations and norms 2.1.7.573–96. Hygienic requirements for the use of wastewater and their sludge for irrigation and fertilization. Moscow: Department of Public Services of Russian Federation, 1997, 54 p (In Russian).
4. Pahnenko, E. N. Sewage sludge and other non-traditional organic fertilizers. Moscow: Laboratory of Sciences, 2007, 311 p (In Russian).
5. Ladonin, V. F., Merzlaya, G. E., Afanasev, R. A. Strategy of use of sewage sludge and composts on their basis in agriculture [Ed. Milashchenko, N. Z.]. Moscow: Agroconsult. 2002, 140 p (In Russian).

6. Sychev, V. G., Merzlaya, G. E., Petrova, G. V. Ecological-agrochemical properties and efficiency of vermi- and biocomposts. Moscow: Pryanishnikov Agrochemistry Research All-Russian Institute. 2007, 276 p (In Russian).

7. State Standard of the Russian Federation R 17.4.3.07–2001. The nature conservancy. The soil. Requirements to the properties of sewage sludge when used as fertilizer. Moscow: Information-publishing center of the Russian Ministry of health. 2001 (In Russian).

8. Program and methodology of research in geographic network of field experiments on integrated application of chemicals in agriculture, Moscow: Pryanishnikov Agro-chemistry Research All-Russian Institute. 1990, 187 p (In Russian).

9. Collection of methods for the determination of heavy metals in soils, in hothouse soil and crop products. [Eds. Ovcharenko, M. M., Kuznetsov, N. V.] Moscow: Depart-ment of Agriculture and Food of Russian Federation, 1998, 97 p (In Russian).

10. Dospehov, B. A. Methodology of field experiment. Moscow: Kolos (Ear in Rus), 1979, 416 p (In Russian).

11. Afanasev, R. A., Merzlaya, G. E. Dynamics of the Mobile Forms of Phosphorus and Potassium in the Soil of Prolonged Trials. Russian Agricultural Sciences. 2013, 39(4), 332–336.

CHAPTER 25

TRANSFORMATION OF MOBILE PHOSPHORUS IN THE SOILS OF AGROECOSYSTEMS DURING PROLONGED TRIALS

RAFAIL A. AFANAS'EV and GENRIETTA YE. MERZLAYA

CONTENTS

ABSTRACT

The authors revealed new regularities of mobile phosphorus dynamics in the soils of different agroecosystems during prolonged systematic fertilizer application. They had shown that mobile phosphorus content in different soils increased only at the first rotations of field crop rotations. Later in spite of positive phosphorus balance the content of its mobile forms didn't increase and even tended to decrease as a result of phosphorus transition in stiff state. In case of negative balance mobile phosphorus content in soils was recompensed through slow-moving phosphates supply. Ecological balance at the agroecosystems was maintained due to these processes of phosphorus transformation. This equilibrium prevented losses of phosphorus due to surface and subsurface flows of the element for environment; so risks of eutrophication decreased. Also influence of phosphorous fertilizer on the biodiversity of soil microflora was ascertained. The mobile phosphorus dynamics in different soils, which was revealed during the prolonged field experiments, could be the model of phosphorus fertilizer transformation in condition of agricultural production.

25.1 INTRODUCTION

Mobile phosphorus content in soils at the time of intensive chemicalization was simpliciter connected with the level of artificial and organic fertilizer application. In Russia in 1965 before intensive chemicalization rate of artificial fertilizer (NPK) usage per 1 hectare of arable land per year was on average 20 kg of primary nutrient; while by five-year periods its application increases: in 1966–1970 – up to 28 kg/ha; in 1976–1980 – up to 65 kg/ha; in 1986–1990 – up to 99 kg/ha. While fertilizer (especially phosphorus) application in the country increased the soil fertility improved (particularly mobile phosphorus supply increased). So in period 1971–1990 the tillage share with low content of mobile phosphorus decreases from 52 to 22% while increasing the share of soils that had high and medium phosphorus content [1].

At the same time the increase of mobile phosphorus content in soils is but one effect of phosphorous fertilizer intensive application. A considerable portion of the phosphorus above its carry-over with crop yields transformed into not mobile forms creating the reserve of phosphorus plant nutrition [2–4]. According to [5] during 25 years about 300 kg/ha of phosphorus were applied above its carry-over; the entire amount was stayed in soil. The amount was sufficient to harvest 2 t/ha of grain yield during 25–30 years unless the phosphorous fertilizer application.

Since 1995 the balance of phosphorus in the agriculture of our country tended the pattern: carry-over became exceed apply. That resulted in trend of decrease of mobile phosphorus in arable lands, although the trend was not so evident as was expected earlier. At the same time the regularities of transformation of applied phosphorus fertilizers were studied insufficiently. Outstanding scientist K. E. Ginsburg has drawn attention to this fact: "The absorbing capacity of soils in the case of phosphorus confuses our calculations of increase of mobile phosphorus content in soil because while applying mobile phosphorus unknown but considerable part of these transforms into poorly soluble and hard-to-reach for plants forms" [4, p. 124]. Our paper allows spy out longstanding dynamics and character of transition of mobile soil phosphorus into poorly soluble forms. Our study uses results of prolonged trials.

25.2 MATERIALS AND METHODOLOGY

Experimental technique was based on the materials of prolonged trials and our proper studies. We analyzed the effect of systematic organic and mineral (including phosphorous) fertilizer application in increased rates on change of mobile phosphorus content. Different soils were studied: the soddy gleyic heavy textured loamy soil (Lithuania), the sod-podzol heavy textured loamy soil (Moscow Region), the soddy-podzolic easy-loamy soil (Smolensk Region), the sod-podzol loamy sandy soil (Belarus) and ordinary chernozem (Stavropol).

We calculated the economic balance of phosphorus by principal variants of field experiments taking into account the applied phosphorus content with rotation of field rotations. We detected the influence of systematic long-termed phosphorous fertilizer application on mobile phosphorus content in soils.

The content of mobile forms of phosphorus in different soils was determined by the methods accepted in agrochemistry [6]. The main methods of determination of soluble phosphorus content in different soils were: the method of Egner-Rim -0.04 normal solution of $(CH_3CHOHCOO)_2Ca \cdot 5H_2O$ soil extract at a pH of 3.5–3.7; the method of Kirsanov 0.2 normal solution of HCl $5H_2O$ soil extract; the method of Machigin $-$ 1% solution of $(NH_4)_2CO_3$ extract. The methods of determinations are indicated when depicting the results of studies.

25.3   RESULTS AND DISCUSSION

K.I. Plesyavichius who studied the efficiency of long-termed mineral fertilizer application on the soil [7]; these experiments were realized in Lithuanian Agriculture Research Institute. In the variant $N_{225}P_{324}K_{350}$ applied during the rotation with annual phosphorous fertilizer application on an average in seven fields deployed in nature when the economic balance of phosphorus was 68 kg/ha in the first rotation and 73 kg/ha in the second – the mobile P_2O_5 content (according to Egner-Rim) to the end of rotations amounted to 42 and 43 mg/kg soil (Figure 25.1).

The change of the index compared with their beginning equaled respectively 12 and 1 mg/kg when productivity of crop rotation was for the first period (1961–1968) on an average 44.8 centner of grain units/ha and for the second period (1969–1974) – 46.5 C/ha. So almost around 140 P_2O_5 centner /ha applied above its carry-over during two rotations of field crop rotations to the end of the second rotation transformed in soil in the forms not extractable by according to Egner-Rim.

On our calculation the expenditures of fertilizer phosphorus for increase of mobile phosphorus content in surface soil by 10 mg/kg at the first rotation were 57 kg/ha P_2O_5, at the second rotation – the mathematical procedure above.

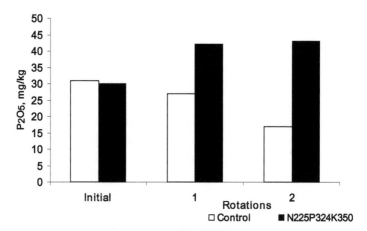

FIGURE 25.1 Dynamics of phosphorus distribution in soddy-gleyic heavy loamy textured dried along rotation of field crop rotations (Lithuania) [Note: Bright columns – control, dark columns – rotation of field rotations].

Also the change of mobile phosphorus content in the soddy-podzolic heavy textured loamy soil of another experiment was enough expressed. This long-termed experiment was carried out at the Central Experimental Station (Pryanishnikov All-Russian Agrochemistry Research Institute, Moscow Region) [8]. During seven rotations of a four course rotation at 28-year-long regular application of fertilizers maximum mobile phosphorus content (according to Kirsanov) was determined at the fourth rotation. Then the content was decreased (Figure 25.2), though the positive balance of phosphorus for 28-year-long experiments was 2700 kg/ha.

The expenditure of fertilizer phosphorus applied above its carry-over for increase of mobile phosphorus content in surface soil by 10mg/kg per 1 hectare on an average equaled 119 kg P_2O_5.

In this variant for 12-year-long aftereffect of fertilizers winter wheat yield (on the average for tree rotations) was 17.2 centner per hectare as compared with 12.0 centner per hectare in the control variant. So due to formerly applied fertilizers preceding increment was 5 centner of winter wheat per hectare.

During 30-year-long regular application of mineral fertilizers for all the crops except perennials to the soddy-podzolic easy loamy soil (Smolensk Region) in the case of the mineral system of fertilization $N_{990}P_{990}K_{990}$ was applied at the first rotation of crop rotation (1979–1989), $N_{450}P_{450}K_{450}$ was

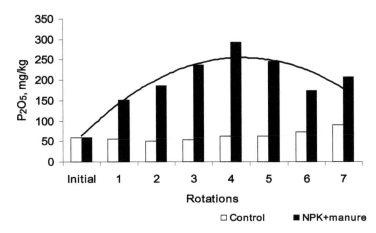

FIGURE 25.2 Dynamics of phosphorus distribution in soddy-podzolic heavy textured-loamy soil along rotation of field crop rotations (Moscow Region) [Note: Bright columns – control; dark columns – rotation of field rotations].

applied both at the second (1990–1995) and the third (1996–2001) rotations and $N_{405}P_{405}K_{405}$ was applied at the fourth rotation (2002–2008).

For the first two rotations of crop rotation, with balance of phosphorus of 943 kg/ha, the content of mobile phosphorus in soil increased from 149 to 210 mg/kg, hence, the value of increment was 61 mg/kg. By the end of the fourth rotation, with balance of phosphorus of 523 kg/ha in sum for the third and the fourth rotations, even a decrease in the mobile phosphorus content in the arable layer was observed: from 210 to 174 mg/kg, for example, by 36 mg/kg (Table 25.1, Figure 25.3).

TABLE 25.1   Influence of Fertilizers on Productivity of Crop Rotation and Mobile Phosphorus Content in the Soddy-Podzolic Easy Loamy Soil (Smolensk Region)

Index	Control without fertilizers	NPK	Manure	NPK+manure
1–2 rotations (17 years)				
Productivity in average year, centners of g.u. per hectare	24.0	34.4	28.6	34.6
Applied P_2O_5, kg/ha	—	1440	340	1780
Carry–over P_2O_5, kg/ha	369	497	422	500
Balance P_2O_5, kg/ha	–369	+943	–82	+1280
P_2O_5 content in the soil in the beginning of the rotations, mg/kg	170	149	143	166
P_2O_5 content in the soil in the end of the rotations, mg/kg	65	210	85	185
Variation of P_2O_5 content in soil, mg/kg	–105	+61	–58	+ 19
3–4 rotations (30 years)				
Productivity in average year, centners of g.u. per hectare	20.1	29.6	26.0	26.0
Applied P_2O_5, kg/ha	—	855	252	1107
Carry–over P_2O_5, kg/ha	222	332	287	300
Balance P_2O_5, kg/ha	–222	+523	–35	+807
P_2O_5 content in the soil in the beginning of the rotations, mg/kg	65	210	85	185
P_2O_5 content in the soil in the end of the rotations, mg/kg	56	174	160	213
Variation of P_2O_5 content in soil, mg/kg	–9	–36	+75	+28

Note: g.u. – grain unit, equivalent 1 kg of wheat grains.

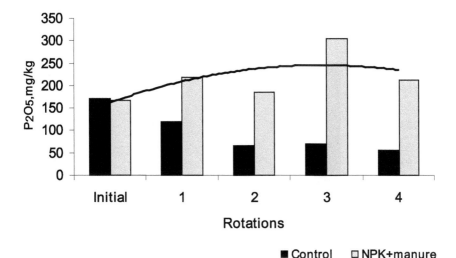

FIGURE 25.3   Mobile phosphorus content in the soddy-podzolic easy loamy soil along rotation of field crop rotations (Smolensk Region) [Note: Bright columns – control, dark columns – rotation of field rotations].

In the subsurface soil mobile phosphorus content (P_2O_5) in the control variant of the experiment in the end of the fourth rotation (2008) compared with middle of the first rotation (1983) decreased from 1.8 to 1.2 t/ha at the expense of element carry-over by yield (Figure 25.4).

In the variant with application of organic-mineral system for this period mobile phosphorus supply in the subsurface soil (20–100 cm) remained at the level 1.9 t/ha, for example, it was impossible to notice appreciable change the mobile phosphorus content in the subsurface soil. This records an exhaustive transformation of there migratory phosphorus into slow-moving forms if not to take into slow-moving forms its migration outside the limits of controlled soil layer.

Studies of the Stavropol Scientific Research Institute of Agriculture [9] showed that in conditions of ordinary loamy chernozem annual fertilizer phosphorus application for crops of a six coarse rotation rate of $N_{120}P_{90}K_{120}$ first rotation of crop rotation ensured positive phosphorus balance (along Machigin) 270 kg/ha (Table 25.2, Figure 25.5).

Mobile phosphorus content in subsurface soil in the end of the rotation increased to 26 mg/kg or 13 mg/kg compared with the beginning of the rotation. During the second and third rotations at the same rate and

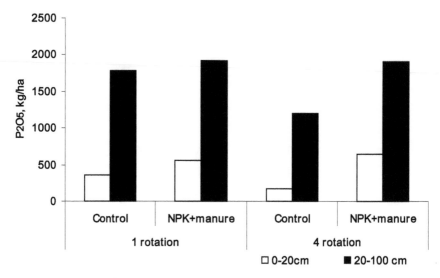

FIGURE 25.4 Mobile phosphorus supply content in the soddy-podzolic easy loamy soil along layer of profile (Smolensk Region) [Note: Bright columns – 0–20 cm, dark columns – 20–100 cm].

the same total phosphorus balance 542 kg/ha mobile phosphorus content in the soil of the variant increased to 52 mg/kg, for example, 26 mg/kg compared with the beginning of the second rotation. But to the end of fifth rotation of crop rotation i.e12 years after the end of the third rotation in spite of systematic mineral fertilizer application at the same rate with phosphorus balance for these years 250 kg/ha mobile phosphorus content in the surface soil increased by only 2 mg/kg.

The application of a higher dose of phosphorus, such as $N_{120}P_{150}K_{120}$, at the balance of phosphorus for the first rotation of 568 kg/ha, with total balance for first three rotations of 1718 kg/ha, and with balance for the fourth and fifth rotations of 609 kg/ha, resulted in the content of mobile phosphorus in the surface layer of 58, 76 and 70 mg/kg, correspondingly. Hence, for the last two rotations of the field crop rotation, the application of 609 kg/ha of phosphoric fertilizers not only has not increased the content of mobile phosphorus in soil, but even had lowered the content by 6 mg/kg as compared to the end of the third rotation. The analysis of mobile phosphorus content not only in the surface layer but in the subsurface soil of soil profile of ordinary chernozem revealed the close dependence of the intensity of its transformation in slow-moving forms on the

TABLE 25.2 Influence of Mineral Fertilizers on Productivity of Crop Rotation and Mobile Phosphorus Content in Ordinary Chernozem (Stavropol Territory)

Index	Control without fertilizers	$N_{120}P_{30}K_{120}$	$N_{120}P_{90}K_{120}$	$N_{120}P_{150}K_{120}$
1 rotation (6 years)				
Productivity in average year, centners of g.u. per hectare	27.7	34.3	35.2	35.7
Applied P_2O_5, kg/ha	–	150	450	750
Carry-over P_2O_5, kg/ha	139	177	180	182
Balance P_2O_5, kg/ha	–139	–27	+270	+568
P_2O_5 content in the soil in the beginning of the rotations, mg/kg	13	13	13	13
P_2O_5 content in the soil in the end of the rotations, mg/kg	13	21	26	58
Variation of P_2O_5 content in soil, mg/kg	0	+8	+ 13	+45
2–3 rotations (18 years)				
Productivity in average year, centners g.u. per hectare	26.5	32.7	34.5	33.3
Applied P_2O_5, kg/ha	–	300	900	1500
Carry-over P_2O_5, kg/ha	277	342	358	350
Balance P_2O_5, kg/ha	–277	–42	+542	+ 1150
P_2O_5 content in the soil in the beginning of the rotations, mg/kg		21	26	58
P_2O_5 content in the soil in the end of the rotations, mg/kg	12	28	52	76
Variation of P_2O_5 content in soil, mg/kg	–1	+7	+26	+18
4–5 rotations (30years)				
Productivity in average year, centners of g.u. per hectare	19.0	23.2	24.5	24.6
Applied P_2O_5, kg/ha	–	180	540	900

TABLE 25.2   Continued

Index	Control without fertilizers	$N_{120}P_{30}K_{120}$	$N_{120}P_{90}K_{120}$	$N_{120}P_{150}K_{120}$
Carry-over P_2O_5, kg/ha	226	281	290	291
Balance P_2O_5, kg/ha	−226	−101	+250	+609
P_2O_5 content in the soil in the beginning of the rotations, mg/kg	12	28	52	76
P_2O_5 content in the soil in the end of the rotations, mg/kg	20	30	54	70
Variation of P_2O_5 content in soil, mg/kg	+8	+2	+2	-6

Note: centners g.u. – see Table 25.1.

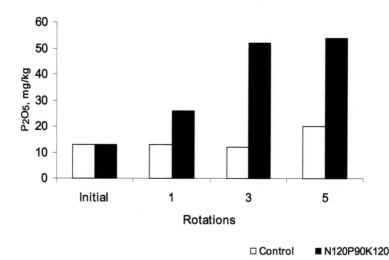

FIGURE 25.5   Mobile phosphorus content in ordinary chernozem (Stavropol Territory) [Note: Bright columns – control, dark columns – $N_{120}P_{90}K_{120}$].

value of positive balance of phosphorus in the agroecosystem. As can be seen from Figure 25.6 with the increase of balance P_2O_5 in the variant and $N_{120}P_{90}K_{120}$ and content of mobile phosphorus transformed in slow-moving state increased. On the whole during five rotations of crop rotation more than 1000 kg/ha P_2O_5 transformed in slow-moving forms.

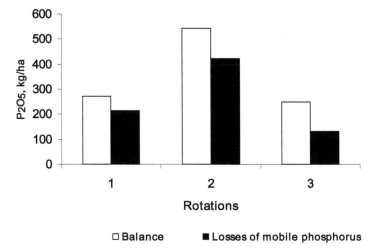

FIGURE 25.6 Dependence of the mobile phosphorus transformation in slow-moving forms (losses) in 1-meter-long layer of ordinary chernozem (the variant $N_{120}P_{90}K_{120}$) on the value of balance P_2O_5 in rotations of crop rotation (Stavropol Territory) [Note: Bright columns –balance, dark columns – losses of mobile phosphorus].

According to the research of Pryanishnikov All-Russian Scientific Research Institute of Agrochemistry together with Lomonosov Moscow state University unilateral use of mineral fertilizer decreased the total number of microorganisms in soil and Shannon index biodiversity (Table 25.3).

Manure application increased the total number of microorganisms, but decreased Index biodiversity. With use of organic-mineral fertilizer

TABLE 25.3 Influence of Fertilizers on the Content of Microorganisms in the Sod-Podzol Easy Loamy Soil (Smolensk Region)

Index	Control without fertilizers	NPK	Manure	NPK+manure
Proteobacteria, cells/gram $\times 10^6$	13.6	17.1	22.0	20.3
Actinobacteria, cells/gram $\times 10^6$	19.3	13.4	18.1	13.3
Firmicutes, cells/gram $\times 10^6$	11.9	5.1	20.4	22.5
Bacteroidetes, cells/gram $\times 10^6$	2.0	2.5	5.9	3,7
The total number of microorganisms, cells/gram $\times 10^6$	46.8	38.1	66.4	59.8
Index biodiversity	5.0	4.7	4.6	4.9

system increase of the total number of microorganisms almost without its biodiversity reducing was to be observed. Reductions in the size of microorganisms, which were marked in the table for the most part, occurred at the expense of Actinobacteria, and increase – at the expense of Proteobacteria and Bacteroidetes. In conditions of fertilizer aftereffect the number of proteolytic bacteria increased in 1.5–2 times. These bacteria (e.g., *Pseudomonas fluorescens, Pseudomonas putida, Brevundimonas vesicularis*) actively transformed organic phosphorus [10]. Generally soil microbial community was represented by more than 40 species belonging to 34 genera. The high number of microorganisms and their diversity indicate a sufficient degree of cultivation of investigated soddy-podzolic soil.

In light soils dynamics of mobile phosphorus according to increase of positive phosphorus balance in agrocoenoses in contrast to clay and heavy loamy can go on increasing. This can be seen on mobile phosphorus dynamics in loamy sandy soil of the long-termed experiment that was carried out at the Agricultural Experimental Station (Grodno, Belarus) [11]. Figure 25.7 showed that mobile phosphorus content (according to Kirsanov) at the expense of systematic mineral fertilizer application increased from

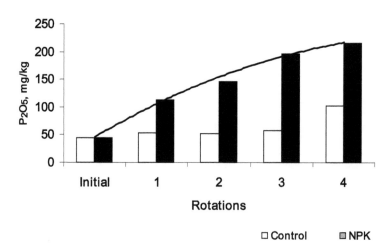

FIGURE 25.7 Mobile phosphorus dynamics in soddy-podzol loamy sandy soil along rotation of field crop rotations [Note: Bright columns – control, dark columns – NPK].

44 mg/kg in the beginning of the first rotation to 216 in the end of the fourth rotation of a four coarse rotation.

This result may be explained by smaller phosphatic capacity of loamy sandy soils in comparison with heavier soils [4] and less active transition of mobile phosphorus in the slow-moving. Nevertheless, the expenditure of phosphoric fertilizers for the increase of 10 mg/kg in mobile phosphorus content in soil raised even in case of the increase in the concentration of mobile forms. In the third rotation of crop rotation, the expenditure was 22 kg of P_2O_5/ha, and already in the fourth rotation the expenditure raised to 56 kg/ha.

The analysis of the long-term dynamics of the mobile forms of phosphorus in long field experiments with clay and loam soils showed the regularity its transition after a time in slow-moving forms with a decrease in concentration of mobile forms. Depending on agrochemical qualities of soils time constraints of the transition matured in different periods: in the soddy gleyic heavy textured loamy soil (Lithuania) – after 7 years (in the second rotation of crop rotation), in the soddy-podzoic easy-loamy soil (Smolensk Region) – after 17 years (in the third-fourth rotations of crop rotation), and in the ordinary chernozem after 18 years (in the fourth-fifth rotations of crop rotation.

Inasmuch as soil is bio inert system which has formed under the influence of biological factors it is inherent the function of conservation and transformation of substances [12, 13] and it reacts to soluble phosphorous fertilizer application according to Le Chatelier principle: If a system at equilibrium experiences a change then the equilibrium shifts to partially counter-act the imposed change [14]. In different soils these functions are manifested in accordance with root natural causes. In soddy-podzolic soils with increased iron and aluminum compounds content applied phosphorous fertilizers transform in phosphate sesquioxides oxides while in carbonate-enriched chernozems and chestnut soils the function of phosphorus conservation manifests itself in emergence of phosphates with different basicity including sparing soluble compounds, for example, apatite. Obviously the specific reasons determined a small increase or even a decrease of moving phosphorus content in the last rotations of the indicated crop rotations were due to increase of intensity of phosphorous fertilizer transformation in sparing soluble forms from the first rotations to following rotations of crop rotations owing to strengthening reaction of soil as a system to the redundant water-soluble phosphorus fertilizer application which was increased summarized over rotations.

From the above is follows that traditional soil test, for example, moving phosphorus determination as index of its effective fertility didn't create an adequate representation of its natural fertility.

There are reports [15] that content of 600–800 kg residual phosphates in the soil per 1 hectare ensures almost all of the soils its optimal phosphate status. So for an objective assessment of the phosphate level it is expedient to take into account not only data on the mobile phosphate content in the soil but also supplies of slow-moving forms which would be used by crops in future. This is important for economic estimation of arable lands. The estimation allows take into account its natural fertility.

25.4   CONCLUSIONS

1. Prolonged systematic fertilizer application with rates of phosphorus, which exceed carry-over of the element by crop yields increases mobile phosphorus content in soils of agroecosystem in the beginning of intensive fertilizer application only. In the further due to phosphorus transition in sedimentary forms mobile phosphorus content in soils increases very slightly or even has a tendency to decrease. Intensity of mobile phosphorus transition in slow-moving forms largely depends on soil type: the lowest intensity of the transformation is characteristic of light soils in particular loamy sandy; the intensity also depends on the duration of the interaction of mobile phosphates with soil and on the value of positive phosphorus balance.

2. Transition of applied phosphorus in sedimentary phosphates of soil as a result of its immobilization ensures phosphorus plant nutrition during for a number of years even upon termination of phosphorus fertilizer application. The processes in many respects predetermine crop production level including the grain production observed in our country in the years after perestroika.

3. Fertilizer use in general increases soil fertility, enriches their with soil microflora. Unilateral fertilizer application decreases diversity

of microorganisms, while use of combined organic-mineral fertilizer contributes to its conservation. For all that many species of protolithic bacteria influence on processes of phosphorus transformation in conditions of aftereffect fertilizer in agroecosystems.

KEYWORDS

- agrochemistry
- eutrophication
- fertilizers
- soil

REFERENCES

1. Sychev, V. G., Mineev, V. G. Role of Pryanishnikov All-Russian Scientific Research Institute of Agrochemistry in solving complex problems of agriculture chemicalization. Fertility. 2011, №3, 2–4 (In Russian).
2. Cook, J. W. Soil fertility regulation/Moscow. Kolos (Ear in Rus.). 1970, 520 p (In Russian).
3. Black, K. A. Plant and soil/Moscow. Kolos (Ear in Rus.). 1973, 501 p (In Russian).
4. Ginsburg, K. E. Phosphorus of general soil types in USSR. Moscow. Kolos (Ear in Rus.). 1981, 542 p (In Russian).
5. Sychev, V. G., Shafran, S. A. Agrochemical properties of soils and mineral fertilizer efficiency/Moscow. Pryanishnikov All-Russian Scientific Research Institute of Agrochemistry, 2013, 296 p (In Russian).
6. Agrochemical methods of investigation of soils. Moscow. Nauka (Science in Rus.). 1975, 656 p (In Russian).
7. Plesyavichius, K. I. Fertilizer application systems comparison in heavy soils. The results of long-termed researchers with fertilizers in regions of the country. Proceedings of Pryanishnikov All-Russian Scientific Research Institute of Agrochemistry. 1982, Issue 12, 4–82 (In Russian).
8. Efremov, V. F. Study of the role of organic manure organic matter in raising sod-podzol soil fertility. The results of long-termed researchers in the system of geographical network of experiments with fertilizers in Russian Federation/Moscow. Pryanishnikov All-Russian Scientific Research Institute of Agrochemistry. 2011, 47–71 (In Russian).

9. Shustikova, E. P., Shapovalova, N. N. Productivity of ordinary chernozems while long-termed systematic mineral fertilizer applying. The results of long-termed researchers in the system of geographical network of experiments with fertilizers in Russian Federation. Moscow. Pryanishnikov All-Russian Scientific Research Institute of Agrochemistry. 2011, Issue 1, 331–351 (In Russian).

10. Merzlaya, G. E., Verkhovtseva, N. V., Seliverstova, O. M., Makshakova, O. V., Voloshin, S. P. Interconnection of the microbiological indices of soddy-podzolic soil on application of fertilizer over long period of time. Problems of agrochemistry and ecology. 2012, №2, 18–25 (in Russian)

11. Shuglya, Z. M. Fertilizer system in crop rotation. The results of long-termed researchers with fertilizers in regions of the country. Proceedings of Pryanishnikov All-Russian Scientific Research Institute of Agrochemistry. 1982, Issue II. 94–118 (In Russian).

12. Williams, V. R. The collected works. Moscow, Selkhozizdat (Agricultural Edition). 1951, V.6. 576 p (In Russian).

13. Shein, E. V., Milanovsky, E. Yu. Spatial heterogeneity of properties on different hierarchic levels as a basis of soils structure and functions.. Scale effects in the study of soils. Moscow. Publishing House of Moscow State University. 2001, 47–61 (In Russian).

14. Glinka, N. L. General chemistry. Leningrad. Chemistry, 18th edition. 1976, 728 p (In Russian).

15. Fertilizers, their properties and how to use them [Ed. Korenkov]. Moscow. Kolos (Ear in Rus.). 1982, 415 p (In Russian).

CHAPTER 26

ACCOUNTING WITHIN-FIELD VARIABILITY OF SOIL FERTILITY TO OPTIMIZE DIFFERENTIATED FERTILIZER APPLICATION

RAFAIL A. AFANAS'EV

CONTENTS

ABSTRACT

This chapter states regularities of the within-field variation of soil fertility, which are important for variable rate fertilizer application under conditions of precision agrotechnologies inclusive, the technologies limiting agroeconomic efficiency. As is well known the usual (traditional) fertilizer practice stipulates their application taking into account-averaged indices of soil fertility: mobile plant food elements

(N, P, K etc.) content in the plow layer. At the same time a part of the plants gets excess of mineral nutrition, and other part – its deficiency. That results in shortage of agricultural products, its deterioration and also the pollution of the environment and the soil with the excesses of agrochemicals in overfertilized plots. In the last decades traditional technologies give place to high-precision agro-technologies with differentiated fertilizer application taking into account within-field heterogeneity of soil fertility. There are several constraints for widespread adoption of high-precision agro-technologies inclusive an underestimation of the character of within-field variability of soil quality. Our investigations reveal eight regularities of the within-field variation of agrochemical indices, which characterize soil fertility in arable soils. These regularities would allow more seriously estimate the efficiency of variable rate fertilizer application under conditions of precision agrotechnologies.

26.1 INTRODUCTION

The main procedures of precision agriculture include differentiated agrochemical application when taking into account within-field variability of fertility and crop status. The agrochemicals are mineral and organic fertilizers, amendments and other inputs. The usual (traditional) fertilizer practice utilizes average rate of fertilizer usage for an individual field. As a rule both procedures use soil analysis when calculating optimal rate of fertilizer. Users employing the traditional technology calculate the optimal rate by averaging results of soil analysis of the whole field; the alternative technology prescribes averaging the results for every within-field contour. Since now there are not visual contours of within-field boundaries all the processes of differentiated application of agrochemicals use satellite navigation system (GPS – Global Positioning System). There are many ways of admeasurement of these contours, which are differentiated one from another by fertility level.

These ways are small-grid sampling and the agrochemical analysis of these samples throughout an entire field with later geostatistic data processing; preliminary yield, electroconductivity or landscape scanning; remote (aerospace) sensing of earth surface. Sampling and analyzing are carried out within the bounds of these contours formerly a priori distinguished and created by some way for calculating differentiated fertilizer and amendment rate and their application realized off-line. For differentiated nitrogen additional fertilizing of vegetating crops they use the photometric methods of nitrogenous nutrition diagnostics by biomass green color or determine elasticity of herbage by crop-meter. The diagnostic devices of these technologies are coordinated with fertilizer applicator, which works on-line. This is the general scheme of preparation and carrying out of procedures for differentiated application of agrochemicals under conditions of precision agriculture.

Accordance with the usual standpoint of ordinary agrochemists they hope for increasing returns under conditions of differentiated application as compared with application by averaged rate because of different fertility of individual plots. Due to this additional expenditures caused by placing of hard- and dataware of high- precision agro-technologies might be compensated. However, the experience of many agriculturalists shows necessity elaborating new ways of decision the problem. In particularly investigations performed in the USA (state Idaho) [1] over a 30-year period with conventional and variable rate nitrogen fertilizer application (data obtained from a seed potato operation) indicated discouraging result: variable rate nitrogen application was found to be unprofitable for the field when compared uniform nitrogen application since the total costs associated with variable rate fertilizer application outweighed the benefits obtained from maintaining the optimal plant available nitrogen levels. This is not unique information concerning the theme. Besides that there are a certain traditional character of agriculture and a sluggishness of landusers thinking. As a result in the last decade there is the decline of interest in the differentiated application of fertilizers. The decline could be explained by the cyclical nature of the development of new technologies [2].

Our investigations show that for practical use of differentiated application of fertilizers it is important more accurately take into account features of within-field variability of soil fertility and mineral nutrition including the features limiting prospective efficiency.

26.2   MATERIALS AND METHODS

The main methods of the investigations were statistical and graph-analytical analysis and the generalizations of the agrochemical characteristic of the plow layer of sod-podzolic clay loam soil. Site: Testing area of the Central Experiment Station of the Pryanishnikov All-Russian Agrochemistry Research Institute (Moscow region). Testing area was a part of cropping rotation field that comprised about 4 ha (200×200 m^2). The part included 400–10 × 10-m square plots on which 400 composite soil samples were taken and analyzed. Taking soil samples and their agrochemical analysis were carried out in accordance with the methods used in the agrochemical service and Russian scientific institutes. Humus (organic carbon) content was analyzed by Tiurins method (oxidation of soil organic matter in $K_2Cr_2O_7 + H_2SO_4$), mobile potassium and mobile phosphorus – by Kirsanovs method (0.2 normal HCl-soil extract), easy-mobile potassium and easy-mobile phosphorus – in low saline $CaCl_2$-extract, easy-hydrolysable nitrogen – in 0.5 normal H_2SO_4, N-NO_3 – in H_2O-extract, pH – in saline suspension (1 normal KCl) [3].

26.3   RESULTS AND DISCUSSION

We revealed eight regularities concerned within-field variability of agrochemical indices. The regularities would be important for technologies of precision agriculture. In the first place it was found that the soil reaction as well as humus and mobile phosphorus and potassium content more or less corresponded to a normal distribution.

In fact plots with the relatively average values of the easy-hydrolysable nitrogen, mobile phosphorus and potassium content occupied more than half of the area, while the number of plots with bulk and minority of the content was noticeably less (Figures 26.1–26.3). Thus 350 plots contained 101–250 mg/kg mobile P_2O_5 in their soil, while plots with lower and higher values of these indices occupied minimum square. Similar results characterized territorial distribution of sites with different easy-hydrolysable nitrogen and mobile potassium.

The second feature of soil spatial structure as used here is that the maximum characteristic of agrochemical indices variability (easy-hydrolysable

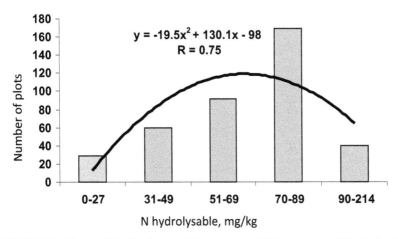

FIGURE 26.1   Curve of the distribution of the number of 400 plots versus N hydrolysable content in the CES testing area soil.

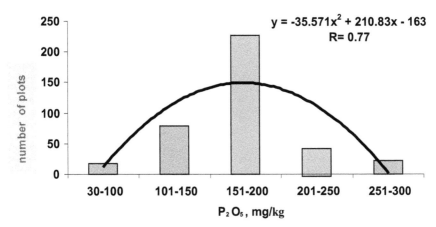

FIGURE 26.2   Curve of the distribution of the number of 400 plots versus mobile phosphorus (P_2O_5) content in the CES testing area soil.

nitrogen, mobile potassium and mobile phosphorus) was found in plots that had relatively smaller and larger values of the content while decreasing the variability in average interval. Trends of the variability sometimes even have negative meaning in average interval. (Figures 26.4–26.6). When investigating the plots, which had the average values of mobile phosphorus content we found, that coefficient of variability of these indices was smaller than 5%, while it exceeded 15–25% for the plots with

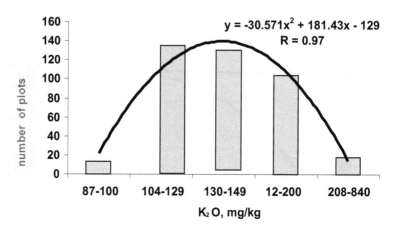

FIGURE 26.3 Curve of the distribution of the number of 400 plots versus mobile potassium (K_2O) content in the CES testing area soil.

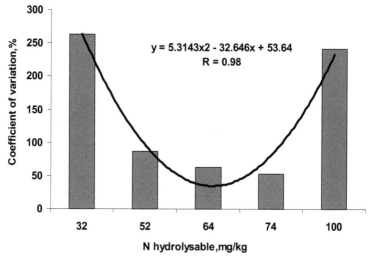

FIGURE 26.4 Variability of N hydrolysable content in soils of the 400 plots of agricultural test area.

marginal indices. Coefficient of correlation between the theoretical calculations and the facts (R) was 0. 89.

According Table 26.1, variability of within-group humus as well as mobile P and K considerably (ten times as large) increased with increasing elementary plot area from 0.1 to 4 ha. It is necessary to take into account

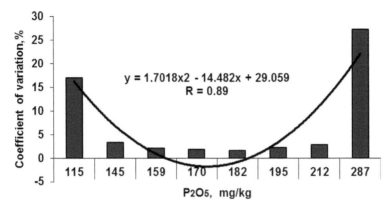

FIGURE 26.5 Variability of mobile phosphorus content in soils of the 400 plots of agricultural test site.

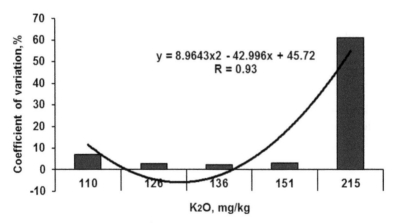

FIGURE 26.6 Variability of mobile potassium content in soils of the 400 plots of agricultural test site

this regularity when developing high agrotechnologies finding a compromise between the expediency of distinguishing fertility contours with minimum intracontour variability of soil fertility for increasing of fertilizer efficiency, on the one hand, and the minimum number of such contours of the field for reducing the cost of taking and analyzing soil samples, on the other. The most acceptable is selection by methods of geostatistics in one plot as a rule no more than 5–6 contours with different level of soil fertility. Thus it is necessary use every agrochemical index according to general area of a field.

TABLE 26.1   Dependence of Variability of Soil Agrochemical Indices on the Area of the Plots Being Averaged

Number of plots	Plot area, ha	Coefficients of variation of agrochemical characteristics, $V\%$		
		Humus	P_2O_5	K_2O
40	0.1	0.8	2.1	2.4
20	0.2	1.6	3.5	4.6
8	0.5	3.2	7.3	10.0
4	1.0	5.6	12.6	17.1
2	2.0	10.0	20.8	27.6
1	4.0	18.6	30.7	41.2

The fourth regularity of spatial heterogeneity of the soil cover consists in the smooth, gradual transition from the maxi mum values of the agrochemical indices to smaller and vice versa from smaller to maximum (Figure 26.7).

Line of trend that describes the change of mobile phosphorus content of 200m transect of the agricultural test site in limits 170–276 mg/kg P_2O_5 approximates the facts with coefficient of correlation R = 0.88. Consideration of this characteristic of the soil structure is very important from a practical viewpoint since when designing and creating machines for variable rate agrochemicals application it allows providing for a

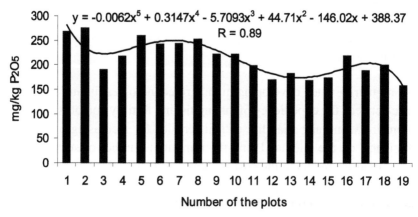

FIGURE 26.7   Mobile phosphorus content in soil of 200m transect of the agricultural test site.

comparatively smooth change of their rates during travel over the field. This facilitates both designing and operating processes of fertilizer and amendment application under production conditions.

The fifth regularity of within-field variability of agrochemical indices in this context is the dependence of the level of soil fertility level on the meso- and microtopography of fields, which to a considerable extent causes all the above-mentioned regularities. Soils of one topographic location influence surrounding soils by leaching, nutrient transfer and deposition of chemical components. Under conditions of the testing area N-NO$_3$ content in soil of elevations prevails over N-NH$_4$ content, whereas in depressions ammonium-nitrogen content exceeds nitrate level due to greater anaerobiosis and reduction of nitrates migrating with subsurface and intrasol flow into depressions.

Greater mobile phosphorus content was also discovered in depressions (Figure 26.8).

This regularity allows the use of the results of topographic survey for preliminary revealing of fertility contours as elementary plots for soil sampling and agrochemical analyzing for the purpose of decreasing agrochemical soil analyzing costs compared to traditional grid sampling.

The sixth regularity of spatial heterogeneity of soil fertility represents noncoincidence of boundaries of various agrochemicals contours among themselves. The regularity was confirmed by the result: the coefficients of correlation between various agrochemical indices of intrafield contours

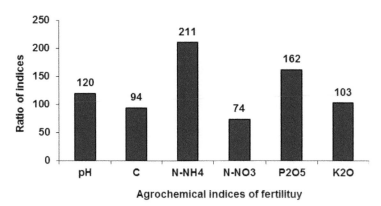

FIGURE 26.8   Ratio of soil agrochemical indices of depressions of the test site vice versa elevations.

were insignificant (Table 26.2). Although there are similar tendencies of the spatial distribution of soil agrochemical characteristics between its separate parts in the limits of a field due peculiarity of the landscape corresponding to the sixth regularity nevertheless considerable differences of fertility of different intrafield parts are obvious. Therefore, for each within-field territorial contour different rates and ratios of nutrients in fertilizer corresponding to the agrochemical characteristics of this contour are necessary. Taking into account this thesis of the technology of variable rate fertilizer application it is required to use machines with simultaneous application of different types of fertilizers or repeated passes of machines adapted for applying one type of fertilizer.

It is necessary to refer greater values of variability of easy-mobile plant nutrients as compared with the values of less mobile nutrients to next (the seventh) regularity. As it is seen in the Figure 26.9, coefficients of variation of easy-mobile N, P, and K are half or twice as much again as the values of corresponding mobile nutrients. But for all that crop yield to a greater extent depends on an easy-mobile nutrients (in particular on phosphorus) than on a less mobile nutrients (Figure 26.10). Taking this regularity into account is very important for increase of agroeconomic efficiency of variable rate fertilizer application because it allows more punctually to take into consideration plant-nutrient need. In other words during the agrochemical analysis of soils side by side with traditional indices of its nutrient (mobile N, P, K) supply it is important to determine easy-mobile nutrients and to use these values while calculating the optimal fertilizer rates. This would increase fertilizer application efficiency and avoid losses of the fertilizers.

TABLE 26.2 Coefficients of Correlation of Soil Agrochemical Indices At the Agricultural Test Site

Index	Humus, %	pH	P_2O_5		K_2O
			mg/kg	mg/L	mg/kg
pH	0.17				
P_2O_5, mg/kg	0.9	0.23			
P_2O_5, mg/L	0.35	0.37	0.45		
K_2O, mg/kg	0	0.12	0.48	0.46	
K_2O, mg/L	0.07	0	0.3	0.3	0.75

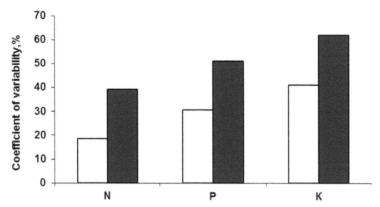

FIGURE 26.9 Variability of mobile (the left row of columns) and easy-mobile (the right row) NPK in soils of the agricultural test site.

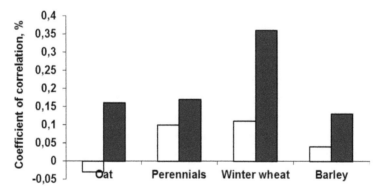

FIGURE 26.10 Influence of mobile (the left row of columns) and easy-mobile (the right row) phosphorus in soils of the agricultural test site on the yield of field crop.

At the same time one ought to take into consideration the eighth regularity that was revealed by our investigation: although crop yield depends on easy mobile nutrients greater than it depends on less mobile forms this yield varies to a lesser degree than proper mobile forms. In particular as it as seen in the Figure 26.11, the values of coefficients of variation of easy-mobile phosphorus (V = 52%) and potassium (V = 62%) in soils of the 400 plots of agricultural test site to a marked degree surpass coefficients of variation of annual grasses hay yield (V = 41%). With lack of nutrition plants increase root development, they can increase solubility of hard-to-reach soil nutrients, biological activity of soil. The lack of nutrition also

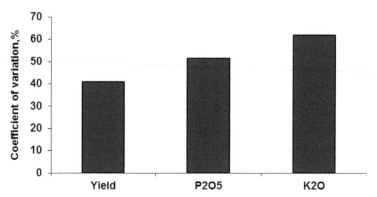

FIGURE 26.11   Variability of annual grasses hay yield and easy-mobile phosphorus and potassium in soils of the agricultural test site.

activates the functioning of soil microflora. Particularly it is known that activity of soil phosphatase sent off by microorganisms increases when content of mobile soil phosphorus decreases and vice versa [4, 5]. Many phosphororganic soil substances (phytin, phospholipids, nucleoproteids) are drawn into exchange of microorganisms with subsequent dephosphorylation and transition of phosphorus into soluble compounds. There are microorganisms in soil that capable transform hard-to-reach compounds of potassium whose exometabolites promote the transition of potassium into soil solution [6]. The activity of nitrogen fixers also increases with decrease of easy-mobile nitrogen content in soil. It is important take into account this regularity when estimating the efficiency of variable rate fertilizer application because to a certain degree differences between variants of traditional and differentiated fertilizer application become smooth through ecological flexibility of plants and also the flexibility of soil microflora which responds to dynamics of soil nutrients.

Of course the ability has their limits. For example studies of differentiated application of nitrogen fertilizer in Tyumen region of the Russian Federation where soils were leached Chernozemic, and crops – spring wheat, showed that application of limited rates of nitrogen counting on grain yield 2 t/ha variable rate application of nitrogen fertilizer didn't advantage first of all economically as compared with application by averaged rate. The plants at the expense of soil nitrogen equalized to a certain extent the difference in low rates of fertilizers applied in different parts of a field. However, with increasing of rates of nitrogen counting on grain yield 3–4 t/ha differences in nitrogen supply of plants because of its variable rate application resulted in essential economic efficiency [7]. So it is possible to come to a conclusion: if the differences between intrafield agrochemical indices or differentiated fertilizer rates are low one cannot expect essential efficiency of their application because of physiologic flexibility of plants.

The regularities of intrafield variability of soil fertility, demonstrated by the example of the testing area of the Central Experiment Station of the Pryanishnikov All-Russian Agrochemistry Research Institute, are characteristic for other types and subtypes of soils where we carried out similar investigations [8]. That indicates their similarity and expedience of utilization when developing high-precision agro technologies of variable rate fertilizer application under different soil-climatic conditions.

26.4 CONCLUSIONS

1. The regularities of intrafield variability of agrochemical indices which are general for all zonal soil types and subtypes may become theoretical grounds for more rational use of agrochemicals when developing technologies of variable rate fertilizer application. We found several distinguishing features of intrafield variability of soil fertility, which is necessary to take into account when trying, increase efficiency of variable rate fertilizer and amendment application. These features are the dependence of agrochemical properties of soil on the meso- and microtopography of fields, smooth shape of conjugacy of soil contours with maximum and minimum values of agrochemical indices, territorial noncoincidence of contours regarding main agrochemical indices, greater values of variability of easy-mobile plant nutrients as compared with the values of less mobile nutrients, dependence of variability on the area of distinguished intrafield contours.

2. Intensity of intrafield heterogeneity of soil fertility has especial significance for choice of technology of fertilizer application (traditional or differentiated). There are two regularities which cause the intensity: firstly – as a rule the ratio of plots with marginal values of

agrochemical characteristics to plots with their average characteristics is low, secondly – soils of marginal plots have the maximum value of agrochemical variability. Basically these regularities of soil intrafield fertility by and large reduce the efficiency of differentiated application of fertilizers.

3. Leveling of the difference in nutrient supply of plants in different parts of a field which is caused to a certain extent by ecologic flexibility of plants also influences on reduction of efficiency and so expediency of differentiation of fertilizer rate. It is necessary to define levels and criteria of soil fertility of agricultural lands for purpose of isolation of areas which are very promising for technologies of differentiated agrochemical application; bearing in mind that the technologies require essential supplemental expenditures as compared with traditional fertilizer application by averaged rate.

KEYWORDS

- **agrotechnologies**
- **hydrolysable**
- **intrafield heterogeneity**
- **precision agriculture**
- **soil**

REFERENCES

1. Watcins, K. B., Yao-chi Lu, Wen-yuan Huang. Economic and environmental feasibility of variable nitrogen fertilizer application with carry-over effects. Journal of Agricultural and Resource Economics. 1998, Vol. 23, №2, 401–426.
2. McGuire, J. Technology coordinator spatial Ag systems. Ohio Geospatial Technologies Conference for Agriculture and Natural Resources (March 24–26. 2003). Columbus. Ohio. 1–42.
3. Agrochemical methods of soil research/Moscow. Nauka (*Science*). 1975, 656 p (In Russian).

4. Rumyantseva, I. V., Devyatova, T. A., Afanas'ev, R. A., Merzlaya, G.Ye. Proceedings of Voronezh State University scientific session. Voronezh. Voronezh State University. 2011, 45–59 (In Russian).
5. Sychev, V. G., Listova, M. P., Derzhavin, L. M. Phosphate regime of soils for agricultural purposes. Bulletin of Geographic network of experiments with fertilizers. Issue 11/Moscow. Pryanishnikov Agrochemistry Research All-Russian Institute. 2011, 64 p (In Russian).
6. Merzlaya, G. Ye., Verkhovtseva, N. V., Seliverstova, O. M., Makshakova, O. V., Voloshin, S. P. Interconnection of the microbiological indices of derno-podzolic soil on application of fertilizer over long period of time. Problems of Agrochemistry and Ecology. 2012, №2, 18–25 (in Russian)
7. Abramov, N. V., Abramov, O. N., Semizorov, S. A., Cherstobitov, S. V. Precision agriculture as a part of resource-saving technologies of crop cultivation. Geoinformatic technologies in agriculture. Proceedings of the International Conference. Orenburg. Publishing Centre of Orenburg State Agrarian University. 2013, 30–40 (In Russian).
8. Afanas'ev, R. A. Regularities of Intrafield Variability of Soil Fertility Indices. Journal of the Russian Agricultural Sciences. 2012, Vol. 38, 36–39.

CHAPTER 27

GAS DISCHARGE VISUALIZATION OF SELECTION SAMPLES OF *TRIFOLIUMPRATENSE* L.

VICTORIA A. BELYAYEVA

CONTENTS

ABSTRACT

The article discusses the results of a qualitative assessment of selection samples of red clover (*Trifolium pratense L.*), as well as the impact of pre-sowing treatment of clover with low intensity X-ray irradiation using a new physical method of investigation – gas discharge visualization. The plants, leaf blades of which have a high intensity of luminescence, differ with largest percentage of sugars. The GDV-bioelectrography allows in short terms to produce a selection of samples of red clover by sugar content, as well as to assess the impact of X-rays on the vitality of clover plants derived from irradiated seeds.

27.1   INTRODUCTION

An important direction of selective practice, together with the research of agrotechnical methods, is the selection of clover on set quality parameters [1]. Traditionally, qualitative assessment of selection samples produced by biochemical methods, disadvantage of which is the complexity and duration of consecutive operations for getting results, and above all – the availability of specific reagents and equipment. An important factor is also the extreme workload of biochemical laboratories during vegetation season, which further increases the time spent on research and decreases efficiency of the method while assessing selection samples.

In this connection, for a more profound and informative study of selection samples of red clover and further experimental material selection, is necessary to use alternative methods of plant objects assessment, in particular – the method of gas discharge visualization (GDV). It is based on Kirlian effect and is a computer recording and subsequent analysis of gas discharge luminescence of biological objects, placed in the electromagnetic field of high tension [2]. The main source of GDV-gram is a gas discharge occurring near the surface of the examined biological due to the interaction of the electromagnetic field with the object of study, in which there is an object surface emission of charged particles involved in the initiation of the initial phases of the gas discharge. GDV-grams are the obtained digital computer images of gas discharge luminescence, arising around the object of study by means of GDV-camera. Condition of biological object is characterized by physiological processes, including the decisive role of the physic-chemical and emission processes, which depend on the structural and emission properties, and also, the total impedance of the object, as well as impedance of the surface of subjects sites. Summary information extracted from the characteristics of luminescence, which is presented as a group of spatially portioned areas of different brightness. Conclusion is given not by studying the anatomical structures of the organism, but on the basis of conformal transformations and mathematical evaluation of multiparameter images [2, 3]. Due to the increasingly widespread of GDV method, researchers from different countries are starting to use it to investigate the impact of environmental factors on biological objects. The attractiveness of this method lies primarily in its sensitivity that sets high requirements for the selection of biological samples, as well as stability of the laboratory conditions and the measurement algorithm to prevent the occurrence of additional errors due to the non-homogeneity of the sample, and other factors that may bring about lowering the accuracy of the result.

GDV-bioelectrogram analysis allows defining parameters of the studied objects: glow area, entropy, form coefficient, fractality, and intensity of luminescence, length of isoline, medium radius of isoline. Bioelectrogram parameters allow characterizing the level of energy homeostasis of the organism as a whole. In particular, the use of this method in the selection of plants will allow selecting with a sufficient probability the best hybrids immediately after crossing, according to the seeds evaluation [4].

This method has been used by researchers in assessing the degree of infestation rough elm (*Ulmus glabra Huds.*) with phytopathogenic fungus (*Graphium ulmi Schw.*), wheat grain with fungi of the genus *Fusarium* [5]. According to the characteristics of the corona discharge viability of seeds can be assessed, diseased and healthy grapes, leaves, fruits; apple can be distinguished, differentiated cultivars of plants [6–9]. Since vitality (viability) of bioobject, its functioning at an optimal level depend on the level of energy supply and its ability to maintain energy homeostasis in a relatively constancy level, the evaluation of these characteristics of the object allows to judge about its functional activity, which is reflected in the comparability of data obtained from using different techniques [3]. First of all, the GDV technology can be used as a method of comparative evaluation of the general condition of plant organisms. Essential arguments in favor of gas discharge visualization are: simplicity of application, convenience of processing and storage of information, efficiency receiving and processing data.

27.2 MATERIALS AND METHODOLOGY

In the first experiment by GDV-bioelectrography we investigated leaf blades of four cultivars of red clover: Minskiy Mutant, Yaskravy, Ustodlivy, Daryal and one local native sample – Dargavski (altitude 1600 m above sea level). Clover cultivars Minskiy Mutant, Ustodlivy, Yaskravy were introduced from the Republic of Belarus. Clover cultivar Daryal was zoned in the republic since 1993 and is commonly used in experiments on the selection of this culture as a standard. It was obtained by artificial hybridization.

For the experiment 20 plants of red clover in stooling phase with different sections of each experimental plot were selected by the three most developed leaves, and in the budding phase from 25–30 plants selected three most developed leaves from 4–5 internodes. At the stage of budding it was determined the percentage of sugar in the samples of clover by standard biochemical methods.

This method was also used in the second experiment to explore the possibility of its using in estimation of presowing treatment efficiency of seeds of red clover cultivar Daryal and synthetic population Syn-316, irradiated with X-rays at doses of 240, 560, 720, 800 cGy (cultivar Daryal) and 320, 640 cGy (population Syn-316) at a constant exposure dose of 80 cGy/s. In this experiment energy of germination of irradiated seeds of clover (the third day of the experiment) and hardness of the seeds (the seventh day of the experiment) were evaluated. We sowed the irradiated seeds of red clover, and in the stooling phase investigated leaf blades of these plants by GDV-bioelectrography.

Filming of the bioelectrograms implemented on the apparatus "BEO GDV-Camera" in static mode (exposure 1s). Samples of the leaf blades were filmed by series of 10 files. Subsequent processing of obtained parameters of bioelectrograms (intensity of luminescence, form coefficient, entropy of isoline, fractality of isoline, medium radius of isoline, length of isoline) was performed with using software packages "GDV Scientific Laboratory" and "Statistica 6.0." Also it was performed correlation analysis by Spearman between sugar content in samples of red clover and the parameters of GDV-bioelectrogram as well as analysis of variance (ANOVA).

27.3   RESULTS AND DISCUSSION

The analysis of parameters of bioelectrograms in the first experiment has discovered significant differences in the intensity of luminescence from 121.33 relative units (rel. un.) (cultivar Yaskravy) to 131.74 rel. un. (cultivar Ustodlivy), the form coefficient of 35.33 rel. un. (native sample Dargavski) to 54.28 rel. u (cultivar Minskiy Mutant) and the entropy of isoline from 0.259 rel. un. (cultivar Minskiy Mutant) to 0,449 rel. un. (native sample Dargavski) in stooling phase. Spread of the fractality of isoline in the samples while the stooling phase is insignificant.

To investigate the dynamic of the bioelectrogram parameters the most informative indicators were: the intensity of luminescence, form coefficient, fractality of isoline. From phase of stooling to budding phase negative trend parameters were observed: the intensity of luminescence, fractality of isoline. Thus, the reduce levels of these parameters to the budding phase was 24.4% and 0.8%, respectively (cultivar Daryal), 17.8% and 1.2% (native sample Dargavski), 15.4% and 2.2% (cultivar Minskiy Mutant), 12,6% and 1.26% (cultivar Ustodlivy), 0.7% and 2.3% (cultivar Yaskravy). Value of the form coefficient parameter, which characterizing the unevenness, irregularity of the outer contour luminescence contrast, tended to increase at 31.7%; 31.4%; 9%; 21.3% respectively for the first four cultivars (Figure 27.1).

It was found out that the average intensity of luminescence of clover leaf blades was slightly higher in the stooling stage than in the budding stage (Figure 27.2), which is probably due to high sugar content, as well

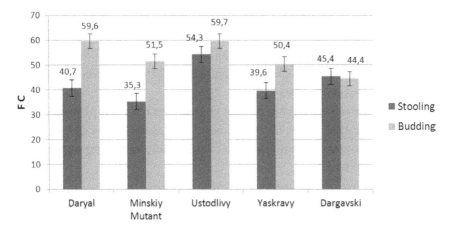

FIGURE 27.1 Dynamics of the form coefficient of red clover samples by phases of development (along the axis X – names of red clover samples, along the axis Y – values of form coefficient).

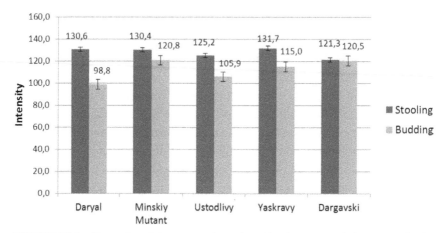

FIGURE 27.2 Dynamics of the average intensity of luminescence of clover samples by phases of development (along the axis X – names of red clover samples, along the axis Y – values of the intensity of luminescence).

as high content of antioxidants: vitamins, flavonoids and other bioactive compounds [10]. Maximum intensity of luminescence in the stooling phase was observed in tetraploid cultivar Ustodlivy – 131.7 rel. un.

Plants with a high content of soluble sugars are valuable as sources of new cultivars of red clover, differentiating by high fodder value as dependent of sugars feed flavoring taste.

Equally important is also selection for increasing seed production, as an indicator of constant renewal of the meadow, and increase of the sugar content in the nectar is correlated with its quality indicators and provides a higher insemination of clover inflorescences of ehntomofile culture by attracting insect pollinators [11].

Content of sugars is an important characteristic for the plant itself, as their concentration affect on the osmotic pressure of the cell sap, they regulate the content of high molecular substances, protecting plants from freezing in winter [1].

We have performed analysis of the correlation between the parameters of GDV-bioelectrograms and sugar content in the samples of clover. It showed that there is a reliable close positive relationship between the intensity of luminescence and the sugar content in the green mass of red clover in the budding stage (Table 27.1).

TABLE 27.1 Correlation Relationship Between the Average Intensity of Luminescence of Leaf Blades of Red Clover and Sugar Content in the Green Mass While the Budding Stage (p <0.5)

Sample	Intensity of luminescence (in relative units)	Content of sugar, (%)	R
Daryal	98.8	2.81	0.75
Dargavski	120.8	4.08	0.72
Minskiy Mutant	105.9	3.61	0.77
Ustodlivy	115.0	3.34	0.69
Yaskravy	120.5	3.98	0.76

Note: R – correlation coefficient.

Our investigations have shown, plants, intensity of luminescence of which was 100–120 relative units, had, according to biochemical analyzes, the largest percentage of sugars. This is due, apparently, to the fact that the whole course of anabolism and catabolism in the plant organism occurs through transformation of carbohydrates. Precisely carbohydrates and, first of all, sugars are related to the energetic component of life. Intensity of luminescence characterizes the level of energy homeostasis of organism, its bioactivity. It is also known that sugar, crude protein and nitrates are associated with protein metabolism in plant organisms [12]. It is important the fact that the capacity for the biosynthesis of amino acids in plant cells in nitrogen assimilation is determined by quantitative sugar content. In studies of other authors it was showed that in clover plants there is a correlation between the content of sugar and protein, also between protein and nitrates in cloudy weather. There is a direct correlation between the levels of sugar and carotene in main mowing in warm sunny weather and between sugar and dry matter in the aftermath in cloudy weather [13]. It is obvious that with a lack of time to conduct biochemical analyzes it is possible to predict the content of the nutrients associated with sugar in the samples of clover. Consequently, using GDV research of leaf blades of red clover and determining the intensity of luminescence, we can give not only a rapid assessment of the rank of sugar content in the samples, but also to evaluate presumably the content of carotene, dry matter, protein, nitrates, which significantly speed up the selection process for increase the quality of products [14, 15].

In the second experiment while the study of the indicators of viability (energy of germination, hard seeds) of seed samples of red clover cultivar Daryal, irradiated with X-rays at doses of 240–800 cGy, we found that the maximum vigor was marked by irradiation at a dose of 800 cGy – 35%, in this variant of the experiment quantity of hard seeds also was minimal – 14% (Figure 27.3).

The study of similar parameters in red clover seeds of the synthetic population Syn-316, irradiated at doses of 320, 640 cGy, showed that the maximum energy of germination (64.7% versus 38.2% in the control) was observed in the variant of experiment with irradiation at a dose of 640 cGy. This dose is also optimal for reducing percentage of hard of seeds (10.0% versus 19.2% in the control) (Figure 27.4).

It is obvious that during the irradiation of seeds of red clover of cultivar Daryal at a dose of 800 cGy and seeds of synthetic population Syn-316 at a dose of 640 cGy, the effect of radiation hormesis is observed, inducing positive physiological processes in them.

Analysis of bioelectrogram parameters of leaf blades in plants of red clover, grown from irradiated seeds, revealed differences in all investigated GDV indices between control and irradiated samples for both cultivar Daryal, and synthetic population Syn-316 (Table 27.2 and 27.3).

It was found that intensity of luminescence of leaf blades, characterizing the degree of energy supply of biological object, was maximal in

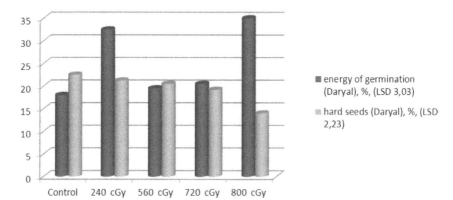

FIGURE 27.3 Energy of germination and hard seeds of samples of red clover of cultivar Daryal after X-ray irradiation (along the axis X – X-ray radiation absorbed dose, along the axis Y – percentages of energy of germination and hard seeds).

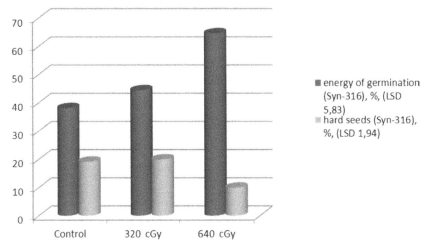

FIGURE 27.4 Energy of germination and hard seeds of the samples of red clover synthetic population Syn-316 after X-ray irradiation (along the axis X – X-ray absorbed radiation dose, along the axis Y – percentages of energy of germination and hard seeds).

plants of cultivar Daryal, grown from seeds irradiated at a dose of 800 cGy and was 31.69 rel. un.

The research of plants of cultivar Daryal, grown from seeds irradiated at a dose of 800 cGy, showed that they have higher values of bioelectrogram parameters, compared with the control and the rest of the variants:

TABLE 27.2   Bioelectrogram Parameters of Leaf Blades of Cultivar Daryal

Dozes	Intensity of luminescence ($p = 2.839$-e-005)	Form coefficient ($p = 5.569$ e-004)	Entropy of isoline ($p = 3.644$ e-005)	Medium radius of isoline ($p = 3.553$ e-005)	Length of isoline ($p = 6.416$ e-005)
Control	27.97	78.28	0.7041	1.159	551.8
Daryal (240 cGy)	31.06	76.84	0.6265	1.348	625.8
Daryal (560 cGy)	24.27	81.4	0.8305	1.125	558.3
Daryal (720 cGy)	22.89	79.04	0.6698	1.106	531.4
Daryal (800 cGy)	31.69	74.89	0.5112	1.400	644.6

Note. Values of significance levels (p) are given for the Kruskal-Wallis test.

TABLE 27.3   Bioelectrogram Parameters of Leaf Blades of Synthetic Population Syn-316

Dozes	Intensity of luminescence ($p = 0.002$)	Form coefficient ($p = 0.008$)	Entropy of isoline ($p = 0.001$)	Medium radius of isoline ($p = 0.001$)	Length of isoline ($p = 0.001$)
Control	27.12	69.25	0.9791	1.490	627.2
Syn-316 (320 cGy)	29.54	73.16	0.8035	1.383	618.1
Syn-316 (640 cGy)	31.41	68.01	0.6828	1.769	732.5

Note. Values of significance levels (p) are given for the Kruskal-Wallis test

medium radius of isoline – 1.4 rel.un., length of isoline – 644.6 rel. un., that obliquely indicates a broader and more powerful crown luminescence of leaf blades.

Parameter values: form coefficient (characterizes irregularity of luminescence contour), entropy of isoline (characterizes the degree of ordering of the biosystem), by contrast, were lower after irradiation at a dose of 800 cGy, which is an indicator of more uniform nature of the luminescence and the relative balance of energy homeostasis of plants. In our opinion, this indicates significant adaptation reserves of these plants, the high activity of functional systems that supports the equilibrium state, and increases productivity.

While the study of bioelectrogram of synthetic population Syn-316 leaf blades it was found that intensity of luminescence of leaf blades, grown from irradiated seeds, higher than in control, both in absorbed radiation at a dose of 320 cGy and 640 cGy – 29.54 and 31.41 rel. un. respectively. In variant with an absorbed radiation dose of 640 cGy the form coefficient (68.01 rel. un.) and entropy of isoline (0.6828 rel. un.) at Syn-316 samples was lower than in the other variants, it shows the balance of energy homeostasis in plants obtained from the seeds, irradiated at the indicated dose. In contrast, the medium radius of isoline (1.769 rel. un.) and length of isoline (732.5 rel. un.), characterizing the extent of the object corona luminescence, while irradiation at a dose of 640 cGy were certainly higher than the same indicators found in the control and in the variant with irradiation at a dose of 320 cGy. Consequently, crown of luminescence in samples of Syn-316, irradiated at a dose of 640 cGy, is wider and leveled. That can be regarded as a confirmation of increased vitality and energy supply of these plants.

27.4 CONCLUSIONS

The selection of red clover plants, distinguished by higher contents of sugars may be carried out not by chemical, but physical method with bioelectrogram registration of leaf blades and release samples with high sugar content at the maximum luminescence intensity ($\geq$ 100–120 rel. un.). Proposed method allows examining majority of samples in a short period, to identify the most valuable plants and to select them for further breeding.

Considering the known correlation between quantitative sugar content and some accompanied biochemical components in the plant raw material of red clover, it is possible to forecast their content in the samples.

Parameters of bioelectrogram (intensity of luminescence, medium radius of isoline and length of isoline) distinguish by maximum values, but form coefficient and entropy of isoline – by minimal values, in the research of leaf blades of red clover plants, grown from seeds irradiated at doses of 800 cGy (cultivar Daryal) and of 640 cGy (synthetic population Syn-316), that can be regarded as functioning at an optimal level of adaptation.

The obtained data correlate with the experimental results and testify that maximum energy of germination and minimal content of hard seeds are fixed after X-ray irradiation of seeds of cultivar Daryal at a dose of 800 cGy and synthetic population Syn-316 at a dose of 640 cGy.

The research of bioelectrogram parameters allows to establish indirectly the degree of influence of ionizing radiation on the functional activity of clover plants, having a direct impact on the level of productivity, and to pick the optimal doses of radiation.

KEYWORDS

- bioelectrogram
- luminescence
- red clover
- X-ray irradiation

REFERENCES

1. Krukovskaya, O. V. Biochemical evaluation of samples and selection of source material to create clover cultivars with improved quality of crude protein. Abstr. cand. diss. Moscow, 1990, 22 p (In Russian).
2. Korotkov, K. G. Basics of GDV. St. Petersburg: Publisher of St. Petersburg State Institute of Precision Mechanics and Optics (Technical University), 2001, 360 p (In Russian).
3. Korotkov, K. G. Principles of GDV analysis. St. Petersburg: "Renome," 2007, 286 p (In Russian).
4. Melnychuk, A. D., Lazarevich, S. S., Latypov, A. Z. Application of Gas Discharge Visualization in plant breeding. Proc. of the III International Scientific Congress on GDV Bioelectrography: "Science, Information, Spirit"/St. Petersburg: Publisher of St. Petersburg State Institute of Precision Mechanics and Optics (Technical University), 1999, 32–33 (In Russian).
5. Priyatkin, N. S., Korotkov, K. G., Kuzemkin, V. A., Dorofeyeva, T. B., Investigation of the influence of environment on the condition of the plants on the basis of GDV. News of high schools. Instrument engineering, 2006, Vol.49. №2, 67–72 (In Russian).
6. Čater, M., Batič, F. Determination of Seed Vitality by High Frequency Electrophotography. Austria; Horn, 1998, Phyton, Annales Rei Botanicae. Vol. 38(2). 225–237. (in English).
7. Kononenko, I., Sadikov, A. Vitality of plants through coronas of fruits and leaves. Proceedings of the VI International Scientific Congress on GDV Bioelectrography: "Science, Information, Spirit". St. Petersburg: Publisher of St. Petersburg State Institute of Precision Mechanics and Optics (Technical University), 2002, 45–46. (in English)
8. Sadikov, A., Kononenko, I. Latest Experiments with GDV Technique in Agronomy. Proc. of the 6-th International Multi-Conference Information Society. Slovenia; Ljubljana, 2003, 110–113. (in English)
9. Skočaj, D., Kononenko, I., Tomaži, I., Korošec-Koruza, Z. Classification of grapevine cultivars using Kirlian camera and machine learning. Res. Rep. Biot. Fac. UL. Agriculture. 2000, Vol. 75(1). 133–138. (in English)

10. Samorodova-Bianki, G. B. Biochemical features of clover. Cultural flora of the USSR. Perennial leguminous grasses (clover, deervetches). Moscow, 1993, – Vol. 13. 125–136 (In Russian).

11. Bekuzarova, S. A. Breeding of red clover. – Vladikavkaz: Publisher Gorsky State Agrarian University, 2006, 176 p (In Russian).

12. Izmailov, S. F. Smirnov, A. M. New directions of plant physiology. Moscow: Nauka, 1985, 122–142 (In Russian).

13. Kuchin, I. N. Daily dynamics of nutrients and biologically active substances in clover plants. Proc. of the Gorky Institute of Agriculture: "Intensification of production and use of feed." Gorky: Publisher Gorky Agricultural Institute, 1991, 19–25 (In Russian).

14. Belyayeva, V. A. Comparative characteristic of bioelectrogram parameters of the leaf blades of *Trifoliumpratense* L. in various phases of development. Medico-Biological Bulletin. 2007, Vol.7. Iss. 13. 210–213 (In Russian).

15. Bekuzarova, S. A., Belyayeva, V. A., Kharchenko, A. Y. Method of selecting plants of clover with high sugar content. Patent №2380885. Published 10.02.2010. Bull. №4, (In Russian).

CHAPTER 28

APPLICATION *GALEGA ORIENTALIS* LAM. FOR SOLVING PROBLEMS OF REDUCTION THE COST OF FORAGE

IGOR Y. KUZNETSOV

CONTENTS

ABSTRACT

The results studies show that the best conditions for the growth and development of plants of a *Galega orientalis*, intended for production of green mass and haylage, formed in mixed crops with *Melilotus officinalis* in the ratio 50+50% from the rate in one-species planting at cleaning cover crops barley in the phase of the output of the plants in the receiver and the level of mineral nutrition on the planned productivity of hay 7 and 9 t/ha. Is thus reduced contamination of crops, increases the collection feed from 1 ha and more profitable crops. To improve the palatability of green mass and improve the quality of harvested forage (hay, haylage, silage) a *Galega orientalis* should cultivate in a mixture with *Phleum pratense* in the ratio 50+50% from the rate in one-species planting. The research results can be widely used in agricultural production, as *Galega orientalis* allows you to get the cheaper product for a long period of time (up to 10 and more years) in comparison with traditional perennial leguminous grasses and annual crops, different low-cost energy saving technology of cultivation.

28.1   INTRODUCTION

The interest towards the cultivation of *Galega orientalis* recent years has grown significantly and culture is gradually becoming a central link in the feed production, because of its highly productive longevity, forage advantages and sustainable harvest of the seeds. Research conducted in the Voronezh region, the Urals, the Volga region and other regions of Russia showed the promise of wide introduction in the production of *Galega orientalis* as a valuable forage plants [1–4]. Review of scientific information indicates that a *Galega orientalis* gives high yield in Kazakhstan, Ukraine, Finland, Belarus, Lithuania, Latvia and Estonia, and other countries of the world [5–10].

The *Galega orientalis* is a good honey plant. Cultural bees willing to visit it during flowering. The productivity of nectar plants not inferior to sainfoin. In addition to bloom *Galega orientalis* begins very early, right after flowering gardens. The results of the research showed that the number of nectar allocated one flower *Galega orientalis* for 48–72 hours isolation of 0.55–0.82 mg, and the total productivity of nectar plants – 154–231 kg/ha.

According to N.M. Semenova [11–12], on the experimental field of the Chelyabinsk scientific research institute on average over 8 years of testing, since 1983 year, the yield of green mass of *Galega orientalis* when two mowing using amounted to 41.6 t/ha, of dry matter 9.73 t/ha, the yield digestible protein – 1.5 t/ha. The average productivity of 1 hectare of sowing for 14 years was: the content of dry matter – 9.74 t/ha, digestible protein – 1.64 t/ha, of fodder units – 9.38 t/ha, exchange energy – level 113.0 MJ/ha, which is 2.5 times higher than in *Medicago sativa* when two mowing use, in average of 5 years in the conditions [12].

In the conditions of the Perm region [13] yields of green mass of *Galega orientalis* in the first year of life was 14.0 t/ha, on the second or third years of life – 22.5–29.0 t/ha.

In conditions of Leningrad region on the experimental fields of the St. Petersburg Academy of agriculture according to N.A. Donskih [14] the yield of green mass of *Galega orientalis* was within 30–75 t/ha, the yield of dry matter on the average for three years amounted to 11.2 to 13.9 tons/ha. In conditions of experimental humidity plants (1992 year) she declined to 7–8 t/ha of dry matter.

In the Northeastern part of the Volga-Vyatka region, according to L.V. Tchkalov [15], in sample fields Vyatka agricultural institute maximum yield of green mass of *Galega orientalis* obtained at early terms of sowing (15 and 30 may) and amounted to 48 t/ha.

Late planting dates (July 15) yield of green mass decreased to 6.7 t/ha. In the first year of life productivity depended on the number of seedings, in the second and subsequent years it has leveled off and was in the range from 50.8 to 64.8 tons/ha. Seeding with row spacing of 45 cm provided the highest yield of green mass both in the first and in the second year of use. When a non-coated growing yield of green mass in the second year of life *Galega orientalis* amounted to 52.8 t/ha, and in the version with the joint sown with barley she declined to 21.5–28.0 t/ha.

In the Central Chernozem region in the Belgorod oblast [16] the yield of green mass of *Galega orientalis* for the second year of life at the beginning of flowering was 35 tons/ha.

According to G.S. Kuznetsov [17], the experimental field of Sverdlovsk agricultural institute, in average, during 1978–1982 years collected 26.3 t/ha for green mass of *Galega orientalis* (with an annual spring frost damage), which is equal to 5 t/ha of dry matter or 1.25 t/ha protein.

Conducted research on the cultivation of *Galega orientalis* in different zones of the Republic of Bashkortostan has shown the expediency of its cultivation. In southern steppe zone, according to S.N. Nadezhkin and I. Y. Kuznetsov [18], the average for the years 1981–1984 dry matter yield of *Galega orientalis* in one-species planting on the experimental field of the Bashkir agricultural institute amounted are 5.36 t/ha, for the years

1987–1993 – 7.7 tons/ha, 1995–2007 years – 7.3 tons/ha. This is undoubtedly a promising low-cost culture.

In experiments S.N. Nadezhkin and V. A. Zaitseva [19] found that crops *Galega orientalis* mixed planting grasses in a ratio of components by 50% in productivity exceeded single-species and were less cluttered. According to A.N. Kshnikatkina [20] and V.A. Varlamov [21], the increased share of the bean component from 45 to 75% and reduction of grain to 40% growth caused the yields of green mass 3.4–4.2 t/ha.

The problem of joint growth of *Galega orientalis* with cereal and perennial leguminous grasses, that have different rates of growth and development requires further decisions. Of particular interest is the study of mixed crops *Galega orientalis* with perennial cereal grasses (*Bromopsis inermis, Festuca pratensis, Phleum pratense*) and perennial leguminous grasses (*Melilotus albus* and *Melilotus officinalis, Trifolium pratense*), especially with *Melilotus officinalis* at different ratios of rates of application components and their productive longevity. *Melilotus officinalis* and *Melilotus albus*, with the rapid pace of development, in mixtures with *Galega orientalis*, successful struggle with weeds in the first years of his life [22–26].

Among researchers there is no consensus on the use of mineral fertilizers in one-species and mixed crops *Galega orientalis*. Not fully explored aftereffect of mineral fertilizers. In Belarus on sod-podzol and peaty soils researchers A.S. Meerovskiy, N.E. Bokhan, A.I. Meleshko [27] considered potassium is a major element in increasing the yield of *Galega orientalis*. On the use of nitrogen fertilizer on crops *Galega orientalis* literature is contradictory information. Researchers N.M. Semenova and O.F. Slepec [28] note that the *Galega orientalis* house does not require nitrogen fertilizer and responds to their inclusion in the negative. However, A.F. Panova [29] considers the possible application of the nitrogen at a dose of 30 kg/ha in spring for an extra feeding.

As for phosphoric-potash fertilizers, the information of N.G. Alcova [30] say about the increased production of *Galega orientalis* with increasing insertion norms of phosphorus and potassium. So for example, if the intake of phosphorus from 60 kg/ha; and potassium, with 60 and 90 kg/ha to 30 kg/ha yield of *Galega orientalis* has increased by 21.1 and 20.9%. A. M. Saharov [31] also considered that a dose of making phosphoric-potash fertilizers should be 100–120 and 120 and 150 kg/ha.

However, I.P. Trynenkov and A.A. Korchagin [32] believe that the intake of phosphorus and potassium to 120 to 240 kg /ha do not provide a reliable increase in the crop yield, compared with doses of phosphorus 60, potassium 60–120 kg/ha.

28.2 MATERIAL AND METHODOLOGY

Based on the analysis of scientific literature and controversial issues on the technology of cultivation of *Galega orientalis* affected by scientists, us in 2000–2010 years on the experimental field of the Bashkir state agrarian university were founded and conducted the long-term field experiments on study of formation of optimum density of crops *Galega orientalis* in one-species and mixed planting with grasses, in one-species and mixed seeding with *Melilotus officinalis* on different background aftereffect of mineral nutrition with the planned yield of 5, 7 and 9 t/ha of hay.

Tasks research included:

- to identify the optimal structure of the grass *Galega orientalis* one-species and mixed planting with grasses;
- to identify the optimal structure of the grass *Galega orientalis* in one-species and mixed seeding with *Melilotus officinalis* with different ratio of seeding rate components and levels of mineral nutrition on the planned yield
- to define the indicators of photosynthetic activity and productivity of crops *Galega orientalis* depending on methods of cultivation.

On agro-climatic zoning of the territory of the experienced fields belongs to a relatively warm, with an average moisture area. Climatic conditions are characterized by large continental with sharp fluctuations of the annual and diurnal variations of weather, unstable and uneven distribution of precipitation, sharp change of air temperature and quick transition from the harsh winters to hot summer, dry air and richness of solar energy.

Southern forest-steppe belongs to the zone of insufficient humidity. The sum of effective temperatures is 2100–2300°C. Annual rainfall 475–575 mm. Rainfall distribution is extremely uneven. Hydrothermal coefficient is 1.0 to 1.2. The arrival of photosynthetic active radiation ranges from 1900 to 2860 kcal/ha.

Research on the formation of the optimal density of crops *Galega orientalis* in its pure form or in mixtures with grasses and *Trifolium officinalis* was conducted with a variety of *Galega orientalis* Gale on leached chernozem of medium and heavy loam particle size distribution by the statement of the stationary field experiments. Power humus horizon was 45–50 cm, total stocks of moisture in the meter layer of soil reached 300–350 mm. Humus content in the topsoil was, on average 8–9%, total nitrogen – 0.5%, phosphorus – 0.2%, potassium – 0.7%.

According to standard techniques were conducted the following studies: phenological observations, density of stalks, the dynamics of the linear growth, temperature, water and nutrient regime of soil structure and botanical composition of the grass, a symbiotic apparatus, the leaf surface, photosynthetic potential yield of green mass, hay, dry matter, the chemical composition of plants with use of methods of state testing [33–36].

28.3   RESULTS AND DISCUSSION

One of the crucial conditions of formation of highly productive plantations of the *Galega orientalis* for 10 or more years of use is the optimum seeding. Attaching exclusive importance in 1992–1997 years at the chair of plant growing, forages production and fruit and vegetable growing Bashkir state agrarian university studied the reaction of the *Galega orientalis* for a wide range of seeding rate under different sowing methods for finite density of standing of plants in order to identify the optimal structure of the grass in the first years of life, and the effects of different harvesting time cover crops, joint seeding with *Trifolium pratense* and *Melilotus officinalis* at different levels of mineral nutrition with the purpose of formation of high grass [37].

The maximum yield of green mass of the *Galega orientalis* house in an average of 6 years of use was obtained by a member seeding with 15 cm row spacing and placement of seeds in rows across 2.5 cm is based on a finite density of standing of plants 2.6 million pieces per 1 hectare and totaled 28.8 t/ha The profitability of cultivation of crops were within 315–340%. Wide-row crops yielded yield ordinary crops in the first years of life. In wide-row sowing with row spacing of 45 cm and the placement of seeds in a row in 1.5, 2.0 and 2.5 cm at the end the stand density 0.9; 1.1 and 1.5 million units/ha seal grass marked the second year of life, and the alignment of yield with an ordinary planting – only in the fourth year of life. Collection digestible protein was higher in the member seeding [37].

When studying the formation of crops *Galega orientalis* with leguminous plants (for the production of silage) in various periods of time cleaning cover crops and different backgrounds mineral nutrition is the most productive in the first years of life obtained by planting it in mixture with *Trifolium pratense* and *Melilotus officinalis*. In mixed crops with legumes grasses first cut was due to growing accompanying components that competed with *Galega orientalis*. In the second mowing, due to more intensive growth of the *Galega orientalis*, its share in planting *Trifolium pratense* increased, and in the mixture with *Melilotus officinalis* dominated.

The highest productivity of *Galega orientalis* (1995–1997 years) marked the introduction of mineral fertilizers on the planned productivity of hay 9 t/ha were obtained 40.8 t/ha of green mass with the control of 37.5. Mineral fertilizers for the planned yield of hay 7 and 9 t/ha has increased the density of the grass and reduced contamination of sowing. The lowest contamination observed in the rate of *Galega orientalis* with *Trifolium pratense*.

The research results G.G. Zainetdinov [38] for the study of the formation of the grass *Galega orientalis* in one-species and mixed crops with legumes components, and in particular with *Melilotus officinalis*, relative norms of seeding 50+50% for different terms of cleaning cover crops of barley and levels mineral nutrition, showed a high responsiveness of *Galega orientalis* on entering of mineral fertilizers.

In the work [38] for 7 years of enjoyment of the highest hay yield was obtained on the perennial leguminous grasses in barley in the phase of plants in a pipe with mineral fertilizers for the planned harvest 9 t/ha and constituted in one-species planting of the *Galega orientalis* house of 7.85 t/ha, and when it is sown with the *Melilotus officinalis* 8.45 t/ha (Table 28.1).

The introduction of mineral fertilizers on the planned productivity of hay 7 and 5 t/ha of the program receive sung in pure sowing of *Galega orientalis* in the early period of harvesting barley average for years of use was 106.97 and 134.77%, while sowing with clover – 108.83 and

TABLE 28.1   The Impact of the Level of Mineral Nutrition and Harvesting of the Barley Crop Yield One-Species and Mixed Planting of *Galega orientalis*

The term of harvest	Planned yield hay, t/ha	*Galega orientalis* in one-species planting			*Galega orientalis* in the mixture with *Melilotus officinalis*		
		Green weight	Dry substance	Hay	Green weight	Dry substance	Hay
Exit	5	30.7	5.61	6.73	32.7	6.01	7.22
plants	7	34.1	6.24	7.49	35.3	6.46	7.61
in the tube	9	35.7	6.53	7.85	38.4	7.03	8.45
Milk-wax	5	28.9	5.29	6.35	31.0	5.69	6.83
ripeness	7	33.3	6.09	7.32	33.9	6.16	7.40
grain	9	34.3	6.28	7.55	36.3	6.60	7.97
Wax	5	29.2	5.33	6.43	29.4	5.38	6.46
ripeness	7	32.0	5.84	7.02	33.0	6.05	7.27
grain	9	33.2	6.07	7.29	34.9	6.38	7.67

*According to Nadezhkin, Zainetdinov [38] on average, over the years 1997–2003.

144.40%. Actual yield of hay were respectively 7.49 and of 6.73; to 7.61 and 7.22 t/ha.

For the period of 7 years of cultivation [38] *Galega orientalis* humus content in the arable horizon of leached chernozem increased by 0.38% from the source for 6.81 and made up 7.19%. The accumulation of humus when entering of mineral fertilizers on the planned yield of 7 and 9 tons of hay from 1 ha occurred more intensively and amounted to 0.44 and 0.47%.

In recent years more and more attention to study and introduction in manufacture attract joint crops *Galega orientalis* with perennial cereal grasses, since the latter contribute to improving the balance of nutrients and eating green mass [39].

In our experiments yield mixtures *Galega orientalis* + *Phleum pratense* by 50% from the rate in one-species planting was over nine years of use 26.6 t/ha of green mass with the control (clean sowing *Galega orientalis*) 24.8 tons/ha or exceeded by 9.7%.

On the seventh year of use of the grass, the share of *Phleum pratense* was to cut 35 and 8%. The share of participation of *Galega orientalis* mixtures ninth year (2000 year) use was 80–88%, cereal component – 12–19%, forbs – 1%. This marked the seal of the grass. *Festuca pratensis* and *Bromopsis inermis* showed high competitiveness in all years of use. On the seventh year of use, the share *Bromopsis inermis* in the first cut was 54%, *Festuca pratensis* – 70%, and the percentage of their participation in the second harvests has increased respectively by 15 and 9%. For the ninth year of use of the equity participation of *Festuca pratensis* decreased to 46%, *Bromopsis inermis* – up to 15%. The productivity of grass mixtures *Galega orientalis* + *Festuca pratensis* was 21.6 t/ha of green mass, mixtures *Galega orientalis* + *Bromopsis inermis* – 22.0 t/ha.

With age, one-species and mixed planting of *Galega orientalis* with grasses (for 10–15 years of use with grass) marked reduction in the productivity of crops, compared with the state of crops on 8–9 years of use (Table 28.2).

A mixture of *Galega orientalis* with *Phleum pratense* on 50% from the rate was 15 years (1992–2006 years,) use the greatest productivity, making in average for the year 28.4 t/ha of green mass, with single-species planting of *Galega orientalis* 27.0 tons/ha. The inclusion in the grass mixture of *Galega orientalis, Festuca pratensis,* and *Bromopsis inermis* 50% led to the formation of the productivity of crops at the level of 21.6 t/ha and 23.9 t/ha of green mass, respectively, lower than the productivity of single-species planting of *Galega orientalis* by 14.5% and 8.3%.

The basis for conducting experiments (2003–2007 years) [40] on the study of the formation of the optimal density of crops *Galega orientalis* pure and mixed seeding with *Melilotus officinalis* on different backgrounds mineral nutrition with the planned yield of 5, 7 and 9 t/ha of hay was held earlier experiments M.H. Kiraev (1997–1999) [41], and G.G. Zainetdinov (2000–2003) [42]. Is of considerable interest for producers of agricultural products the term of use of the crop on years, since the longer-term herbage can generate high productivity, the more economical it is the sowing.

According to the results of our studies [40] found that mineral fertilizers had a positive impact on growth processes one-species and mixed planting of *Galega orientalis* on all variants. With the improvement of food regime of soil plant height was increased.

According to our research, the beginning of vegetation of plants by years of use was noted mainly in the third decade of April. *Galega orientalis* early forms yield of green mass, which can be used to feed animals

TABLE 28.2 Comparative productivity of *Galega orientalis* and its mixtures with grasses (the experimental field of the Bashkir state agrarian university, t/ha, 1992–2006 years).

Culture	The productivity of green mass, t/ha									Deviation from the control	
	Years of use										
	first – seventh (1992–1998)	eighth – tenth (1999–2001)	eleventh (2002)	twelfth (2003)	thirteenth (2004)	fourteenth (2005)	fifteenth (2006)	the amount of the harvest in 15 years	in an average year	t/ha	%
Galega orientalis	20.4	39.0	34.2	28.2	30.1	28.5	24.0	405.0	27.0	–	–
Galega orientalis + Festuca pratensis	15.5	32.7	26.4	24.5	25.3	23.7	18.1	324.6	21.6	–5.4	–14.5
Galega orientalis + Bromopsis inermis	17.8	34.3	27.5	26.4	28.6	26.3	22.4	358.6	23.9	–3.1	–8.3
Galega orientalis + Phleum pratense	23.1	38.3	35.1	29.1	30.7	29.6	26.1	427.3	28.4	1.4	+3.7

with 20–25 may, 15–20 June (the first cut) and in the aftermath from 10 to 20 July (second cut).

The most intensive growth in above-ground biomass is observed in the second and third ten days of may. Significant difference in the timing of spring flowering options unchecked. The offensive phase of budding-the beginning of flowering took place from 30 may to 12 June. The offensive phase of budding in pure and mixed crops plants *Galega orientalis* with *Melilotus officinalis* was noted at the 49th day after regrowth. The beginning of formation of the yield of the second cut in the years of use grass took place on 9–12 day. The develop of plants of *Melilotus officinalis* from the beginning of the growth phase of budding in the years of studies lasted 39–44 days, *Galega orientalis* 41–47 days. The results of our research showed that over a period of 11 years of cultivation of *Galega orientalis* (1997–2007) the humus content in the arable horizon of leached chernozem increased by 0.45% −6.79% and amounted 7.14%.

In our experiments 2004–2007 years leaf surface *Galega orientalis* planting it in its pure form was before the first cut 118.4–152.5 thousand m²/ha, before the second – 86.9–135.2 thousand m²/ha, mixed with *Melilotus officinalis* (total area of leaves), respectively 122.4–161.9 and 95.1–131.0 thousand m²/ha (Table 28.3).

On average 11 years of use (1997–2007 years) before the first cut was for leaf surface level 125.6–139.9 thousand m²/ha, before the second – the 98.9 to 127.9 thousand m²/ha, mixed with *Melilotus officinalis*, respectively 127.1–148.3 and 114.1–129.8 thousand m²/ha The maximum size of the leaves when all terms of cleaning cover crops of barley was for in versions with application of mineral fertilizers for crop 9 tons of hay with 1 hectare. Photosynthetic potential of pure and mixed crops *Galega orientalis* with *Melilotus officinalis* in the years of studies ranged from 1.7 to 4.0 million m²* days/ha.

Photosynthetic potential of pure and mixed crops *Galega orientalis* depending on the level of mineral nutrition and the period of harvesting cover crops barley average for 1997–2007 years is defined by the following equation:

$$Z = -18163.211 + 72.953*x + 360.4*y + 0.122*x*x - 0.74*x*y - 1.72*y*y,$$

where Z – photosynthetic potential crops, million m²* days/ha; x – the level of mineral nutrition of plants (5, 7 and 9 t/ha); y – harvesting

TABLE 28.3 The Impact of Harvesting Barley and Levels of Mineral Nutrition on the Area of Leaf Surface Plants *Galega orientalis* Planting it in One-Species Planting and Mixed With Yellow Sweet Clover

| Culture | Planned yield hay, t/ha | Harvesting of barley | Years of use | | | | | | In an average 11 years (1997–2007) | |
| | | | First-seventh (1997–2003) | | Eighth–tenth (2004–2006) | | Eleventh (2007) | | | |
			1 cut	2 cut	1 cut	2 cut	1 cut	2 cut	1 cut	2 cut
Galega orientalis in one-species planting	5	1	124.2	112.1	127.7	110.3	133.2	118.5	128.1	112.3
		2	127.9	95.4	129.6	101.2	131.0	95.7	129.5	98.9
		3	130.2	125.5	125.4	105.8	121.6	112.5	125.6	111.0
	7	1	134.7	135.9	136.8	120.3	137.2	123.6	136.5	124.1
		2	134.1	139.7	131.2	118.9	135.5	128.7	132.6	125.0
		3	123.2	142.1	132.2	116.9	128.8	131.6	129.7	124.8
	9	1	145.3	131.3	138.9	125.2	137.6	128.0	139.9	127.0
		2	144.9	126.5	137.8	125.9	139.1	135.2	139.5	127.9
		3	128.3	142.1	132.1	119.2	133.8	126.4	131.7	125.3

TABLE 28.3 Continued

| Culture | Planned yield hay, t/ha | Harvesting of barley | Years of use | | | | | | In an average 11 years (1997–2007) | |
| | | | First–seventh (1997–2003) | | Eighth–tenth (2004–2006) | | Eleventh (2007) | | | |
			1 cut	2 cut	1 cut	2 cut	1 cut	2 cut	1 cut	2 cut
Galega orientalis in the mixture with *Melilotus officinalis*	5	1	129.6	118.0	130.2	112.7	131.0	116.1	130.2	114.4
		2	131.2	128.7	128.9	116.1	129.7	118.0	129.5	119.0
		3	129.7	125.1	127.1	110.1	124.3	115.1	127.1	114.1
	7	1	144.2	124.9	138.4	121.2	139.2	121.5	139.7	122.0
		2	147.1	117.3	136.0	121.4	130.5	127.1	137.1	121.7
		3	149.2	112.1	140.2	116.0	149.7	110.6	143.9	114.1
	9	1	154.4	129.7	145.5	128.8	150.6	131.0	148.3	129.4
		2	147.8	134.9	148.5	126.8	147.2	129.2	148.1	128.9
		3	147.5	140.1	141.6	126.5	138.4	129.6	142.1	129.8

Note. Harvesting barley: 1 – exit plants in the tube; 2 – milk –wax ripeness grain; 3 – wax ripeness grain

*Experimental field of the Bashkir state agrarian university, thousand m²/ha, 1997–2007 years.

barley (exit plants in the tube, milky-wax the ripe grain, wax ripeness of grain).

In our experiments on the program receiving the planned yield of hay influenced pure and mixed crops *Galega orientalis*, harvesting cover crops of barley, levels of mineral nutrition and its effects, the age of the grass. We observed the increase of productivity of hay with age grass. The average for the eleven areas with the greatest hay yield was obtained on the perennial leguminous grasses in barley in the phase of plants in the tube with the introduction of mineral fertilizers on the planned productivity of hay 9 t/ha and was in pure sowing *Galega orientalis* 7.73 t/ha, and when it is sown with *Melilotus officinalis* 8.18 t/ha (Table 28.4).

This is consistent with the implementation of the program of receipt of hay on 85.8 and 90.8%. The introduction of mineral fertilizers on the planned productivity of hay 7 and 5 t/ha of program execution receipt of hay in one-species planting of *Galega orientalis* during the early period of harvesting barley average for years of use has made 106.4 and 135.4%, while sowing with *Melilotus officinalis* – 106.7 and 143,8%. Actual yield of hay were respectively 7.45 and 6.77; 7.47 and 7.19 t/ha.

When harvesting barley in the stage of milky-wax and wax ripeness grain yield of hay in one-species planting of *Galega orientalis* was at an average of 3.50 and 5.48%, mixed cropping in the 3.91 and 7.18% less in comparison with its cleaning in the phase of the output of the plants in the tube and amounted respectively 7.06 and 6.93; 7.32 and 7.10 t/ha.

Analysis of energy efficiency showed that the highest values of energy coefficient and coefficient of energy efficiency at all levels of mineral nutrition are achieved when cleaning cover crops of barley in the phase of the output of the plants in the tube. Cleaning of barley on the stage of milky-wax and wax ripeness grain led to a decrease of the energy factor in one-species planting of *Galega orientalis* on 10.14–14.80 and 11.99–18.43%, coefficient of energy efficiency – 4.16–15.07 and 12.87–18.14%. Maximum power efficiency and energy efficiency ratio was noted in sowing the *Galega orientalis* + *Melilotus officinalis* is when harvesting barley in the phase of the output of the plants in the tube with the level of the planned yield of hay 5 t/ha and was 15.69 and 7.70.

The costs of the total energy of mineral fertilizers on the planned productivity of hay 7 and 9 t/ha reduced performance of energy coefficient

TABLE 28.4 Hay Yield and Program Execution Receive the Planned Harvest of *Galega orientalis* in One-Species Planting and Mixed with *Melilotus officinalis*, Depending on the Terms of Harvest of Barley and Levels of Mineral Nutrition

Planned yield hay, t/ha	Harvesting of barley	Hay yield, t/ha							% program execution
		Years of use					In an average 11 years (1997–2007)		
		first – seventh (1997–2003 years)	eighth (2004 year)	ninth (2005 year)	tenth (2006 year)	eleventh (2007 year)	the amount	average per year	
5	1	6.73	6.95	6.25	7.01	7.15	74.51	6.77	135.4
	2	6.35	6.60	6.30	7.15	7.06	71.59	6.50	130.0
	3	6.43	5.75	6.11	7.19	6.63	70.73	6.43	128.6
7	1	7.49	7.51	7.29	7.80	7.04	81.97	7.45	106.4
	2	7.32	7.30	6.26	7.67	6.98	79.46	7.22	103.1
	3	7.02	7.04	7.31	7.61	6.87	77.98	7.08	101.1
9	1	7.85	7.80	7.34	7.91	7.01	85.04	7.73	85.8
	2	7.55	6.90	7.33	7.94	7.10	82.15	7.46	82.8
	3	7.29	7.45	7.15	7.65	7.05	80.37	7.30	81.1

TABLE 28.4 Continued

Planned yield hay, t/ha	Harvesting of barley	Hay yield, t/ha							% program execution
		Years of use					In an average 11 years (1997–2007)		
		first – seventh (1997–2003 years)	eighth (2004 year)	ninth (2005 year)	tenth (2006 year)	eleventh (2007 year)	the amount	average per year	
5	1	7.22	7,25	7.14	7.24	6.95	79.12	7.19	143.8
	2	6.83	6,45	6.85	6.35	6.86	74.35	6.75	135.0
	3	6,46	6.21	6.40	7.00	6.80	71.66	6.51	130.2
7	1	7,61	7.31	6.67	7.85	7.10	82.26	7.47	106.7
	2	7.40	6.55	7.30	7.79	7.01	80,46	7.31	104.4
	3	7.27	7.01	6.65	7.83	7.24	79.65	7.24	103.4
9	1	8.45	8.40	6.30	8.16	8.03	90.05	8.18	90.8
	2	7.97	7.65	7.31	8.20	7.94	86.93	7.90	87.7
	3	7.67	7.75	6.25	8.41	7.15	83.31	7.57	84.1

Note. Harvesting barley: 1 – exit plants in the tube; 2 – milk-wax ripeness grain; 3 – wax ripeness grain

*Experimental field of the Bashkir state agrarian university, 1997–2007 years.

and coefficient of energy efficiency in comparison with the control. The increment of gross energy in one-species and mixed crops of *Galega orientalis* to raising the level of mineral nutrition increased. In one-species planting of *Galega orientalis* the introduction of mineral fertilizers on the planned productivity of hay 7 and 9 t/ha increment of gross energy was greater than in control average harvesting date of barley on 7266.27 and 12403.20 MJ/ha and was 80768.18 and 85905.11 MJ/ha. In mixed planting *Galega orientalis + Melilotus officinalis* increment of gross energy exceeded the control, respectively 5519.54 and 9300.64 MJ/ha and was 77033.30 and 74866.61; 82543.49 and 86324.59 MJ/ha on average for years of use, the greatest increment of gross energy was seeded *Galega orientalis + Melilotus officinalis* cleaning cover crops of barley in the phase of the output of the plants in the phone and the application of mineral fertilizers on the planned productivity of hay 9 t/ha and was 99964.78 MJ/ha, which exceeded the control on 18765.32 MJ/ha or of 23.11%.

Thus, in the year of sowing of perennial leguminous grasses maximum value of pro forma net income and the increment of gross energy were obtained by harvesting of barley in a phase of wax ripeness of grain and making the calculated doses of mineral fertilizers on the planned grain yield of 2.5 t/ha

Profitability was 71.81%, power efficiency and energy efficiency ratio respectively 4.43 and 2.63. The increment of gross energy exceeded the control 11592.4 MJ/ha.

28.4 CONCLUSIONS

1. On the formation of crops *Galega orientalis* and its productivity is influenced by the inclusion in the sowing of cereals and legumes components, harvesting time cover crops, mineral fertilizers and their effects.

2. In the abundance of moisture in the soil is preferable to use cleaning cover crops sowing under. In sowing under crops sowing norm of seeding of *Galega orientalis* increase by 15%, and the coating culture decrease by 25–30% from recommended.

3. To increase consumption of green mass and improve the quality of harvested forage (hay, haylage, silage) a *Galega orientalis* should

cultivate in a mixture with *Phleum pratense* in the ratio 50+50% of seed mixtures in one-species planting.

4. The best conditions for the growth and development of plants *Galega orientalis*, intended for production of silage, formed in mixed his crops with *Trifolium pratense* and *Melilotus officinalis* when cleaning cover crops of barley in the phase of the output of the plants in the receiver and the level of mineral nutrition on the planned productivity of hay 7 and 9 t/ha. Is thus reduced contamination of crops, increases the collection feed from 1 ha and improves profitability.

5. On the basis of the conducted research it can be concluded that the highest yield of green mass, collecting the dry matter and hay are achieved in the cultivation of *Galega orientalis* mixed with *Melilotus officinalis* in the ratio 50 + 50% of seed mixtures in one-species planting, harvesting of barley in the phase of the exit plants in the tube and the application of mineral fertilizers on the planned productivity of hay 9 t/ha.

6. In the years of use grass mineral fertilizers on the planned productivity of hay 7 and 9 t/ha was economically justified. Most pro forma net income from 1 ha is obtained in the cultivation of *Galega orientalis* in one-species planting and mixed with *Melilotus officinalis* when cleaning cover crops of barley in the phase of the output of the plants in the phone and the application of mineral fertilizers on the planned productivity of hay 7 t/ha The maximum increment of gross energy was provided in sowing the *Galega orientalis* + *Melilotus officinalis* is with cleaning cover crops of barley in the phase of the exit plants in the tube and the application of mineral fertilizers on the planned productivity of hay 9 t/ha. Energy efficiency ratio was 6.78.

KEYWORDS

- feeding value
- forage
- forages grasses
- perennial grasses
- protein

REFERENCES

1. Kshnikatkina, A. N., Gushchina, A. V. Methods of cultivation of *Galega orientalis*. the *Galega orientalis* problems of cultivation and use: Abstracts. Chelyabinsk. Publishing house of the Chelyabinsk agricultural research institute. 54 (In Russian).
2. Nadeikin, S. N., Kiraev, M. H. Eastern Galega (*Galega orientalis*). Ufa. Publishing house of the Bashkir state agrarian university, 2001, 49 p (In Russian).
3. Kuznetsov, Y. U. Basic directions of development of fodder production in the Republic of Bashkortostan. Recommendations for conducting spring field works 2013 in the Republic of Bashkortostan. Ufa. Publishing house of the Ministry of agriculture of the Republic of Bashkortostan, 2013, 51–56 (In Russian).
4. Bushueva, V. I. Biochemical Evaluation of the Varieties Samples of the Red Clover and *Galega orientalis*. V. I. Bushueva. Ecological consequences of Increasing Crop Productivity. Plant Breeding and Biotic Diversity. [Eds, A.Iv. Opalko, L. I. Weisfeld et al.] Apple Academic Press Inc. 2014, 287–295.
5. Barbakadze, L. N. The nutritional value of the *Galega orientalis* and the efficiency of its use in the diets of cattle: abstract of thesis for PhD in Agr. Sci. Saransk, 1986.18 P (In Russian).
6. Belous, N. V., Bespalova, L.E, Bulakh, T. M. Efficiency of cultivation of Eastern Galega green fodder and seeds in irrigated conditions of the South of Ukraine. the *Galega orientalis* problems of cultivation and use: Abstracts. Chelyabinsk. Publishing house of the Chelyabinsk agricultural research institutes. 1991, 45–46 (In Russian).
7. Lukashov, V. N., Ostrovsky, M. S., Shiyanov, G. N., Zhumagulov, J. J. *Galega orientalis* in the South-Eastern Kazakhstan. *Galega orientalis* problems of cultivation and use: Abstracts. Chelyabinsk. Publishing house of the Chelyabinsk agricultural research Institute. 1991, C. 51–52 (In Russian).
8. Varis, E. Goatsrue (*Galega orientalis,* L.) a potential pasture legume for temperate conditions. Journal of the Agricultural Sciences in Finland. 1986, Vol. 58. 83–100.
9. Kausanen, P. Goatsrue (*Galega orientalis,* L.) – new persistent forage legume. Efficient Grassland Farm. Berks. 1983, 294–295.
10. Raig, H. A. About the use of the new fodder crops – *Galega orientalis*/The way of solving the problem of fodder protein in Belarus, Lithuania, Latvia and Estonia. Zhodino. 1984, C. 74–77 (In Russian).
11. Semenova, N. M. Prospects of the introduction of *Galega orientalis* in the Urals. the *Galega orientalis* problems of cultivation and use. Theses of reports. Chelyabinsk. Publishing house of the Chelyabinsk agricultural research Institute. 1991, 24–26 (In Russian).
12. Semenova, N. M. Productive longevity resource-saving culture – *Galega orientalis* in conditions of the Chelyabinsk region. The Introduction of nonconventional and rare agricultural plants. Materials of all-Russian research-and-production conference. Penza. Publishing house of the Penza State Agricultural Academy. 1998, Vol. 2. 126–127 (In Russian).
13. Bugreev, V. A., Voloshin, V. A., Osheva, G. M. Culture of the big opportunities/ Kormoproizvodstvo (Grassland). 2006, №6, 28–29 (In Russian).

14. Donskih, N. A. Scientific substantiation of methods of creation of long-cutting grass in the North-West of Russia. Abstract of thesis for a DSc. in Agr. Sci. St. Petersburg – Pushkin. 1998, 40s (In Russian).
15. Tuchkalov, L. V. Cultivation of *Galega orientalis* in farms of Kirov region. The *Galega orientalis* problems of cultivation and use. Theses of reports. Chelyabinsk. Publishing house of the Chelyabinsk agricultural research institute. 55–56 (In Russian).
16. Sereda, P. A. The cultivation of *Galega orientalis* in the Belgorod region. The *Galega orientalis* problems of cultivation and use. Theses of reports. Chelyabinsk. Publishing house of the Chelyabinsk agricultural research institute. 42–43 (In Russian).
17. Kuznetsov, G. S. *Galega orientalis* in the Sverdlovsk region. The *Galega orientalis* problems of cultivation and use: Abstracts. Chelyabinsk. Publishing house of the Chelyabinsk agricultural research institute. 55–56 (In Russian).
18. Nadeikin, S. N., Kuznetsov, Y. U. In new technologies for cultivation of *Galega orientalis*. Resource-saving technologies of cultivation of agricultural crops in Bashkortostan. Ufa. Publishing house of the Bashkir state agrarian university, 2007, 76–79 (In Russian).
19. Nadeikin, S. N., Zaitseva, V. A. Mixed crops of Eastern *Galega* with grasses. Quality of crop production and methods of its improvement: materials of the regional scientific conference. Publishing house of the Bashkir state agrarian university, 1998, 197–200 (In Russian).
20. Varlamov, V. A. Formation of stable legument-cereal grass on leached chernozem of forest-steppe zone of the Volga region. Abstract of thesis for PhD in Agr. Sc. Penza. 2000, 24 p (In Russian).
21. Kshnikatkina, A. N. Formation of highly productive agrophytocenoses of new fodder crops in forest-steppe of Volga region: author's abstract of dissertation for the DSc. in Agr. Sci. Kinel. 2000, 44 p (In Russian).
22. Zubarev, Y.A Comparative bioenergy value of mixed crops *Galega orientalis* with perennial leguments and grasses. Quality of crop production and methods of its improvement: Materials of regional scientific conference. Ufa. Publishing house of the Bashkir state agrarian university, 1998, 172–174 (In Russian).
23. Ievlev, N. I. Initial stages of ontogenesis *Galega orientalis* Lam. in the subzone of middle taiga. Introduction of nonconventional and rare agricultural plants. Materials of III International scientific conference. Penza. Publishing house of the Penza state agricultural academy. 2000, Vol. 1. 127–129 (In Russian).
24. Kiraev, M. H. Formation of highly productive crops (*Galega orientalis* Lam.) for feed purpose in southern forest-steppe of the Republic of Bashkortostan. Abstract of thesis for PhD in Agr. Sc. 1999, S.43–44 (In Russian).
25. Leontiev, I. P., Bikbulatov, S. G. *Galega orientalis* Lam. – reserve protein. Kormoproizvodstvo (Grassland). 1997, №3, C. 21–23 (In Russian).
26. Sagirova, R. A. Cultivation technology of Eastern Galega in connection with introduction in Priangarye. Introduction of nonconventional and rare agricultural plants: proceedings of III International scientific-production conference. Penza. Publishing house of the Penza state agricultural academy. 2006, Vol. 2. 109–110 (In Russian).
27. Meyerovsky, A. S., Bokhan, N. E., Meleshko, A. I. Influence of fertilizers on the productivity of Eastern Galega in BSSR. *Galega orientalis* problems of cultivation

and use: Theses of reports. Chelyabinsk. Publishing house of the Chelyabinsk agricultural research institute. 73–74 (In Russian).

28. Semenova NM, Slepec, O. F. Response to fertilizer Galega orientalis on leached chernozem of the South Urals. the Achievements of agrarian science in practice Ural agriculture. Collection of scientific works. Chelyabinsk. Publishing house of the Chelyabinsk agricultural research institute. 1995, 98–109 (In Russian).

29. Panova, A. F. Influence of spring mineral fertilization on productivity of Eastern Galega. Technology of cultivation of agricultural crops in conditions of Nonchernozem. Collection of scientific works. Saransk. Publishing house of the Mordovian state University. 1981, 84–91 (In Russian).

30. Alcove, N. G. *Galega orientalis* Lam. for fodder. Sheep-breeding. 1988, №5, 28–29.

31. Sagarov, A. M. *Galega orientalis*. Agriculture of Russia. 1986, №7, 24.

32. Trynenkov, I. P., Korchagin, A. A. Quality and yield of green mass, seeds of *Galega orientalis* depending on the levels of supply. Application spirit bards and fertilizers to increase crop yields. Collection of articles. Ivanovo. Publishing house of the Ivanovo agricultural academy of sciences. 1996, 138–143 (In Russian).

33. Methodology the state of crops testing. Moscow. Kolos. 1971, Issue 3. 30–33 (In Russian).

34. Methodics of field and vegetation experiments with fertilizers and herbicides. Moscow. The all-union academy of agricultural sciences, 1990, 174 (In Russian).

35. Methodical instructions on conducting field experiments with forage crops. Moscow, Publishing house: Moscow agricultural academy. 1995, 175p (In Russian).

36. Nichiporovich, A. A., Strogonova, L. E., Chmore, S. N., Vlasova, M. P. The photosynthetic activity of plants in crops. Moscow. Publishing house of Academy of Sciences USSR. 1961, 52–84 (In Russian).

37. Nadeikin, S. N., Kiraev, M. H. Productivity *Galega orientalis* depending on methods of cultivation. Introduction of nonconventional and rare agricultural plants. Materials of all-Russian research-and-production conference. Penza. Publishing house of the Penza state agricultural academy. 1998, Vol. 4, 85–86 (In Russian).

38. Nadeikin, S. N., Zainetdinov, G. G. Forming grass *Galega orientalis* in the first years of life. Materials of international scientific-practical conference "Ways of increasing the efficiency of agriculture in conditions of Russia's joining the World Trade Organization." Ufa. Publishing house of the Bashkir state agrarian university. 2003, 167–169 (In Russian).

39. Kuznetsov, Y. U., Nadeikin, S. N. The role of perennial leguminous grasses in the intensification of fodder production. Modern farming systems: experience, problems, and prospects. Materials of international Scientific-Practical Conference devoted to the 80th Anniversary from the Birthday of academician, V. I. Morozov. Ulyanovsk. Publishing house of the Ulyanovsk state agricultural academy. 2011, 199–207 (In Russian).

40. Nadeikin, S. N., Kuznetsov, Y. U. *Galega orientalis* to feed and seeds/Ufa. Publishing house of the Bashkir state agrarian university. 2008, 144 p (In Russian).

41. Kiraev, M. H. The formation of highly productive crops *Galega orientalis* for feed purposes in southern forest-steppe of the Republic of Bashkortostan: Thesis abstract of thesis for PhD in Agr. Sci. Ufa, 1999, 16 P (In Russian).

42. Zainetdinov, G. G. Methods of forming grass *Galega orientalis* in the Eastern steppe region: abstract of thesis for PhD in Agr. Sci. Ufa, 2003, 16 p (In Russian).

CHAPTER 29

THE EFFECT OF AROMATIC PLANTS ON THE INCIDENCE AND THE DEVELOPMENT OF MALIGNANT TUMORS

VALERY N. EROKHIN, TAMARA A. MISHARINA,
ELENA B. BURLAKOVA, and ANNA V. KREMENTSOVA

CONTENTS

ABSTRACT

The effect of different doses of savory essential oil on the development of spontaneous leukemia was studied on mice. The drug efficiency was determined from the survival curves, animal life spans, and the incidence of leukemia. The savory essential oil in low doses added with drinking water (150 ng/mL) or with feed (2,5 µg/g) increased the average lifetime of mice by 20–35%. The low doses of essential oil from this aromatic plant seems promising as a prophylactic agents.

29.1 INTRODUCTION

The problem of malignant diseases prevention and the search for drugs decelerating or arresting the development of malignant neoplasms is very urgent. Many products of plant origin, for example, herbs, spices, and their extracts, possess wide biological activity, including antioxidant and pharmacological ones [1, 2]. The addition of quercetin as quercetin-aglucon, rutin, and also products containing these compounds (dried apples and onions) in the feed of mice has increased the content of reduced glutathione and decreased the content of oxidized glutathione and mixed disulfide protein- glutathione in the animal liver. The antioxidant activity of plant flavonoids is caused by their ability to inhibit prooxidant enzymes, give complexes with the cations of iron and copper, and catch radicals of oxygen and nitrogen being the donor of hydrogen [3]. These compounds are applied in small doses; they have low toxicity and are recommended for usage for decrease in the risk of disease caused by increased oxidation of cell components. Synthetic antioxidant from the class of hindered phenols [β-(4-hydroxy-3,5-ditretbutylphenyl)propionic acid (phenozan)] also shows significant antitumor activity both in low and very-low doses when administered into the organism of leukemic mice [4].

Among natural antioxidants of plant origin, an impotent place is taken by essential oils, which are a mixture of volatile compounds isolated from spice aromatic plants. The presence of antioxidant properties in many essential oils, including ones that do not contain phenol derivatives, has been proved in model experiments [5]. It has been shown that lemon essential oil and its separate components inhibit oxidation of human low-density lipoproteins in vitro with efficiency close to the efficiency of synthetic phenol antioxidants [6]. Thymol, carvacrol, eugenol, and their derivatives have shown a dose dependent decrease of mitochondrial activity of cancer cells [7].

The goal of this work was to study the effect of the summer savory (*Satureja hortensis* L.) essential oil in low and ultra-low doses in drinking water or food on the life span and development of spontaneous leukemia in AKR mice in the course of their entire lives.

Savory essential oil used in the work contained 0.5–1.7% of each of the following monoterpene carbohydrates (α-thujene, α-pinene, camphene, β-pinene, β-myrcene, sabinene, α-phellandrene, α-terpinene), 2.1% of γ-cymene, 14.8% of γ-terpinene, 2.8% of bornyl acetate, 18.1% of thymol, 37.8% of carvacrol, and 4% of caryophyllene. A high content of thymol, carvacrol, and γ-terpinene was responded for the antioxidant activity of the oil [5, 8]. It was revealed earlier that the addition of thyme oil (1200 mg per 1 kg of mass) in to rat feed increased the general antioxidant status of the animals and kept a high level of polyunsaturated fatty acids in cell membranes during the process of their aging [9]. It should be noted that savory oil doses in our work were by a factor of 100 lower than in the study by Youdim and Deans [9].

Thyme and savory essential oils have a close content of the main components; that is why we hoped that the oil of savory, which is successfully grown in central Russia, would also possess biological activity.

Our research line agrees with the current trend in cancer prevention by efficient low toxicity compounds including antioxidants [10]. Many drugs at low and ultra-low doses demonstrate activities comparable to those at therapeutic doses [11], while their toxicity is much lower.

29.2   MATERIALS AND METHODOLOGY

Experiments were carried out on AKR mice at the age of 3–4 months. The model of AKR mice is interesting since spontaneous leukemia is observed. Note that mouse spontaneous leukemia is most close to human leukemia by the origin, clinical presentation, and morphological properties [12]. Previously, we carried out a detailed kinetic study of the development of different leukemia forms in AKR mice [13]. The kinetic curves of hematological indices (leukocyte count, count of blood corpuscles that can increase the number of leukocytes, and count of undifferentiated (leukemia) cells) were plotted and NMR spectra were obtained for lymphocytes, which allow distinguishing cells at different stages of differentiation.

In savory experiments mice of the first experimental group got drinking water, in which the essential oil of summer savory *Satureja hortensis* L. (Lionel Hitchen Ltd., Great Britain) was added, and standard laboratory feed ad libitum. The content of the essential oil in drinking water was 0.15 mg in 1l. This water was placed into waterers in a sufficient amount. Mice of the second experimental group got pure drinking water and food, into witch the essential oil was added. For obtaining this feed, 0.5 g of savory essential oil was mixed with 100 g of glucose; 50 g of this mixture was added to 100 g of feed and mixed until equal distribution. Consequently, 1g of feed contained about 2.5 µg of savory essential oil.

The experiment was performed for 17 months until the natural death of the last animal.

The life span was recorded from the available dates of animal birth and death. The weight of organs replaces with leukemia cells, thymus and spleen, was determined within one day after death. Leukemia was identified in dead animals by increased weight of thymus (more than 50 mg) and spleen (more than 200 mg). The development of leukemia was evaluated from the life span of affected animals and the leukemia incidence in control animals and animals administered drugs in studied.

The survival (proportion of survived animals vs. age) curves were plotted from the life span data. In order to quantify the drug impact on the leukemia, the Gompertz function was used for nonlinear approximation of the survival curves:

$$S(t) = e\left(-\frac{h_0}{\gamma}(e^{\gamma t} - 1)\right)$$

where S(t) is the proportion of animals that survived to age t; and h_0 and γ are parameters of the function. The process rate largely depends on the γ value. These parameters were calculated using the Gauss–Newton least squares method. In addition to these parameters was calculated the average lifetime.

29.3   RESULTS AND DISCUSSION

Figure 29.1a represents the survival curves of the AKR line of mice in the control and in the case o administration of savory essential oil with drinking water, and Figure 29.2 represents the analogous data for the case

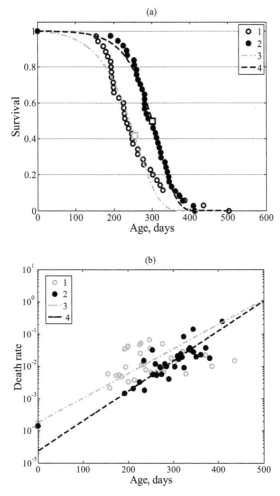

FIGURE 29.1 (a) Survival curves of AKR line mice in the control (1) and in the case of addition of savory oil into drinking water (2). Approximate Gompertz function in the control (3) and in the experiment (4). Selected values of the average lifespan are marked by squares. (b) Death rate of AKR line mice in the control (1) and in the case of addition of savory oil into drinking water (2).

of administration of the essential oil with feed. It is seen that the essential oil in both methods of administration shows remarkable antileukemic action: the survival curves of experimental groups of mice are significantly shifted to the right in comparison with the control. The difference

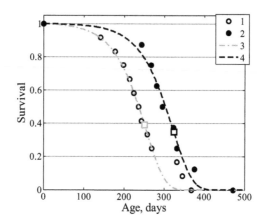

FIGURE 29.2   Survival curves of AKR line mice in the control (1) and in the case of addition of savory oil to feed (2). Approximate Gompertz function in the control (3) and in the experiment (4). Selected values of the average lifespan are marked by squares.

in the date of the beginning of animal mortality is marked in both cases: in the control grope it began after 120 days, in the experimental groups it began after 200–250 days. Figure 29.1b show the curves of death rate of the animals. It is seen that the initial level of mortality in the control group is higher than in the group obtaining savory essential oil. All the survival curves were fitted by the Gompertz function. The kinetic parameters of these curves are presented in the Table 29.1, which shows that the value of the parameter h_0, determining the latent period, in the experiment is lower than in the control.

TABLE 29.1   The Parameters of the Course of Spontaneou Leukosis in the AKR Line of Mice in the Control and in the Case of Consumption of Savory Essential Oil With Drinking Water and Food

Index	Control	Savory
Consumption with water in a dose of 1.5 µg/day		
h_0	1.783×10^{-4}	2.369×10^{-5}
γ	0.0176	0.0214
Average lifetime	254±9	301±8
Consumption with food in a dose of 50 µg/day		
h_0	1.052×10^{-4}	2.962×10^{-5}
γ	0.0209	0.0203
Average lifetime	250±20	325±17

The obtained data give evidence that the constant consumption of savory essential oil significantly increased the latent period (Figures 29.1 and 29.2). It is possible that the consumption of savory essential oil has prophylactic action putting of the terms of contraction of leukemia and mass animal mortality. Thus, the average lifetime of mice increases by 47 days (20%) in the case of the consumption of savory essential oil with drinking water and by 75 days (30%) in the case of its consumption with food in comparison with the control.

29.1 CONCLUSIONS

It was revealed that savory essential oil in low doses in the case of different ways of long-term action increased the length of the latent period (the date of the beginning of leukemic progress) and the average lifetime of mice with spontaneous leucosis.

The obtained results allow us to consider that the usage of containing antioxidants essential oils of aromatic plants (particular savory) in small doses is promising for treatment and prophylactic aims.

KEYWORDS

- antioxidants
- essential oil
- leukemia
- low doses
- savory

REFERENCES

1. Lampe, J. W. Spicing up a Vegetarian Diet: Chemopreventive Effects of Phytochemicals. Am. J. Clin. Nutr., 2003, vol. 78, 579S–583S.
2. Dragland, S., Senoo, H., Wake, K., et al., Several Culinary and Medicinal Herbs Are Important Sources of Dietary Antioxidants,. J. Nutr., 2003, vol. 133, 1286–1290.

3. Meyers, K. J., Rudolf, J. L., and Mitchell, A. E. Influence of dietary quercetin on glutathione redox status in mice. J. Agric. Food Chem., 2008, vol.56, no.3, 830–836.

4. Erokhin, V. N., Krementsova, A. V., Semenov, V. A., and Burlakova, E. B. Effect of Antioxidant β-(4-hydroxy-3,5-ditretbutylphenyl)propionic acid (phenozan) on the Development of Malignant Neoplasms. Biol. Bull, 2007, vol. 34, №5, 485–491.

5. Misharina, T. A., Terenina, M. B., and Krikunova, N. I. Antioxidant Properties of Essential Oils. Applied Biochemistry and Microbiology, 2009, vol. 45, no. 6, 642–647 (In Russian).

6. Takahasi, Y., Inaba, N., Kuwahara, S., and Kuki, W. Antioxidative effect of citrus essential oil components on human low-density lipoprotein in vitro. Biosci. Biotechnol. Biochem., 2003, vol. 67, no. 1, 195–197.

7. Mastelic, J., Jercovic, I., Blazevic, I. et al. Comparative study on the antioxidant and biological activities of carvacrol, thymol, and eugenol derivatives. J. Agric. Food Chem. 2008, vol.56, no.11, p3989–3996.

8. Ruberto, G., Baratta, M. Antioxidant Activity of Selected Essential Oil Components in Two Lipid Model Systems. J. Agricultural Food Chem., 2002, vol. 69, no. 1, 167–174.

9. Youdim, K. A., Deans, S. G., Effect of Thyme Oil and Thymol Dietary Supplementation on the Antioxidant Status and Fatty Acid Composition of the Ageing Rat Brain. British, J. Nutr. 2000, vol. 83, #1, 87–93.

10. Chung, W.-Y., Jung, Y.-J., Surh, Y.-J., et al. Antioxidative and Antitumor Promoting Effects of [6]-Paradol and Its Homologs. Mutat. Res., 2001, vol. 496, # 1–2, 199–206.

11. Burlakova, E. B. Effect of Ultra-Low Doses. Bulletin of the Russian Academy of Sciences, 1994, vol. 64(5), 425–431 (In Russian).

12. Bergolts, V. M., Rumyantsev, N. V. Comparative Pathology and Etiology of Human and Animals/Moscow. Publisher Medicine. 1966, 520 p (In Russian).

13. Erokhin, V. N., Burlakova, E. B., Spontaneous Leukemia-A Model for Studying the Effects of Low and Ultralow Doses of Physical and Physicochemical Factors on Tumorigenesis. Radiation Biology. Radioecology. 2003, vol. 43, no. 2, 237–241 (In Russian).

GLOSSARY

We consciously included some terms in Glossary, which seem to be comprehensible to all. This was done in order to overcome false interpretation of some biological notions, which pronunciation and writing are similar and/or identical to everyday words.

Terminology is a language of science, a component of meta-language. It includes elementary (composed from single word) and composite (combination of words and their equivalents) terms. The word "term" originates from Latin "Terminus," which means the name of the god who protected boundary markers in Roman mythology. On February 23 annually, the Ancient Roman peasants celebrated a festival called the "Terminalia." That is why every term is subordinated to the meaning of the word "limit," for example, it should limit the polysemy and subjectivity and it should be applied in the strictly limited area of meanings. All researchers should use only identical terms for identical concepts.

Terms can form from the words not used in general vocabulary and are usually introduced in the science by replication and/or translation from Latin or any other languages, usually languages – sources of the information, or from words of the native language, which obtain in a particular scientific field a special meaning differing from the everyday one.

The correct use of terms was always and remains now the base of mutual understanding between scientists of close, but different scientific branches. A certain complexity of the terminology is explained namely by that many terms originate from foreign languages. Terms formed from the words of native language are perceived with fewer difficulties, but not always are applied properly.

That is why we recommend becoming acquainted with the glossary not only to young researchers, but to becoming experienced scientists, too, and presenting us their ideas about the terminology improvement, which we will accept with gratitude and which we promise to take into account in the further books. As the example we propose to discuss the use of the term "cultivar."

The diversity of wild plants is determined by spontaneous variability and ecological conditions, which create eliminative factors of natural selection, whereas under the effect of human activity not only artificial diversity of cultivated plants is formed, but also the relationships between particular genotypes change.

Certain plant forms are preserved, others are discarded and new plants with useful characters are created. This represents certain difficulties when classifying cultivated plants. Usually in their taxonomy differences in morphological, anatomic, physiological, geographic, cytogenetic characters are applied, for example, criteria of botanical taxonomy. The main unit of botanical taxonomy is species, which means a group of individuals, characterized by particular, morphological, physiological, ecological and geographic peculiarities, inherent only to them, common phylogenetic origin, possibility of crossing among them, fertility of the posterity and spreading within a particular territory (areal). Plant breeders also apply this concept. However, when working with one, sometimes two or three species it is necessary to distinguish units of intra-species differentiation: subspecies, varieties, sub-varieties, and forms.

Concerning cultivar (variety, sub-variety, form) of cultivated plants, until mid XX century there were differences in the interpretation of this concept, because within one variety sometimes it is possible to find characters of different subspecies and even species. In less domesticated crops the concept of variety is close to such botanical units as subspecies and/or variety; more domesticated plants count hundreds and thousands of cultivars within the same variety. Some of new cultivars were selected from populations formed by inter-taxon crossing as a result of introgression of single genes and splitting in the posterity. Consequently, some progeny, being representatives of one species, can have some characters of another one. Moreover, in some languages, first of all in English, economic concept of cultivar and botanic concept of variety were called by the same term "variety."

In order to give the distinctness to the concept of this notion, International Commission for Nomenclature of Cultivated Plants proposed in 1957 to introduce in the International Code of Nomenclature for cultivated plants the new term cultivar, formed from roots of two English words: cultural and variety. The Code, approved by 14th International Congress of Gardeners, which took place in April 1956 in Nice (France), recommends since January 1 of 1959 to apply for cultivated plants namely this term. In the last edition of the Code, approved in July 2011 in Melbourne (Australia) at the 18th International Botanical Congress, the validity of the term cultivar was confirmed once more. Its content corresponds to following concepts:

- clone — genetically homogeneous group of individuals (which can be sometimes periclinal chimeras), deriving from a single clonal genotype (monogenotypic) as a result of vegetative multiplication or apomixis, for example, in potato, cassava, sweet potato, rubber, mango, avocado, apple, pear, banana, pineapple, strawberry, brambles, grape, peach, cherry, almond, citrus, artichoke, yams, black pepper, olive, fig, pistachio, or edible aroids etc.;
- line — a group of outwardly homogeneous individuals of a common ancestry and more narrowly defined than a strain or variety; in breeding, it refers to any group of genetically uniform individuals formed from the selfing of a common homozygous parent, reproduced by syngenesis and propagating by seeds or spores;
- a group of genetically inhomogeneous individuals of plants having one or more common characters, which allow to distinguish it from other cultivars (variety-population, synthetic, etc.);
- hybrid F_1 — homogeneous group of individuals, which is always restored by crossing two or more selected posterities, lines, clones, simple first generation hybrids F_1.

Anatoly I. Opalko

A

activated sludge
Biological treatment process, it is retrieved from the secondary clarifier as the activated sludge process and consists of microorganisms, non-living organic matter, and inorganic materials.

adaptation
The process of changes of an individual's structure, morphology, and function that makes it better suited to survive in a given environment.

agrocenosis
Association of different organisms forming a closely integrated community of anthropogenic origin (artificial biocenosis).

agroecosystem
A farming ecosystem (US spelling of agrocoenosis).

amylopectin
A starch component which has strongly branched structure and consists of 50,000 –1 million glucose molecules (linkage α-(1,4)-D glycosidic, α-(1,6)-D glycosidic, α-(1,3)-D glycosidic).

amylose
A starch component, which consists of long not branched chains having 1000–6000 glucose molecules (linkage α-(1,4)-D glycosid).

androgenesis
in vitro is a process of microspore development in anther culture or isolated microspore culture in nutrient media resulted in formation of multi cellular structures instead of pollen. These structures produce callus or embryoids, which can regenerate into haploid plantlets.

agroecosystem
A part of the anthropoecosystem as subset of the conventional ecosystem, i.e., a spatially and functionally coherent unit of agricultural activity >>> anthropoecosystem.

alanine (Ala)
A simple aminoacid, one of the early products of photosynthesis. Alanine is formed by transamination when an amino group is donated by glutamine to pyruvic acid. It may be deaminated back to pyruvate for use in the Krebs cycle (or citric acid cycle). An aminoacid present in almost all proteins. The L-isomer is one of the 20 amino acids encoded by the genetic code. Its codons are GCU, GCC, GCA, and GCG. L-Alanine is second only to leucine in rate of occurrence, accounting for 7.8% of the primary structure in a sample of 1.150 proteins.

alien species (Syn. exotic, introduced species, non-indigenous, non-native species)
A species not part of the original flora of a given area, rather, brought by human activity from another geographical region where they evolved or spread naturally. *In cytogenetics*, a species that serves as donor of genomes, chromosomes, or chromosome segments to be transferred to a recipient species or genotypes. We must distinguish between introduced species that

may only occur in cultivation, under domestication or captivity whereas other become established outside their native range and reproduce without human assistance >>> invasive.

alliance (alliancia, all.)
One of the four principle ranks in the hierarchical system of syntaxa, between that of association and order; a plant community of definite floristic composition which presents a uniform physiognomy and which grows in uniform habitat conditions; a rank in a taxonomic hierarchy, treated as equivalent to and being in the rank of order >>> syntaxa >>> syntaxon.

alloploid (Syn. allopolyploid)
A plant that arises after natural or experimental crossing of two or more species or genera; they may contain genomes of the parents in one or more copies >>> allopolyploidy.

allopolyploidy
A type of polyploidy involving the combination of chromosomes from two or more different species. Allopolyploids usually arise from the doubling of chromosomes of a hybrid between species, the doubling often making the hybrid fertile. Many plant species have been derived originally from allopolyploidy, e.g. cultivated wheat: hexaploid bread wheat ($2n=6x=42$) and tetraploid hard wheat ($2n=4x=28$) used for macaroni >>> allopolyploid.

allotetraploid (Syn. amphidiploid)
A plant that is diploid for two genomes, each from a different species >>> allopolyploidy.

amphidiploid (Syn. didiploid, allotetraploid)
Two different diploid chromosome sets present in one cell or organism >>> allopolyploidy.

amphidiploidy
The condition of being amphidiploid >>> allopolyploid >>> allopolyploidy >>> amphidiploid.

aneuploidy
The occurrence of one or more extra or missing chromosomes leading to an unbalanced chromosome complement, or, any chromosome number that is not an exact multiple of the haploid number.

antioxidant
Inhibitor of oxidations, natural or synthetic substances that can retard oxidations.

antioxidant protection (AOP)
The body's defense system against free radicals and the consequences of their effects on the body.

anthropoadaptability
The ability of an individual, taxon or cultivar to satisfy any human needs (utilitarian and aesthetic); the range and extent of reaction is genetically determined.

anthropoecosystem
An ecological system including man with his material and spiritual culture.

apomixis
A sexual reproduction in plants without fertilization or meiosis.

arginine (Arg)
An aliphatic, basic, polar aminoacid that contains the guanido group. The L-form is one of the 20 most common natural amino acids. At the level of molecular genetics, in the structure of the messenger ribonucleic acid mRNA, CGU, CGC, CGA, CGG, AGA, and AGG, are the triplets of nucleotide bases or codons that code for arginine during protein synthesis.

aromatic oils
Are defined as organic compounds multicomponent terpenes, alcohols, aldehydes, ketones and other hydrocarbons produced aromatic plants. Plants or parts thereof, containing essential oils and used to extract it, called aromatic feedstock.

aspartate >>> aspartic acid.

aspartic acid (Asp) (Syn. aminosuccinic acid, aspartate)
An aliphatic, acidic, polar alpha-amino acid ($HO_2CCH[NH_2]CH_2CO_2H$). The carboxylate anion and salts or ester of aspartic acid. The L-isomer of aspartate is one of the 23 proteinogenic amino acids, i.e., the building blocks of proteins. Its codons are GAU and GAC.

B

BAP (6-benzylaminopurine)
Synthetic growth regulator, which belongs to cytokinins.

bedding manure
Mixture of solid and liquid secretions of animals and litter.

biocenosis (US spelling of biocoenosis)
An association of different organisms forming a closely integrated community >>> ecosystem.
The complex of plant, animal, micro-organisms inhabiting the area of land or water body, is characterized by a certain relations between themselves and with abiotic factors of the environment

binomial nomenclature
It is the system of giving a scientific name to an animal or a plant, an outstanding system contributed by Carolus Linnaeus. According to this system, any given animal or plant is given a scientific name consisting of two words. The first word refers to name of the genus while the second word refers to the name of the species. Both the genus and the species are generally given Latin names. For example, Avena sativa for the certain kind plant an onion. >>> International Code of Botanical Nomenclature

biodiversity (biotic diversity)
The existence of a wide variety of species (species diversity), other taxa of plants or other organisms in a natural environment or habitat, or communities within a particular environment (ecological diversity), or of genetic variation within a species (genetic diversity); genetic diversity provides resources for genetic resistance to pests and diseases – not to be confused with biological diversity. The diversity of life in all its manifestations, the degree of variation of life.

bioecofunge-1
The preparation bio-specimen, which was engineered on the base of refined biochemical compounds from the Basidiomycetes fungi and vegetative combinations from plants of genera Polygonaceae, Betulaceae, Cannabaceae, Caprifoliaceae, Scorphulariaceae, Asteraceae.

biological oxygen demand in five days (BOD$_5$)
The amount of oxygen that is used aerobic biochemical oxidation as a result the action of microorganisms and decomposition of labile organic compounds. Is one of the most important criteria for determining the level of water pollution with organic substances, determines the amount of easily oxidized organic pollutants in water.

biological resistance plants
A survival of plants during the period of growth and development from germination to full ripeness.

biotest
The test with the help of herbaceous or woody hosts using inoculation with sap of infected plants.

biotope
An area of habitation with relatively similar conditions providing a living place for a specific assemblage of plants and animals. This area is formed as a result of influence of biocenosis on an ecotope >>> ecosystem >>> ecotope.

biota
Historically established set of plants and animals, united of a common distribution area.

C

callus
Plant tissue consists of undifferentiated cells, which have different morphology and growth characteristics.

catabolism
The set of metabolic pathways that breaks down molecules into smaller units to release energy

cell selection
Selection within a population of genetically different cells *in vitro* by different means and different approaches; selection of cells in their mitotic reproduction and elimination (removal) of the individual physiologically damaged cells, which are the carriers of genetic alterations. Sometimes - the selection of cells having advantages over normal cells. In this case, developed tumors.

chromosome
The microscopic rod-shaped structures that appear in a cell nucleus during cell division, consisting of nucleoprotein arranged into units (genes) that are responsible for the transmission of hereditary characteristics; a DNA-histone protein thread, usually associated with RNA, occurring in the nucleus of a cell; it bears the genes that constitute hereditary material; each species has a constant number of chromosomes.

chromosome conjugation
Joining of homologous chromosomes during meiotic prophase.

cistron
A section of the DNA or RNA molecule that specifies the formation of one polypeptide chain; the functional unit of the hereditary materials; it codes for a specific gene product, either a protein or an RNA.

civilization
Stage of human social development and organization that is considered most advanced; the wild animals and plants domestication gave man a unique opportunity to live a civil (contrary to military) life, without killing anyone in the competition for food.

class >> order >> alliance
These are vegetation unit in the phytosociology. *In bionomenclature*, the principal category of taxa intermediate in rank between phylum (division) and order; one of the four principle ranks in the hierarchical system of syntaxa, above that of order >>> alliance >>> order.

colinearity
The correspondence between the order of nucleotides in a section of DNA (cistron) and the order of amino acids in the polypeptide that the cistron specifies gene.

combined heat and power (CHP)
kind of thermal power plant.

compost
Organic fertilizers produced from the decomposition of organic substances under the effect of microorganisms.

conjugation

A process whereby organisms of identical species, but opposite mating types, pair and exchange genetic material (DNA) gene >>> chromosome conjugation.

chromosomal rearrangement

A type of chromosome abnormality involving a change in the structure of the native chromosome. Usually, the rearrangements are caused by a breakage in the DNA double helices at two different locations, followed by a rejoining of the broken ends to produce a new chromosomal arrangement of genes, different from the gene order of the chromosomes before they were broken.

crop rotation

The alternation of the crop species grown on a field; usually, this is done to reduce the pest and pathogen population or to prevent one-track exhaustion.

cropping rotation

A time-honored process of annual planting rotation crops, which involves changing of different crops that are planting in a given section of field each growing season.

cultivar

A contraction of «cultivated variety» (abbreviated cv.); refers to a crop variety produced by scientific breeding or farmer's selection methods; after International Code of Nomenclature for Cultivated Plant (ICNCP-1995); «cultivar» is synonymous with «Sorte» (German, Ukrainian and Russian), «variety» (English), or "variété" (French).

cysteine (Cys)

An aliphatic, polar alpha-amino acid that contains a sulfhydryl group. A semi-essential amino acid, which means that it can be biosynthesized in humans. A sulfur-containing aminoacid synthesized from methionine and serine. It is involved in the synthesis of biotin. It also acts as a store of sulfur for biosynthesis, and can be broken down to pyruvate.

D

2,4-D (2,4-dichlorophenoxyacetic acid): herbicide induced aberrations of chromosome and possessing strong morphogenetic effect in plant cell and tissue culture in vitro

didiploid (Syn. allotetraploid)
Two different diploid chromosome sets present in one cell or organism.

diploid
A cell with two chromosome sets or an individual with two chromosome sets in each cell; a diploid state is written as "$2n$" to distinguish it from the haploid state "n" >>> allopolyploid >>> allopolyploidy.

DNA sequence
The order of nucleotide bases in the DNA molecule; a succession of nucleotides in a DNA molecule or strand; a succession of any number of nucleotides greater than four is liable to be called a sequence. With regard to its biological function, which may depend on context, a sequence may be sense or anti-sense, and either coding or noncoding. DNA sequences can also contain "junk DNA".

DNA sequencing
The methods and procedures for determining the nucleotide sequence of a DNA fragment and/or chromosome >>> DNA sequence.

E

ecological stability
The precise definition depends on the ecosystem in question, the variable or variables of interest, and the overall context. In the context of conservation ecology, stable populations are often defined as ones that do not go extinct.

ecotope
A relatively homogeneous, spatially-explicit landscape unit that is useful for stratifying landscapes into ecologically distinct features for the measurement and mapping of landscape structure, function and change. Just as ecosystems are defined by the interaction of biotic and abiotic components.

ecosystem
The complex of an ecological community together with a biological component of the environment, which function together as a stable system >>> environment.

edaphic
A nature related to the physical and chemical conditions of the soil. Edaphic qualities may characterize the soil itself, including drainage, texture, or chemical properties such as pH. Edaphic may also characterize organisms, such as plant communities, where it specifies their relationships with soil. Edaphic endemics are plants or animals endemic to areas of a specific soil type. >>>soil

edaphotop
Plot of soil along with part of the lithosphere and hydrosphere, members of the geocenosis.

ELISA (enzyme-linked immunosorbent assay)
A widespread very sensitive method for detecting individual proteins. It makes use of the mechanisms of the immune system. If the immune system recognizes a substance as being foreign, it produces antibodies that attach themselves to the foreign molecule, thereby marking it and a carrier medium will fish it out. This triggers an enzyme-controlled reaction, which leads to a visible colour deposit. They are not to be confused with methods for detecting DNA or DNA sequences.

environment
The sum of biotic and abiotic factors that surround and influence an organism.

F

field germination (ground germination rate)
A measure of the percentage of seeds in a given sample that germinate and produce a seedling under field conditions.

flow cytometry (Syn. FCM, flow cytofluorometries, flow cytophotometry, flow microfluorimetry)
A technique for counting, examining and sorting microscopic particles suspended in a stream of fluid; a technique for identifying and sorting

cells and their components (as DNA) by staining with a fluorescent dye and detecting the fluorescence usually by laser beam illumination.

fundazol (Syn. Agrocit, Benlate, Fungochrom.)
A systemic agricultural fungicide used for control of certain fungal diseases. Active substance: Benomyl (Methyl 1-[(butylamino)carbonyl]-1Hbenzimidazol-2-ylcarbamate, 9CI. Methyl 1-(butylcarbamoyl) benzimidazol-2-ylcarbamate). Very sparingly soluble H_2O; soluble $CHCl_3$; less soluble other org. solvs. Toxic to freshwater fish and aquatic invertebrates.

G

gametophyte
A haploid phase of the life cycle of plants during which gametes are produced by mitosis; it arises from a haploid spore produced by >>> meiosis from a diploid sporophyte.

genome
The total genetic information carried by a single set of chromosomes in a haploid nucleus.

germination (field germination)
The ratio of the number of appeared seedlings by seeds sown in the field, expressed as a percentage.

glutamic acid (Glu)
The carboxylate anions and salts of glutamic acid are known as glutamates. In neuroscience, glutamate is an important neurotransmitter that plays the principal role in neural activation.

An aminoacid ($HOOC(CH_2)_2CH(NH_2)COOH$) involved in purine biosynthesis, occasionally added to plant tissue culture media; it may replace ammonium ions as the nitrogen source; it is of key importance in pollen growth in vitro. Its codons are GAA and GAG.

glycine (Gly)
An amino acetic acid; the simplest alpha amino acid. Having a hydrogen substituent as its side-chain, glycine is the smallest of the 20 amino acids commonly found in proteins, and indeed is the smallest possible. Its codons are GGU, GGC, GGA, GGG of the genetic code.

grain crops
The most important group of cultivated plants are man's basic food product and farm animals, is also the raw material for many industries. Grain crops are subdivided into cereals and legumes. Most cereal crops (wheat, rye, rice, oats, barley, maize, sorghum, millet, broomcorn, panic, paisan, eleusine coracana and others) belongs to the botanical familiy of Gramíнeae, buckwheat - family Polygonaceae; mealy amaranth - Amaranth family. The grain of cereals contains a great deal of carbohydrate and protein, as well as enzymes, B-complex vitamins, PP, and provitamin A. Cereals are raised on all the continents of the earth.

growth inhibitor
Any substance that retards the growth of a plant or plant part; almost any substance will inhibit growth when concentrations are high enough; common inhibitors are abscisic acid and ethylene; other inhibitors, such as phenolics, quinones, terpens, fatty acids, and amino acids affect plants at very low concentrations >>> growth promoter.

growth promoter
A growth substance that stimulates cell division (e.g., cytokinin) or cell elongation (e.g., gibberellin) >>> growth inhibitor

gynoecium
The collective term for the female reproductive organs of a flower, comprising one or more carpels.

H

hermaphrodite
A plant having both female and male reproductive organs in the same flower of the floral receptacle or base of the perianth that surrounds the gynoecium and fruits.

histidine (His)
A basic, polar aminoacid that contains an imidazole group. It is one of the 23 proteinogenic aminoacids. Its codons are CAU and CAC.

hybrid [L. hybrida, the offspring of a tame sow and a wild boar]
(1) Offspring of two parents that differ in one or more inheritable characteristics. (2) Offspring of two different varieties or of two different species.

hybridization
The process of combining different cultivars of organisms to create a hybrid.

hydrothermal coefficient
indicator of natural providing the territory the moisture, characterizing the relation of receipt part of water balance of a precipitation to the maximum size of its account part of an evaporability

hypanthium
A cuplike or tubelike enlargement of the floral receptacle or base of the perianth that surrounds the gynoecium and fruits.

I

International Code of Botanical Nomenclature (ICBN)
Following are some of the major guidelines for scientific naming of plants and animals: 1. Every scientific name should have words either in Latin or be Latinized (i.e., follow Latin grammar). 2. The first word refers to name of the genus and the second word to the name of the species. 3. The name of the genus should start with a capital letter and name of the species with a small letter. 4. Both the names should be printed in italics or else they should be underlined separately. 5. Name of the scientist who first identified and described the species should be abbreviated and written after the species name, preferably in brackets. For example, Homo sapiens Linnaeus is written as Homo sapiens (Linn). The ICN can only be changed by an International Botanical Congress (IBC), with the International Association for Plant Taxonomy providing the supporting infrastructure. This practice is more prevalent in the botanical sciences. Since 2011the ICBN adopted at the International Botanical Congress in Melbourne. For the naming of cultivated plants there is a separate code - the International Code of Nomenclature for Cultivated Plants (IBN).

introduction
Deliberate or accidental relocation of individuals of any species of animals and plants outside the native range into new habitat for them >>> alien species.

invasion
The spreading of a pathogen through tissues of a diseased plant >>> invasive.

invasive (plants or a disease)
Tending to spread prolifically and undesirably or harmfully.

invasiveness
The ability of a plant to spread beyond its introduction site and become established in new locations where it may provide a deleterious effect on organisms already existing there>> invasion >>> invasive.

isoleucine (Ile)
A crystalline aminoacid, $C_6H_{13}O_2$, present in most proteins. It is an essential aminoacid, which means that humans cannot synthesize it, so it must be ingested. Its codons are AUU, AUC and AUA.

L

leucine (Leu)
An aliphatic, non-polar, neutral aminoacid $(HO_2CCH(NH_2)CH_2CH(CH_3)_2$ that, unlike most amino acids, is sparingly soluble in water. Leucine is classified as a hydrophobic aminoacid due to its aliphatic isobutyl side chain. It is encoded by six codons (UUA, UUG, CUU, CUC, CUA, and CUG) and is a major component of the subunits in ferritin, astacin, and other 'buffer' proteins.

leukosis (liukemiya)
Clonal malignant (neoplastic) disease of the hematopoietic system.

LSD_{05}
Least Significant Difference. The minimum difference between means that will result in a "significant" difference at a 5% confidence level. Fisher's Least Significant Difference (LSD) test is one of the post hoc tests. R. A. Fisher proposed this simplest and widely used LSD test in 1935. This method is based on the smallest difference between the two means, which is considered to be significant at a particular level of significance.

lysine (Lys)
Nonessential aminoacid $(HO_2CCH(NH_2)(CH_2)_4NH_2$ found in legumes, whole grains, and nuts. It is an essential aminoacid for humans. Lysine's codons are AAA and AAG. Lysine is a base, as are arginine and histidine. The ε-amino group often participates in hydrogen bonding and as a general base in catalysis.

M

methionine (Met)
This aminoacid is classified as Sulfur containing nonpolar as it has a straight side chain that possess a S-methyl thioether at the γ-carbon $(HO_2CCH(NH_2)$ $CH_2CH_2SCH_3)$; an intermediate in the biosynthesis of cysteine, carnitine, taurine, lecithin, phosphatidylcholine, and other phospholipids; the only one of two aminoacids encoded by a single codon (AUG) in the standard genetic code (tryptophan, encoded by UGG, is the other); the codon AUG is also significant, in that it carries the "start" message for a ribosome that signals the initiation of protein translation from mRNA

microspecies
species described based on minute differences, often used in apomictic taxa such as *Taraxacum* or *Rubus* et al.

mineral nutrition of plants
It is a set of processes of absorption, movement and assimilation by plants of chemical elements obtained from the soil in the form of mineral salt ions.

monophyletic
A group of species that share a common ancestry, being derived from a single inter-breeding population.

mutagen
An agent that increases the mutation rate within an organism or cell; for example, X-rays, gamma-rays, neutrons, or chemicals [base analogues, such as 5-bromouracil, 5-bromodeoxyuridine, 2-aminopurine, 8-ethoxycaffeine, 1.3.7.9-tetramethyl-uric acid, maleic hydrazide; some antibiotics; alkylating agents, such as sulfur mustards (ethyl-2-chloroethyl sulfide), nitrogen mustards (2-chloroethyl-dimethylamine), epoxides (ethylene oxide),

ethylene mine, sulfates, sulfonates, diazoalkanes, nitrosocompounds (N-ethyl-N-nitroso urea); azide (sodium azide); hydroxylamine; nitrous acid; acridines (hydrocyclic dyes), such as acridine orange] etc.

mutagenesis
The process leading to a mutant genotype >>> mutagen.

N

nodule bacteria
It refers to several species of nitrogen-fixing *Rhizobium* bacteria, which form ball-like nodules along legume roots bot. agr.

nullisome
A plant lacking both members of one specific pair of chromosomes.

nullisomy >>> nullisome.

O

outcrossing
Cross-pollination between plants of different genotypes.

P

paraphyletic
A group of species that descended from a common evolutionary ancestor or ancestral group, but not including all the descendant groups.

phenylalanine (Phe)
An aromatic aminoacid $(HO_2CCH(NH_2)CH_2C_6H_5)$; this essential aminoacid is classified as nonpolar because of the hydrophobic nature of the benzyl side chain; the codons for L-phenylalanine are UUU and UUC; it is a white, powdery solid. L-Phenylalanine (LPA) is an electrically neutral amino acid used to biochemically form proteins, coded for by DNA. The codons for L-phenylalanine are UUU and UUC.

phytocenosis
plant community that exists within the same habitat, characterized by a relatively uniform composition, a certain structure and relationships of plants with each other and with the environment.

phytophage >>> polyphagia >>> polyphagous.

phytophagous
Peculiarity of an insect or other invertebrate feeding on plants; >>> phytophage >>> polyphagia.

polyphagia
The habit of certain insects, esp. certain animals, of feeding on many different types of food [New Latin, from Greek, from polyphagos eating much; >>> phagyphagia >>> polyphagous.

polyphyletic
Designating a group of species arbitrarily classified together, some of the members of which have distinct evolutionary histories, not being descended from a common ancestor.

proline (Pro)
A heterocyclic, non-polar aminoacid, which is present in all proteins; the major pathway for proline synthesis, which takes place in the cytoplasm, is from glutamate, through gamma-glutamyl phosphate and glutamyl-gamma-semialdehyde, a two-step reaction that is catalyzed by a single enzyme, D1-pyrroline-5-carboxylate synthetase. Its codons are CCU, CCC, CCA, and CCG. It is not an essential aminoacid, which means that the human body can synthesize it. It is unique among the 20 protein-forming amino acids in that the amine nitrogen is bound to not one but two alkyl groups, thus making it a secondary amine.
prolonged trials – long-timed experiment that continues for many years.

promotoring agent (bio-preparation)
A noncarcinogenic substance that enhances tumor production in a tissue as it has immune-incentive effect.

Q

qualitative trait (qualitative character)

It is determined by a single gene (monogenic, Mendelian traits) and harakterizuyuutsya discrete values - the color, the number of chastey body, etc.

quantitative traits (quantitative characters)
Phenotypes that differ in degree and can be attributed to polygenic effects. This is interaction product of two or more genes, and their environment. Quantitative trait loci (QTLs) are stretches of DNA containing or linked to the genes that underlie a quantitative trait. Mapping regions of the genome that contain genes involved in specifying a quantitative trait is done using molecular tags such as AFLP or, more commonly SNPs . . This is an early step in identifying and sequencing the actual genes underlying trait variation.

There are not genetic systems constant on composition, which governs development of any quantitative trait under all possible growth conditions. There are several approaches to the description of the inheritance of quantitative traits. Aluminum taking at different concentration has different physiological mechanism of toxic action; any plant has different genetic systems determining reaction of the same genotype at different concentration of toxic agent.

R

relevé (vegetation plot data)
This is source data in geobotany and phytosociology, includes information about the plot, first of all the list of plant taxa with the data on their cover.

resistance
The inherent capacity of a host plant to prevent or retard the development of an infectious disease; there are different types of resistance:
- hypersensitivity (infection by the pathogen is prevented by the plant),
- specific resistance (specific races of the pathogen cannot infect the plant),
- nonuniform resistance (the host prevents the establishment of certain races),
- major gene resistance (races of the pathogen are controlled by major genes in the host),

- vertical resistance (host resistance controls one or a certain number of races),
- field resistance (severe injury in the laboratory, but resistance under normal field conditions),
- general resistance (the host is able to resist the development of all races of the pathogen),
- nonspecific resistance (host resistance is not limited to specific races of the pathogen),
- uniform resistance (host resistance is comparable for all races of the pathogen, rather than being good for some races),
- minor gene resistance (host resistance is controlled by a number of genes with small effects),
- horizontal resistance (variation in host resistance is primarily due to differences between varieties and between isolates, rather than to specific variety × isolate interactions).

S

self-fertile
Capable of self-fertilize (producing seed upon self-fertilization); *self-fertility* – autogamy; *self-fertilizing* – the fusion of male and female gametes from the same individual.

self-fertilize >>> self-fertile

serine (Ser)
An aminoacid synthesized from glycerate 3-phosphate. Serine is also a product of photorespiration and other metabolic reactions. It is a component of several phosphoglycerides. It is broken down by removal of the amino group to form pyruvic acid. It is one of the proteinogenic aminoacids. Its codons in the genetic code are UCU, UCC, UCA, UCG, AGU and AGC.

sewage sludge
The residual, semi-solid material that is produced as a by-product during wastewater treatment of industrial or municipal wastewater. The term septage is also referring to sludge from simple wastewater treatment but is connected to simple on-site sanitation systems such as septic tanks.**soil**
The surface layer of the earth's crust, carrying the land cover, populated mikroogranizmami and having fertility. Soil formed continuously changes under the influence of water, air, living organisms, and other factors;

natural formation consisting of genetically related horizons formed as a result of the conversion of the surface layers of the lithosphere under the influence of water, air and living organisms, has fertility. Consists of solid, liquid (soil solution) and gaseous living (soil fauna and flora) parts. Divided into types: podzolic, gray forest, black soils, gray. >>> **edaphic**

solitary
A tree or brushwood existing alone, growing alone.

somaclonal variation
Somatic (vegetative non-sexual) plant cells can be propagated in vitro in an appropriate nutrient medium; according to the composition and conditions, the cells may proliferate in an undifferentiated (disorganized) pattern to form a callus, or in a differentiated (organized) manner to form a plant with a shoot and root; the cells, which multiply by division of the parent somatic cells, are called somaclones and, theoretically, should be genetically identical with the parent; in fact, in vitro cell culture of somatic cells, whether from a leaf, a stem, a root, a shoot, or a cotyledon, frequently generates cells significantly different, genetically, from the parent; during culture, the DNA breaks up and is reassembled in different sequences which give rise to plants different in identifiable characters from the parent; such progeny are called "somaclonal variants" and provide a useful source of genetic variation. This are species not peculiar to analyzed phytoceonosis, their penetration into the phytocenosis promotes destruction of its structure; they include all non-native species and lead to a gradual change of cenosis in which they introduced.

syntaxon (pl. syntaxa) >> class >> order >> alliance
Vegetation unit in the phytosociology.
syntaxonomic classes (sing. class) >> orders (sing. order), >> alliances (sing. alliance), >> associations (sing. association), >> syntaxon >> syntaxa
They include different ranks in syntaxonomic hierarchy, abstract vegetation units in phytosociology.

T

tetraploid
Having four sets of chromosomes in the nucleus.

tolerance
The ability of a plant to endure attack by a pathogen, well as biochemical and physiological adaptation to concentrations toxic ions and other stresses without severe loss of yield.

translocation
A change in the arrangement of genetic material, altering the location of a chromosome segment; the most common forms of translocation are reciprocal, involving the exchange of chromosome segments between two non-homologous chromosomes.

threonine (Thr)
An aliphatic, polar alpha-aminoacid ($HO_2CCH(NH_2)CH(OH)CH_3$); its codons are ACU, ACA, ACC, and ACG. Together with serine, threonine is one of two proteinogenic amino acids bearing an alcohol group (tyrosine is not an alcohol but a phenol, since its hydroxyl group is bonded directly to an aromatic ring, giving it different acid/base and oxidative properties). The threonine residue is susceptible to numerous posttranslational modifications. An aminoacid derived from aspartic acid. It is broken down to form glycine and acetyl CoA. Isoleucine can be synthesized from threonine.

trisomic
A genome that is diploid but that contains an extra chromosome, homologous with one of the existing pairs, so that one kind of chromosome is present in triplicate >>> trisomy.

trisomy
The presence of a single extra chromosome, yielding a total three chromosomes of that particular type instead of a pair >>> trisomic.

type
In the taxonomy (1) (variety) a kind of cultivated plants subdivided into smaller systematic units; (2) element taxon (herbarium specimen, description, picture or taxon of lower rank), which is constantly associated

the name of the taxon; (3) (in botany) - the highest taxonomic unit phytocoenotic classification.

tyrosine or 4-hydroxyphenylalanine (Tyr)
An aromatic, polar alpha-amino acid. Is one of the 22 amino acids that are used by cells to synthesize proteins. Its codons are UAC and UAU.

V

valine (Val)
An essential, aliphatic, nonpolar amino acid ($HO_2CCH(NH_2)CH(CH_3)_2$). L-Valine is one of 20 proteinogenic amino acids. Its codons are GUU, GUC, GUA, and GUG.

variety
A taxonomic category that ranks below subspecies (where present) or species, its members differing from others of the same subspecies or species in minor but permanent or heritable characteristics. Varieties are more often recognized in botany. For a cultivated form of a plant >>> cultivar.

viral infection
The invasion of the tissue of a plant by pathogenic virus/viruses.

INDEX